Fred Kernot

Cambridge IGCSE®

Chemistry

Revision Guide

Roger Norris

CAMBRIDGE
UNIVERSITY PRESS

University Printing House, Cambridge CB2 8BS, United Kingdom

Cambridge University Press is part of the University of Cambridge.

It furthers the University's mission by disseminating knowledge in the pursuit of education, learning and research at the highest international levels of excellence.

www.cambridge.org
Information on this title: education.cambridge.org

First published 2015

Printed in the United Kingdom by Latimer Trend

A catalogue record for this publication is available from the British Library

ISBN 978-1-107-69799-7 Paperback

All questions and answers provided have been written by the author. In examinations, the way marks are awarded may be different.

Table of Contents

How to use this book: a guided tour

Learning outcomes

By the end of this unit you should:

- Understand the importance of purity in substances in everyday life
- Assess the purity of substances from melting and boiling point data
- Describe paper chromatography and interpret simple chromatograms
- Interpret simple chromatograms, including the use of R_f values
- Understand how locating agents are used to make colourless substances visible on a chromatogram
- Describe and explain how solvents, filtration, crystallisation and distillation are used to separate or purify substances
- Suggest suitable purification techniques to obtain a given product

Learning outcomes – set the scene of each chapter, help with navigation through the book and give a reminder of what is important about each topic

Supplement material – indicated by a bold vertical line. This is for students who are taking the Extended syllabus covering the Core and Supplement content

Progress check – check your own knowledge and see how well you are getting on by answering regular questions

Terms – clear and straightforward explanations are provided for the most important words in each topic

Tip – quick suggestions to remind you about key facts and highlight important points

Unit 10 Electrochemistry

- For a molten compound of a metal with a non-metal (a binary compound), the cathode product is always the metal and the anode product is always the non-metal.
- The table shows the electrode products formed during the electrolysis of particular electrolytes.

Electrolyte	Cathode product	Anode product
molten lead(II) bromide	lead	bromine
molten sodium iodide	sodium	iodine
molten zinc chloride	zinc	chlorine
concentrated aqueous sodium chloride	hydrogen	chlorine
concentrated hydrochloric acid	hydrogen	chlorine
dilute sulfuric acid	hydrogen	oxygen

Table 10.01

Progress check

10.01 Why are steel-cored aluminium cables used for overhead power lines? [3]

10.02 What do these terms mean? Electrical insulator, electrolysis, anode, electrolyte. [5]

10.03 Predict the products at the anode and cathode when the following are electrolysed:

a dilute sulfuric acid [2]

b concentrated hydrochloric acid [2]

c molten lithium bromide [2]

d concentrated aqueous sodium chloride. [2]

10.05 Reactions at the electrodes

During electrolysis:

- Electrons move from the power supply to the cathode.
- Positive ions in the electrolyte move to the negative cathode.
- The positive ions accept electrons from the cathode. Metals or hydrogen are formed:

$$Zn^{2+} + 2e^- \rightarrow Zn$$
$$2H^+ + 2e^- \rightarrow H_2$$

- The reaction at the cathode is a reduction reaction because electrons are gained (see Unit 14).
- Electrons move from the anode to the power supply.
- Negative ions in the electrolyte move to the anode.
- If the anode is inert, the negative ions lose electrons to the anode. Non-metals or oxygen are formed:

$$2Br^-(aq) \rightarrow Br_2(aq) + 2e^-$$
$$4OH^-(aq) \rightarrow O_2(g) + 2H_2O(l) + 4e^-$$

- If the anode is a reactive electrode the metal atoms of the anode loose electrons and form positive ions. The positive ions go into solution and the anode becomes smaller:

$$Zn \rightarrow Zn^{2+} + 2e^-$$
$$Ag \rightarrow Ag^+ + e^-$$

- The reaction at the anode is an oxidation reaction because electrons are lost (see Unit 14).
- Equations showing electron loss or gain like this are called **ionic half equations**.

TERMS

Ionic half equation: An equation balanced by electrons which shows either oxidation or reduction.

You can also write the half-equations like this:

$$Zn^{2+} \rightarrow Zn - 2e^-$$
$$2Br^- - 2e^- \rightarrow Br_2$$

You should stick to one method or the other to avoid getting muddled.

68

Worked example 15.01

a What could you do to tell the difference between an acidic and alkaline solution? **[2]**

The best ways involve chemistry not taste! Either the litmus test or the pH test is suitable. Do not suggest indicators other than litmus or universal indicator because if the acid or base is very weak, a colour change may not be seen.

Acids turn blue litmus red **[1]**; alkalis turn red litmus blue **[1]**.

b Name three different compounds you could add to hydrochloric acid to make the salt calcium chloride. Write word equations for the formation of the salt. **[4]**

Since it is calcium chloride you are making, the substances must be calcium compounds. But not calcium metal because that is not a compound. When writing the equations, concentrate on the products. Do not forget that with a carbonate, salt + carbon dioxide + water are formed. And do not forget the arrow and + signs!

The compounds are calcium oxide, calcium hydroxide and calcium carbonate **[1]**.

Worked example – a step by step approach to answering questions, guiding you through from start to finish

Sample answer

Ammonia, NH_3, is a simple covalent molecule. Graphite is a macromolecule with a giant covalent structure. Compare the volatility and electrical conductivity of these two molecules and draw a dot-and-cross diagram for ammonia. **[7]**

Ammonia is volatile and has a low boiling point **[1]**. Graphite is a giant structure so has a high boiling point **[1]**. Ammonia does not conduct **[1]** but graphite does **[1]**.

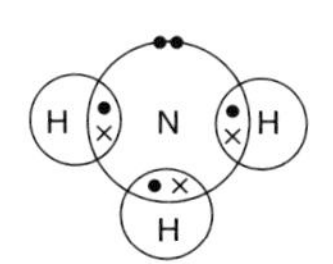

Figure 6.13

Bonding pairs of electrons between each of the three N and H atoms **[2]**.

If this not scored then a pair of bonding electrons between one of the N and H atoms **[1]**.

Lone pair of electrons on the N atom **[1]**.

Sample answer – an example of a question with a good quality answer that is likely to score a high mark.

Exam-style questions

Question 4.01

The electronic structures of the atoms of four elements are:

A 2,8,4 B 2,2 C 2,8 D 2,8,8,1

a Which one of these elements is in Group I of the Periodic Table? Explain your answer. [2]

b Which one of these elements has the lowest proton number? Explain your answer. [2]

c Which one of these elements is in Period 4 of the Periodic Table? Explain your answer. [2]

d Which element is a noble gas? Explain your answer. [2]

e Element A has three naturally-occurring isotopes

i What is meant by the term *isotope*? [1]

ii An isotope of A has a nucleon number of 30. State the number of electrons, protons and neutrons in this isotope. [3]

Question 4.02

Two isotopes of bromine are $^{79}_{35}Br$ and $^{81}_{35}Br$.

a Deduce the number of neutrons in each of these isotopes. [1]

b Bromine has the electronic structure 2,8,18,7.

i Explain how this structure shows that bromine is in Group VII of the Periodic Table. [1]

ii Explain how this structure shows that bromine has a proton number of 35. [2]

c Magnesium reacts with bromine to form magnesium bromide.

i Write the electronic structure for a bromide ion. [1]

ii Write the electronic structure for a magnesium ion, Mg^{2+}. [1]

Unit 4 Atomic structure

Revision checklist

You should be able to:

- State the relative charges and approximate relative masses of protons, neutrons and electrons
- Define nucleon number (mass number) as the total number of protons and neutrons in the nucleus of an atom
- Explain the basis of the Periodic Table in terms of the number of protons
- Explain that isotopes are atoms of the same element with different numbers of neutrons
- Understand that isotopes have the same properties because they have the same number of electrons in their outer shell
- State the two types of isotopes as being radioactive and non-radioactive
- State one medical and one industrial use of radioactive isotopes
- Describe the electronic structure of atoms
- Understand the importance of the noble gas electronic structure

29

Exam-style questions – practice answering exam-style questions and check your answers against those provided at the back of the book

Revision checklist – at the end of each chapter so you can check off the topics as you revise them

Introduction

This book is designed to support students studying the Cambridge IGCSE Chemistry syllabus (0620) from Cambridge International Examinations. The topics in the syllabus have been divided into 30 units and match the syllabus.

The main purpose of this publication is to serve as a revision guide for students. The features of this book outlined above are designed to make learning as effective as possible and to give plenty of opportunity to test yourself and gain confidence before taking examinations. Material indicated by a red line relates to the material that forms the supplementary content of the syllabus.

Practical aspects of chemistry are considered not only in Units 2 and 3 but also throughout the book where relevant. These include questions relevant to practical examinations and to alternatives to practical examinations as well as to coursework.

Particles in motion

Learning outcomes

By the end of this unit you should:

- State the distinguishing properties of solids, liquids and gases
- Describe the structure of solids, liquids and gases in terms of particle separation, arrangement and types of motion
- Be able to describe changes in state
- Explain changes of state in terms of the kinetic theory
- Understand the effect of pressure and temperature on a gas in terms of the kinetic theory
- Describe Brownian motion as evidence for the kinetic theory
- Describe and explain diffusion
- State evidence for Brownian motion
- Know how the rate of diffusion depends on the relative molecular mass of the molecules

1.01 States of matter

The three states of matter are solids, liquids and gases.

Solid	**Liquid**	**Gas**
Figure 1.01	Figure 1.02	Figure 1.03
fixed volume	fixed volume	no definite volume
fixed shape; they keep this shape unless hit	no definite shape	no definite shape
does not flow easily (unless the solid is a powder)	flows easily	spreads out everywhere
cannot be compressed	only compressed a little if at all	can be compressed

Table 1.01

We can explain these bulk properties using the particle theory.

Solid	Liquid	Gas
Figure 1.04	Figure 1.05	Figure 1.06
particles close together	particles close together	particles far apart
particles arranged in a regular pattern	particles arranged randomly	particles arranged randomly
particles vibrate around a fixed point	particles slide over each other randomly and slowly	particles move randomly and rapidly

Table 1.02

It is a common error to think that the particles in liquids are spaced out. They are not. They are close to each other with very little or no space between. They slide over each other and do not have free motion like gases.

Sample answer

Use the kinetic particle theory to explain why a crystal of iodine keeps its particular shape and cannot be compressed but iodine vapour can be compressed and spreads everywhere. **[4]**

The particles in solid iodine are regularly arranged **[1]** so it keeps its shape. They are packed closely together **[1]** so the crystal cannot be compressed. The particles in iodine vapour are far apart **[1]** (as there are no attractive forces between them) **[1]**. When pressure is put on the vapour, the particles can be pushed closer to each other **[1]**.

Progress check

1.01 Describe the three states of matter in terms of shape and volume. [6]

1.02 Describe the difference between solids and liquids in terms of closeness and motion of particles. [4]

1.03 At room temperature and pressure bromine molecules are close together and randomly arranged. Describe the proximity (closeness) and arrangement of bromine molecules in (a) bromine vapour [2] (b) solid bromine. [2]

1.02 Changes of state

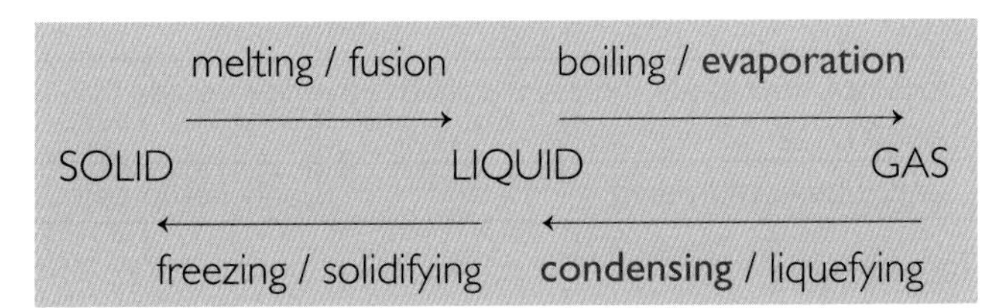

> **TERMS**
>
> **Condensing:** The change in state from a gas to a liquid.
>
> **Evaporation:** The change in state from a liquid to a gas below the boiling point of the liquid.

Under given conditions of temperature and pressure some substances, for example carbon dioxide, sublime. They turn directly from solid to gas or gas to solid.

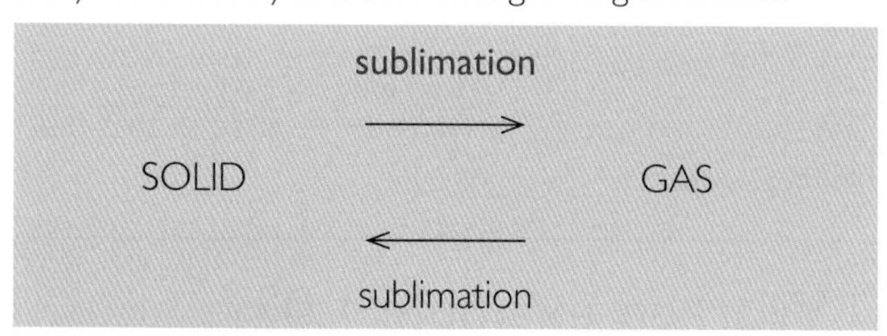

> **TERMS**
>
> **Sublimation:** The change in state directly from solid to gas and/or gas to solid without the liquid state being formed.

1.03 The kinetic particle theory

> **TERMS**
>
> **Kinetic theory:** Particles in solids, liquids and gases behave as hard spheres which are constantly moving from place to place (in liquids and gases) or vibrating (in solids).

- Particles in liquids and gases are in constant motion.
- Particles in a gas are constantly colliding and changing directions. They move randomly.
- As the temperature increases, the particles gain more energy and they move faster.
- The simple **kinetic theory** also assumes that the particles are 'hard' spheres.

We can use the kinetic particle theory to explain many facts, for example:

- Gases can be compressed easily: the particles are far apart. Increasing the pressure on a gas decreases the distance between the particles. So an increase in pressure at constant temperature decreases the volume.

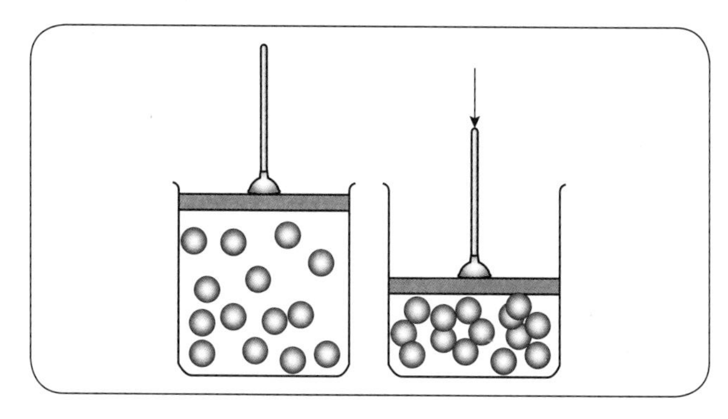

Figure 1.07 Increasing pressure on a gas

- Liquids and solids cannot be compressed easily: the particles are close together or touching. Increasing the pressure has little effect since the particle theory assumes that the particles are 'hard'.
- A gas in a closed container increases in pressure when it is heated. The particles have more energy. They move faster and hit the walls of the container with greater force.
- The volume of gas in a syringe increases when heated at constant pressure: the particles have more energy the higher the temperature, so they hit the walls of the syringe more often pushing the syringe plunger out.

1.04 Explaining changes of state

- An input of energy is needed to melt and boil a substance.
- The energy lessens the forces of attraction between particles when a solid melts or a liquid boils.
- Energy is given out when a substance condenses or freezes.
- This is because the particles are losing some of the energy of their movement (kinetic energy) when condensing or freezing.

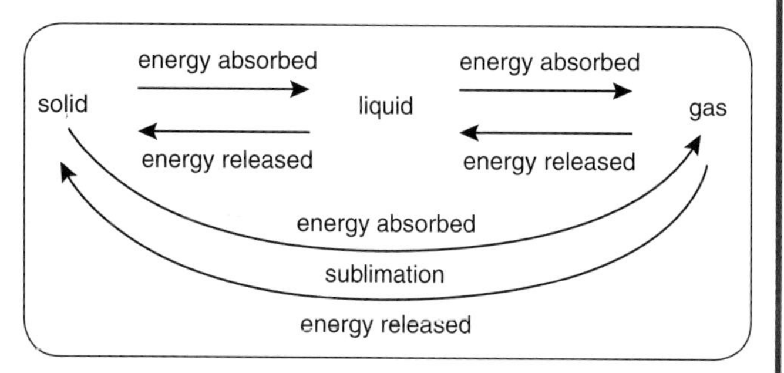

Figure 1.08

> **TIP**
>
> When writing about forces make sure that you:
>
> - refer to attractive forces
> - refer to forces between particles not within particles

Worked example 1.01

Use the kinetic particle theory to explain why a balloon increases in size when the temperature increases. **[5]**

In this type of question your answer should involve a comparison, for example the more energy the higher the temperature (not high temperature). The sequence needed is:
movement → collisions → energy → force → pressure

The particles move faster the higher the temperature **[1]** and collide more frequently with the walls of the balloon **[1]**. The gas particles have more energy the higher the temperature is **[1]** and hit the walls of the balloon with more force **[1]**.

This leads to greater pressure on the wall of the balloon **[1]**.

1.05 Heating and cooling curves

Heating and cooling curves are graphs showing how the temperature changes when a substance is heated or cooled at a steady rate (steady increase or decrease in energy).

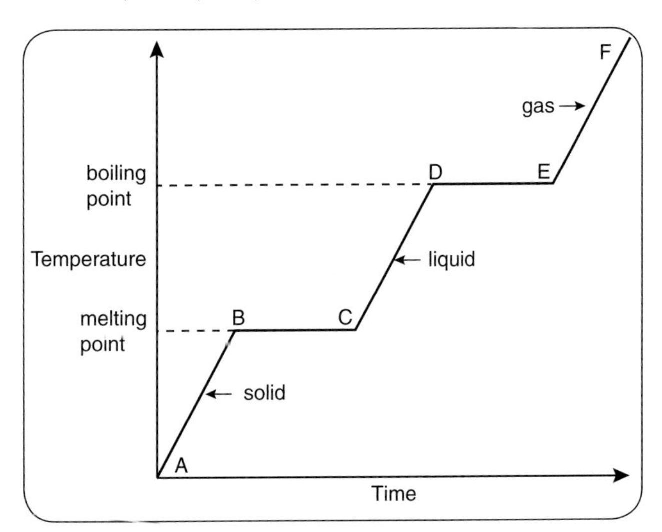

Figure 1.09 A heating curve

A → B: an input of energy increases the temperature of the solid.

B → C: the energy is being used to reduce the attractive forces between the particles in the solid. Liquid as well as solid material is present. The energy does not go in to raise the temperature.

C → D: the energy increases the temperature of the liquid.

D → E: the energy is being used to overcome the attractive forces between the particles in the liquid so that a liquid as well as a gas (vapour) is present.

E → F: the energy increases the temperature of the gas.

The shape of a cooling curve is the reverse of that of a heating curve.

Worked example 1.02

Explain, using ideas of particles and energy, the changes in arrangement and motion of the particles which occurs when zinc melts. **[6]**

You should make sure that the word particles or a type of particle, for example molecule, is mentioned. Start with the solid arrangement and energy:

The particles in solid zinc are regularly arranged **[1]** and only vibrate **[1]**. As temperature increases the zinc particles gain energy **[1]** and the forces between the zinc particles weaken **[1]**.

Then write about the liquid:

In the liquid particles move over each other **[1]** and become irregularly arranged **[1]**.

Progress check

1.04 Explain why liquids cannot be compressed. [2]

1.05 Name these changes of state: (a) gas to liquid [1], (b) liquid to gas below the boiling point [1], (c) solid to gas (without liquid being formed) [1].

1.06 Copy and complete: When liquid sulfur is heated, the temperature of the liquid ___________. At the ___________ point the ___________ is used to overcome the ___________ forces between the ___________ and the temperature remains ___________. (Words to use: attractive; boiling; constant; energy; increases; molecules.) [6]

1.06 Brownian motion

Brownian motion provides evidence for the kinetic particle model.

Examples: 1. the zigzag movement of pollen grains in still water

2. random movement of dust particles in still air

TERMS

Brownian motion: The random movement of small visible particles in a suspension caused by the unequal random bombardment of molecules of liquid or gas on the visible particles.

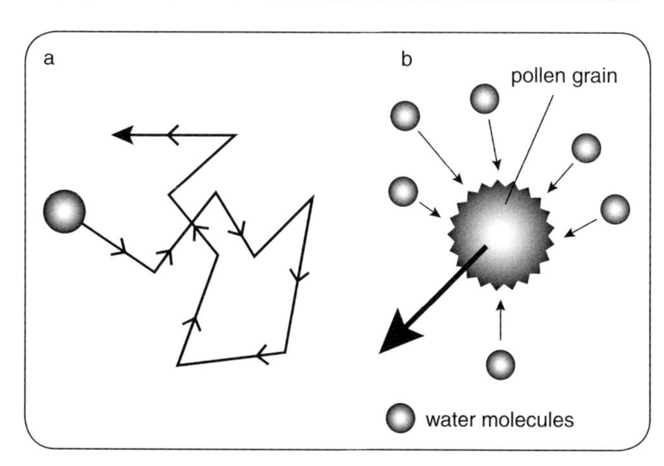

Figure 1.10 **a** Random motion of pollen grain **b** Random bombardment of water molecules on a pollen grain causes movement in the direction →

1.07 Diffusion

TERMS

Diffusion: The spreading movement of one substance into another due to the random motion of the particles.

- **Diffusion** is explained by the kinetic particle theory. Particles collide randomly with each other leading to complete mixing of the particles.
- Diffusion in gases is faster than diffusion in liquids because the particles in gases move faster.
- Diffusion does not happen in solids because the particles are fixed in position.
- The overall movement of particles is from where they are more concentrated to where they are less concentrated.

When writing about diffusion refer to the random movement of the particles. It is a mistake to think that the particles always travel in the direction from high to low concentration.

Diffusion of gases

Figure 1.11 shows the diffusion of bromine vapour in air.

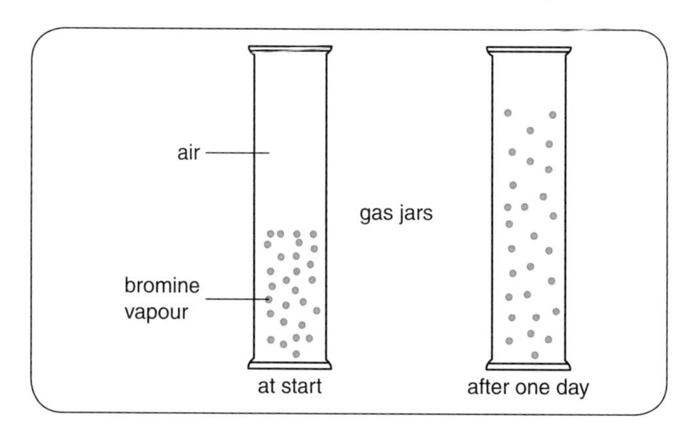

Figure 1.11

Diffusion of liquids

Figure 1.12 shows the diffusion of ink in water.

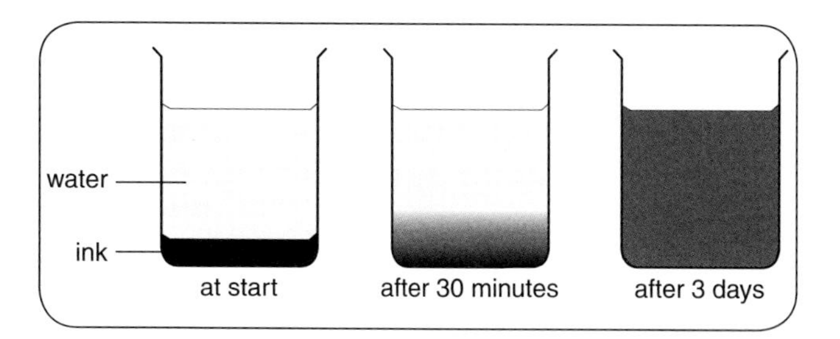

Figure 1.12

Diffusion and relative molecular masses

- The masses of molecules are compared by using relative molecular masses, M_r (see Unit 8).
- The higher the value of M_r the heavier the molecule.
- Molecules with lower M_r move faster than molecules with higher M_r.

Progress check

1.07 Describe the evidence for the kinetic particle theory. [3]

1.08 Define and explain the term diffusion using the kinetic particle theory. [4]

1.09 Explain why hydrogen gas, H_2, diffuses quicker in air than hydrogen sulfide gas, H_2S. [2]

Sample answer

The wall of a rubber balloon is slightly porous. Gases can move through the wall. A balloon contains a mixture of carbon dioxide and helium. Explain why the percentage of carbon dioxide in the balloon increases with time. **[4]**

Molecules with lower relative molecular mass move faster **[1]** by diffusion **[1]** than molecules with higher molecular mass. Helium has a lower molecular mass than carbon dioxide **[1]**. So helium diffuses faster out of the balloon than carbon dioxide **[1]**.

Exam-style questions

Question 1.01

The arrows on the diagram below represent some changes in state.

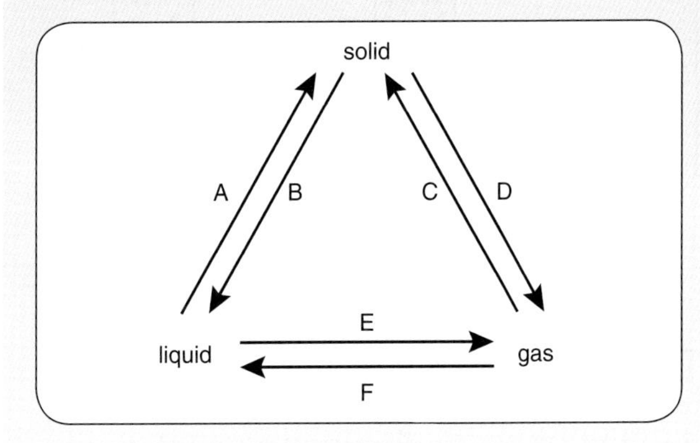

Figure 1.13

a Give the name of the changes B, D and E. [2]

b Describe what is happening to the arrangement and motion of the particles in change A. [4]

c What energy change is taking place in
 i change B [1]
 ii change F? [1]

Question 1.02

Hydrogen chloride, HC*l*, and hydrogen bromide, HBr, are gases which turn blue litmus paper red. A teacher soaked some cotton wool in hydrochloric acid and set up the apparatus shown below. Hydrogen chloride gas evaporated from the hydrochloric acid.

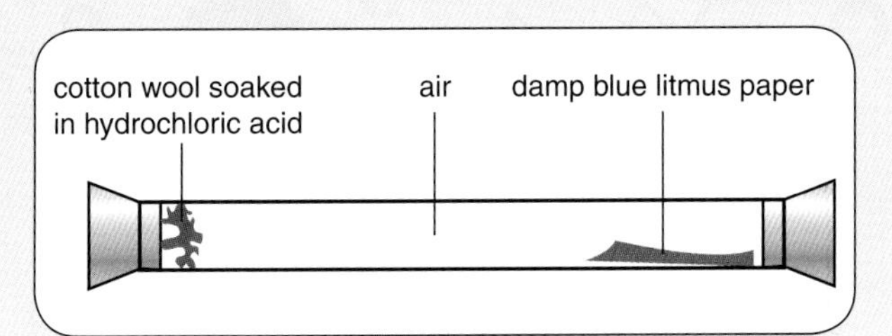

Figure 1.14

Hydrogen chloride gas evaporated from the cotton wool. It was only after two minutes that the litmus paper turned red.

a Use the kinetic particle theory to explain these results. [3]

b The teacher repeated the experiment with hydrobromic acid, which evaporates to produce hydrogen bromide gas. Would the litmus paper turn red, quicker or slower or take the same time? Explain your answer. [3]

c Explain evaporation using the kinetic particle theory. [3]

Question 1.03

Explain using the kinetic particle theory why particles of smoke in still air appear to move in an irregular way. [5]

Revision checklist

You should be able to:

- State the distinguishing properties of solids, liquids and gases
- Describe the structure of solids, liquids and gases in terms of particle separation, arrangement and types of motion
- Describe changes of state in terms of melting, boiling, evaporation, freezing, condensation and sublimation
- Explain changes of state in terms of the kinetic theory
- Describe the temperature and pressure of a gas in terms of the kinetic theory
- Describe Brownian motion as evidence for the kinetic theory
- Explain Brownian motion in terms of particle collisions
- Describe and explain diffusion
- Explain how diffusion depends on relative molecular mass

Experimental chemistry

Learning outcomes

By the end of this unit you should:

- Be able to name and know the use of glassware, for example pipettes, burettes
- Be able to name apparatus for the measurement of time, temperature, mass and volume
- Know that density is the mass of substance in a given volume
- Be able to describe and comment on experimental arrangements
- Know how to record readings, complete tables of data and plot graphs
- Be able to plan investigations, taking into account the control of variables

2.01 Basic laboratory apparatus

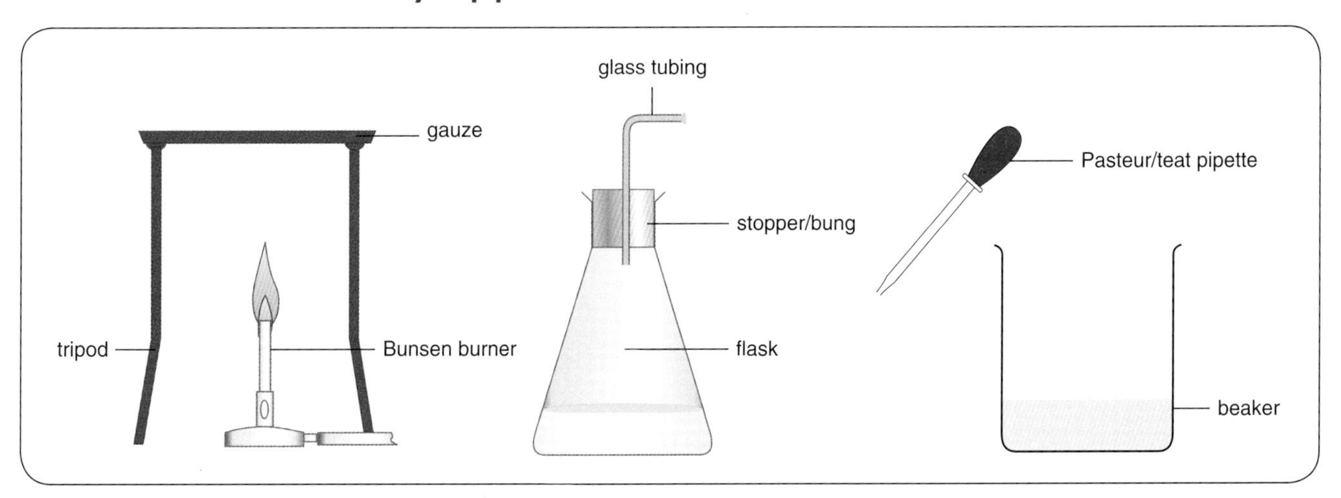

Figure 2.01

You should draw apparatus in cross section not in three dimensions. Make sure that if tubes come out of flasks, they are not cut across by a line on your diagram.

- Flasks are used for carrying out reactions where fairly small amounts of liquid are used.
- Beakers are used to store liquids temporarily or sometimes for carrying out reactions.
- Test tubes are used to carry out reactions where small amounts of liquid are used and not heated.
- Boiling tubes are used to heat small amounts of liquid.

2.02 Recording mass, time and temperature

Mass

- Mass is recorded on a balance. A good balance can give a reading to two decimal places, for example 45.15.
- The unit of mass is the gram (g) or kilogram (kg). 1000 g = 1 kg
- We change g into kg by dividing the mass in g by 1000.

Time

- Time is recorded on a stop clock or stop watch.
- The unit of time is the second (s). For chemical reactions which are slower, we can use minutes (min).

Temperature

- Temperature is measured with a thermometer.
- The unit of temperature is degree Celsius (°C). We can read accurate thermometers to ± 0.1 °C.

2.03 Measuring volumes of liquids

- Volume is measured in centimetres cubed (cm^3) or decimetres cubed (dm^3): $1000\,cm^3 = 1\,dm^3$.
- We can change cm^3 into dm^3 by dividing the volume in cm^3 by 1000. For example, $25\,cm^3 = 25/1000 = 0.025\,dm^3$.
- A burette is used to accurately deliver up to $50\,cm^3$ of liquid. The scale divisions of a burette can be read to the nearest $0.1\,cm^3$.
- A volumetric pipette can deliver a single fixed volume of liquid very accurately, for example $25.0\,cm^3$ of sodium hydroxide solution. Some volumetric pipettes have scale divisions like a burette which can be read accurately.

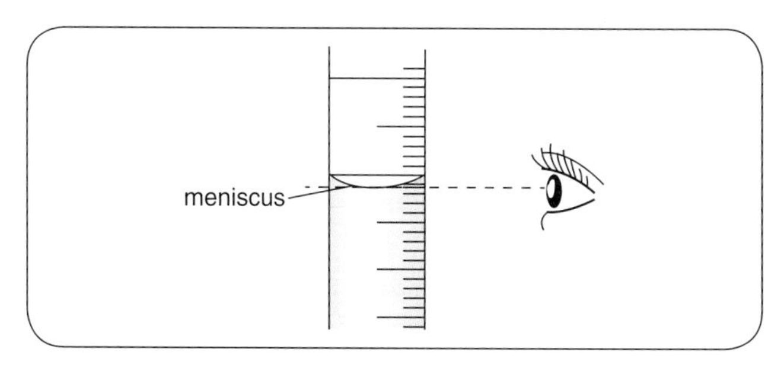

Figure 2.03 When reading a burette, your eye should be level with the bottom of the meniscus

- A volumetric flask is used to make up a solution of a known concentration very accurately. A known mass of solute is dissolved in a small amount of solvent in the flask. The flask is then filled to the graduation mark with solvent.
- A measuring cylinder is used for measuring volumes of solutions where accuracy is not so important. Many measuring cylinders have scale divisions which are only every $2\,cm^3$.

Progress check

2.01 Convert 0.84 kilograms to grams. [1]

2.02 Convert 23 grams to kilograms. [1]

2.03 Convert $36\,cm^3$ to dm^3. [1]

2.04 Convert $0.450\,dm^3$ to cm^3. [1]

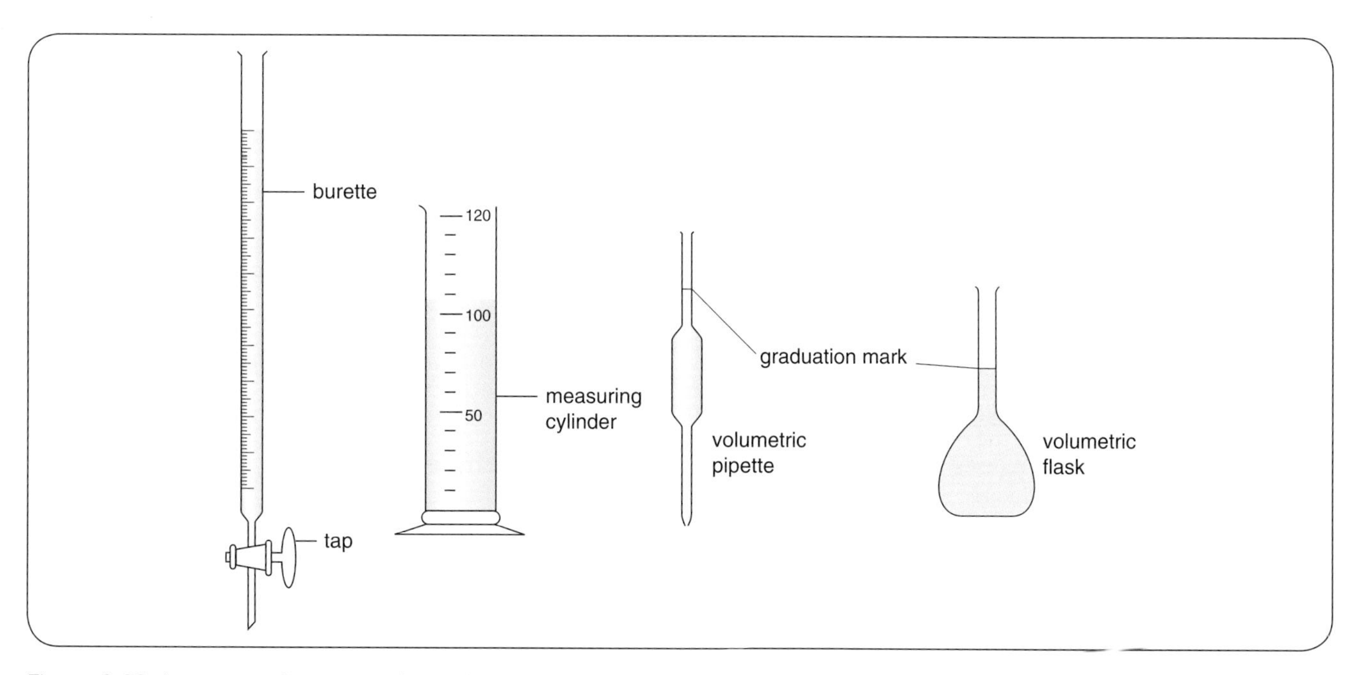

Figure 2.02 Apparatus for measuring volumes of liquids

Sample answer

A student used the apparatus in Figure 2.04 to calculate the concentration of aqueous sodium hydroxide.

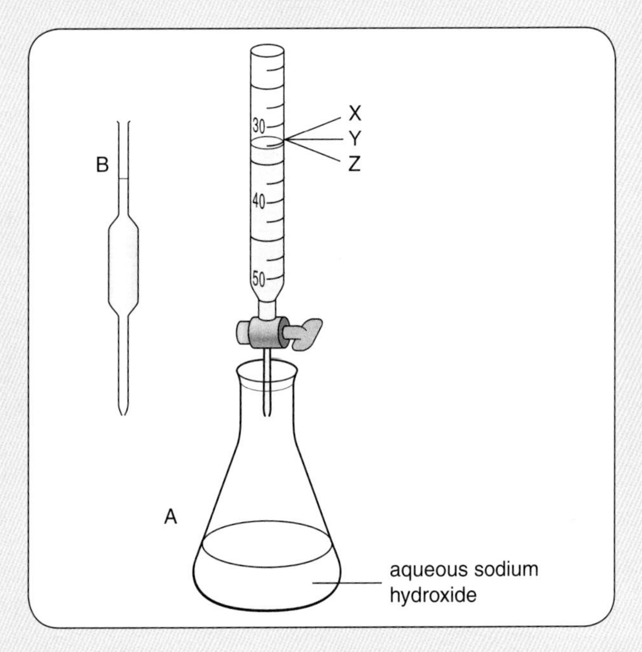

Figure 2.04

a Give the names of the pieces of glassware labelled A and B. **[2]**

A (conical) flask / Erlenmeyer (flask) **[1]**, B volumetric pipette **[1]**.

b In which place, X, Y or Z, should you position your eye to read the burette? Give a reason for your answer. **[2]**

Y **[1]**, your eye should be in line with the bottom of the meniscus **[1]**.

c The diagram shows the initial burette reading and the final burette reading.

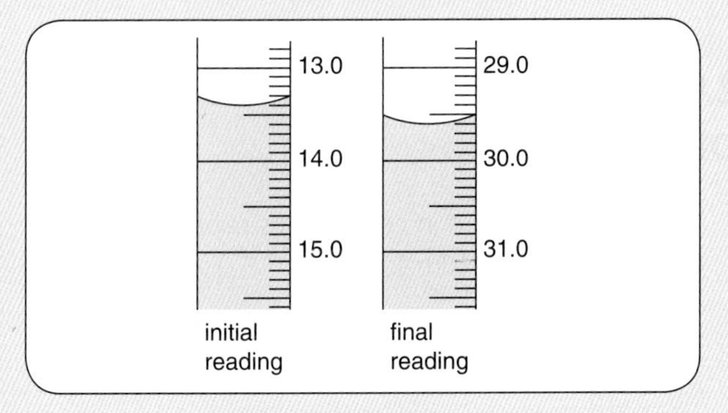

Figure 2.05

Give the initial and final burette readings and deduce the volume of hydrochloric acid which was delivered into the flask. **[3]**

Initial reading = 13.4 (cm^3) **[1]**, final reading = 29.6 (cm^3) **[1]**.

volume = final − initial = 29.6 − 13.4 = 16.2 (cm^3) **[1]**.

Worked example 2.01

When a solution of acid is added to a solution of alkali, the temperature of the mixture increases. A student wants to measure how the temperature varies with the volume of acid added to the alkali.

What equipment should the student use to carry out the experiment accurately and what measurements should be taken? **[8]**

1 *Since temperatures are being measured, heat losses should be minimised:*

Insulated beaker or drinking cup for the alkali **[1]**

2 *Select apparatus for measuring accurate volumes:*

Burette for measuring volumes of acid **[1]**

Fixed volume of alkali put into a beaker measured with a volumetric pipette or burette **[1]**

3 *Carry out the procedure, taking relevant measurements:*

Add known small volume of acid to alkali **[1]**

Stir **[1]**

Record the highest temperature of the solution **[1]**

With an accurate thermometer **[1]**

Repeat by adding further fixed volumes of acid and measuring the temperature **[1]**

2.04 Measuring gas volumes

The volume of a gas produced during a reaction (see Unit 12) can be measured using:

- a gas syringe – **a** in Figure 2.06
- an upturned measuring cylinder (or upturned burette) full of water at the start of the experiment – **b** in Figure 2.06

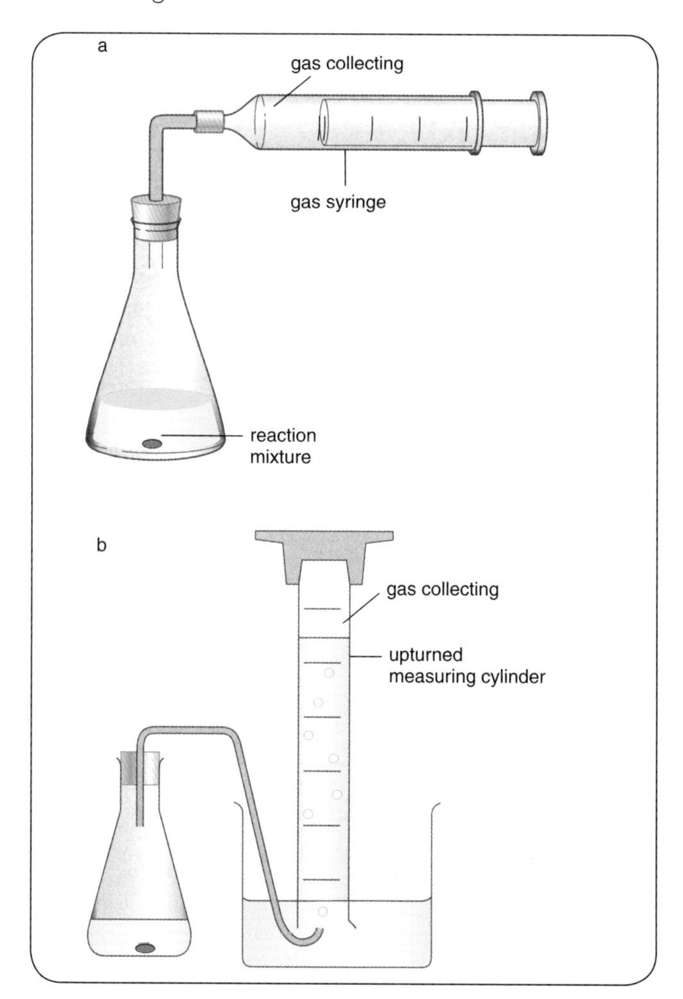

Figure 2.06

Progress check

2.05 What piece of apparatus should you use to measure out 23.4 cm^3 of solution accurately? [1]

2.06 What two different apparatuses you could use for measuring the volume of gases. [2]

2.07 Why would you not use a measuring cylinder to measure 25.0 cm^3 of a solution into a beaker? [2]

2.05 Density

- Density (g/cm^3) $= \dfrac{\text{mass (g)}}{\text{volume (cm}^3\text{)}}$
- Gases with densities that are less than air can be collected by the downward displacement of air – Figure 2.07 **a**.
- Gases with densities that are greater than air can be collected by the upward displacement of air – Figure 2.07 **b**.

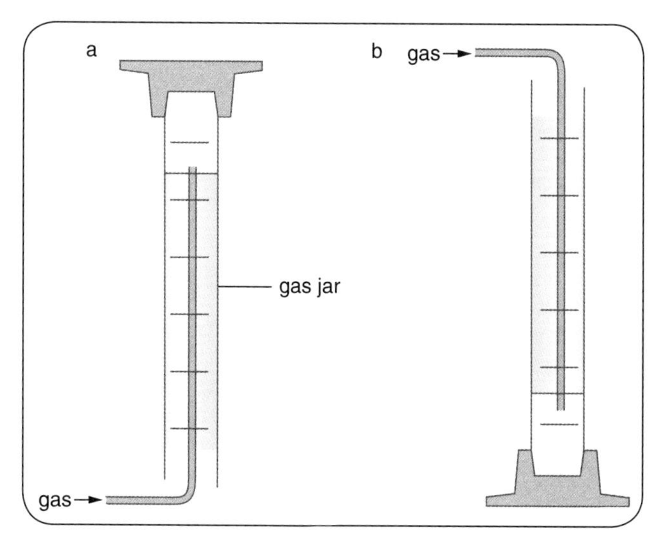

Figure 2.07

- If two liquids do not mix, the less dense one floats on the one that is denser.

It is better not to use the words 'lightweight' or 'light' when referring to density. They do not indicate mass for a given volume.

2.06 Carrying out experiments

When carrying out experiments, we need to decide:

- the apparatus to be used
- the conditions needed, e.g. heat, catalyst
- the measurements to be made
- what variables are involved
- how to make the experiment a fair test
- the accuracy and reliability of the measurements
- safety, for example if harmful gases are given off, use a fume cupboard

Variables and fair testing

A variable is something that is changed in an experiment.

Experiments involving measurements may have several variables.

When calcium carbonate reacts with hydrochloric acid, the volume of gas given off can be measured at various time intervals using the apparatus in Figure 2.06 a.

In this experiment the variables are:

- volume of gas given off
- time
- temperature
- mass of calcium carbonate
- size of calcium carbonate lumps
- concentration and volume of hydrochloric acid

So if we want to find out how volume of gas changes with time:

- We need to keep temperature, the mass of calcium carbonate, the size of the calcium carbonate lumps and the concentration and volume of hydrochloric acid constant (control variables).
- We need to measure the volume of carbon dioxide at different times.
- The variable selected by you (in this case time) is called the independent variable.
- The variable which is measured for each change of the independent variable (in this case the volume of gas) is called the dependent variable.
- The experiment above is a **fair test**.

TERMS

Fair test: An experiment where the independent variable affects the dependent variable and all other variables are controlled.

Accuracy

- Repeat your measurements and take an average, ignoring any **anomalous results**.
- Use instruments with as great an accuracy as possible, e.g. a burette for measuring volumes of liquids accurately instead of a measuring cylinder.
- Read the instruments carefully.
- In an experiment, the overall accuracy depends on the least accurate instrument you are using.

TERMS

Anomalous result: A result or piece of data which does not fit the pattern of the rest of the data.

Progress check

2.08 What is meant by a fair test? [2]

2.09 The table shows how the volume of gas released in a reaction varies with time.

time / s	0	10	20	30	40	50
volume / cm^3	0	5	10	12	20	25

Table 2.01

Which is the anomalous reading? Give a reason for your answer. [3]

Tables and graphs

Tables

independent variable — dependent variable

time/s	volume of gas /cm^3
10	23
20	41

Figure 2.08

Graphs

When drawing graphs:

- plot each point as an ×
- draw lines or curves of best fit
- ignore anomalous points
- the dependent variable is plotted on the y (vertical) axis of the graph

When the points plotted show a curve, do not join one point to the next by a straight line.

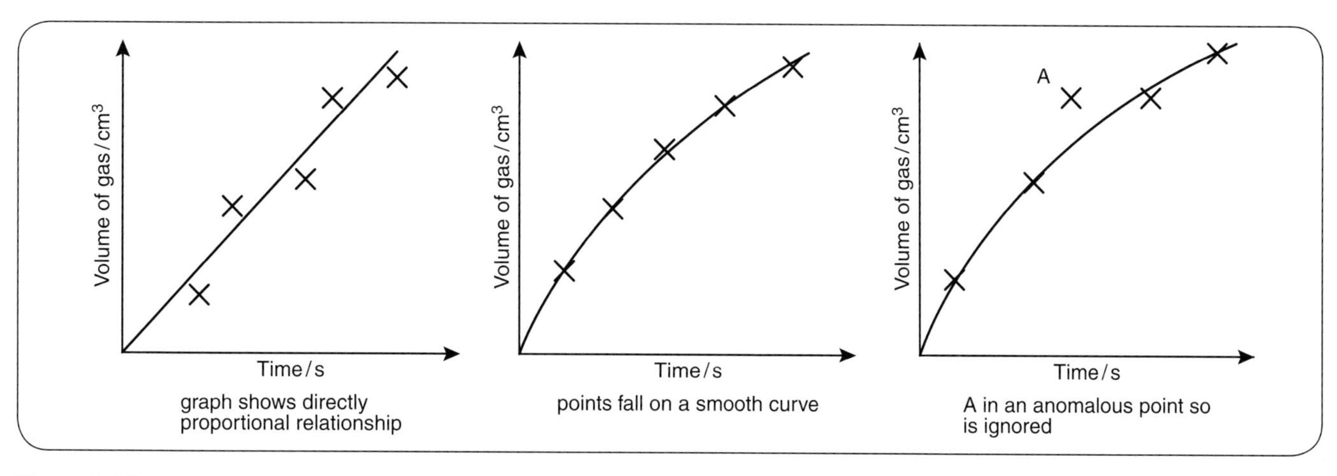

Figure 2.09

Worked example 2.02

When magnesium reacts with hydrochloric acid, hydrogen gas is produced. A student wants to investigate how the volume of gas produced in 20 seconds changes when the temperature changes. Draw the headings of the table that the student should make for the results and list the factors which should be kept constant in this experiment. **[8]**

How to get the answer:

Table:

The quantity chosen by you (independent variable) goes on the left so

Temperature on the left **[1]** with units °C **[1]**

The values you are measuring (dependent variable) goes on the right so

Volume on right **[1]** with units cm^3 **[1]**

Factors kept constant:

The possible variables in an experiment are masses, volumes, concentrations, time, temperature, pressure. You need to select control variables suitable for your experiment. Make sure that you name particular chemicals:

Mass of magnesium **[1]**, size of pieces of magnesium **[1]**

Concentration of hydrochloric acid **[1]**, volume of hydrochloric acid **[1]**

Exam-style questions

Question 2.01

Chlorine is a poisonous gas which is denser than air and soluble in water. A student made chlorine by reacting sodium chloride with concentrated sulfuric acid and collecting the gas using the apparatus in Figure 2.10.

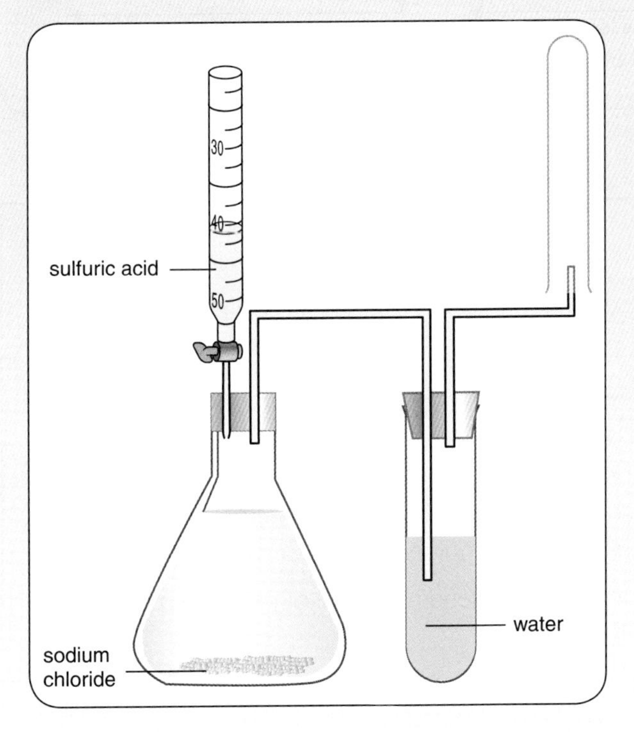

Figure 2.10

a Identify two errors in the diagram. Explain why each is an error. [4]

b Where on the diagram is heat applied? [1]

c Suggest why the experiment should be carried out in a fume cupboard. [1]

d State one hazard of concentrated sulfuric acid. [1]

e State one other safety precaution when carrying out this experiment. [1]

Question 2.02

A student compared the energy released when different fuels were burned in the same spirit burner. The student compared the fuels by measuring the increase in temperature of $100\,cm^3$ of water in a copper can.

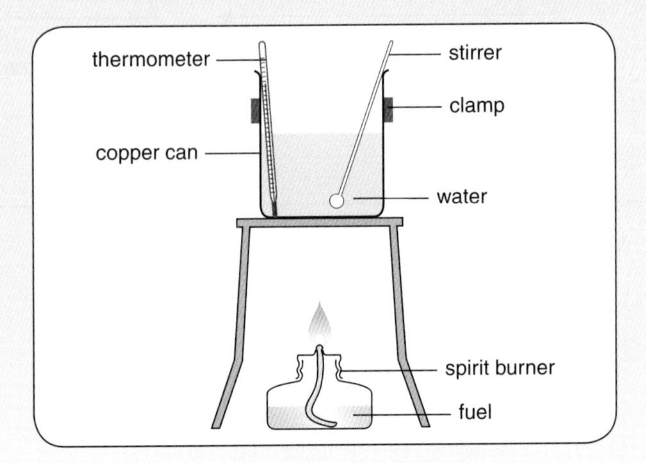

Figure 2.11

a Explain why the student kept the water stirred all the time. [1]

b State two other factors which should be kept constant when comparing the fuels. In each case explain why. [4]

c The student repeated the experiment twice with each fuel. Suggest why. [2]

d What piece of equipment should be used to measure out $100\,cm^3$ of water? Give a reason for your answer. [2]

e The mass of fuel in the spirit burner was weighed to three decimal places. The reading on the thermometer was to the nearest °C. Why was it of little value to give the answer to the calculation to three decimal places? [1]

Revision checklist

You should be able to:

- ☐ Name appropriate apparatus for the measurement of time, temperature, mass and volume
- ☐ Understand the use and accuracy of burettes, pipettes, measuring cylinders and other laboratory glassware
- ☐ Select suitable apparatus for the accuracy required by the experiment
- ☐ Record readings, construct tables of data and plot graphs
- ☐ Understand the importance of controlling particular variables in an experiment
- ☐ Plan an investigation given relevant information

Methods of purification

Learning outcomes

By the end of this unit you should:

- Understand the importance of purity in substances in everyday life
- Assess the purity of substances from melting and boiling point data
- Describe paper chromatography and interpret simple chromatograms
- Interpret simple chromatograms, including the use of R_f values
- Understand how locating agents are used to make colourless substances visible on a chromatogram
- Describe and explain how solvents, filtration, crystallisation and distillation are used to separate or purify substances
- Suggest suitable purification techniques to obtain a given product

3.01 Pure and impure substances

It is important that substances put into medicines and added to food are pure because the impurities may contain substances which are harmful to us.

- A pure substance is a single substance, for example pure sodium chloride is 100% sodium chloride.
- Impure substances contain other substances mixed with them.
- We can tell the difference between a pure and impure substance by differences in melting points and boiling points.

Pure substance	Impure substance
solid has sharp melting point – all the solid melts at same temperature	solid melts over a temperature range and at a lower temperature than the pure solid
liquid has sharp boiling point – all the liquid boils at the same temperature	1. for a solid dissolved in a liquid: liquid boils over a temperature range and at a higher temperature than the pure liquid 2. for a mixture of two liquids: the mixture starts boiling at the boiling point of one of the liquids and rises to the boiling point of the other liquid

Table 3.01

Progress check

3.01 A solid melts between 234–240 °C. Is this solid likely to be pure or impure? Give a reason for your answer. [1]

3.02 Why should zinc oxide used in creams to treat sunburn be pure? [1]

3.03 Explain why salt is spread on roads in icy weather. [3]

3.02 Chromatography

TERMS

Paper Chromatography: The separation of a mixture of soluble compounds using chromatography paper and a solvent.

Paper chromatography is used to separate and purify a mixture of dissolved substances, for example the pigments (colourings) present in food colourings and inks. Figure 3.01 shows how to carry out paper chromatography.

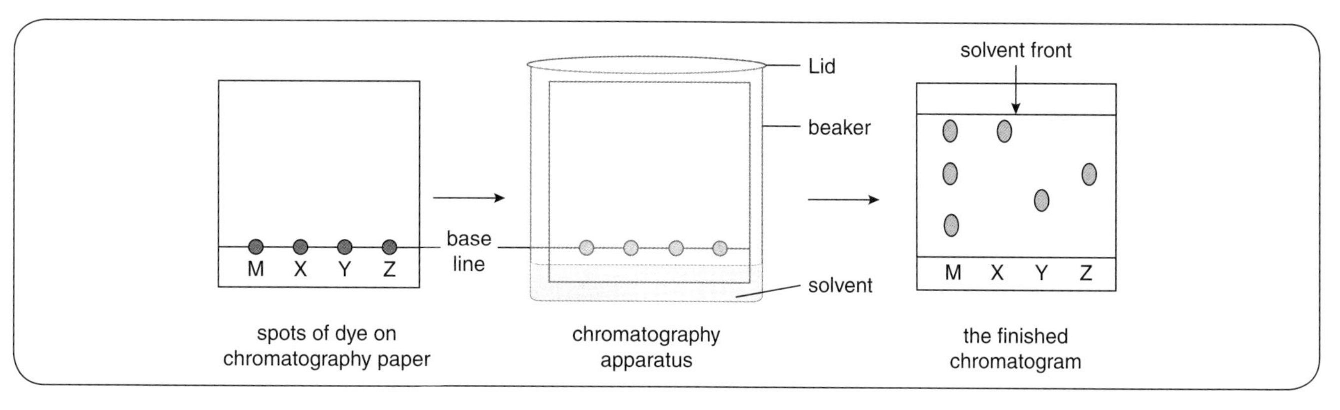

Figure 3.01

- Draw a pencil line on the chromatography paper.
- Put a spot of the concentrated mixture to be separated, M, on the line using a very thin tube.
- Put spots X, Y and Z that you think are in the dye mixture on the base line for comparison.
- Dip the bottom of the paper in the solvent.
- The solvent moves up the paper and allows separation of the mixture.
- Remove the paper when the solvent front is near the top.

The spots on the chromatogram can be compared with those of known dyes. In Figure 3.01, the mixture, M, contains dyes X and Z as well as a third dye.

We can identify the components in the mixture because for a particular solvent they travel a certain distance compared with the solvent front.

TIP

When carrying out chromatography, the solvent level should be below the base line to prevent the mixture washing off into the solvent.

Retention factor (R_f)

TERMS

R_f: In chromatography, the distance moved by a particular substance from the base line divided by the distance moved by the solvent front from the base line.

$$R_f = \frac{\text{distance from base line to centre of spot}}{\text{distance of solvent front from base line}}$$

R_f values can be used to identify compounds or ions because most substances have characteristic R_f values.

Locating agents

TERMS

Locating agent: A substance that reacts with colourless spots on a chromatogram to make them visible as coloured spots.

Many compounds, for example amino acids, are colourless and so cannot be seen on a chromatogram. Spraying the chromatogram with ninhydrin and warming makes the spots appear a purple colour. Ninhydrin is an example of a **locating agent**.

TIP

You do not have to know the names of particular locating agents.

Worked example 3.01

Plant leaves contain pigments called chlorophylls. Describe how you would make a solution of these pigments and use chromatography to separate them. **[8]**

Step 1 is to extract the pigments:

Grind up leaves **[1]**

With solvent **[1]**

Using a mortar and pestle / using a blender **[1]**

Step 2 is to separate the solids from the pigment solution

Filter off the solid remains of the leaves (through glass wool) **[1]**

Step 3 is chromatography of the pigments

Place a small spot of the solution obtained / filtrate on chromatography / filter paper **[1]**

Place the bottom of the paper in solvent / water / alcohol **[1]**

So that the solvent level is below the spot **[1]**

Allow the solvent to run up the paper (and separate the pigments) **[1]**

Progress check

3.04 In chromatography, why is the base line drawn in pencil and not in ink? [1]

3.05 Draw a diagram of the apparatus used in chromatography. In your diagram include the position of a spot of mixture on the base line and the level of the solvent. [3]

3.06 After chromatography, a spot of dye is 18 cm from the base line. The solvent front is 45 cm from the base line. Deduce the R_f value of the dye. [1]

3.03 Purification by use of a solvent

Liquids which mix with each other are said to be miscible.

Liquids which do not mix are immiscible.

Some substances dissolve in water. Others dissolve better in organic solvents like hexane. We can use these differences in solubility to separate two solutes dissolved in a solvent.

Example:

The solvents hexane and water are immiscible. Iodine is more soluble in hexane than in water. Sodium chloride (salt) is more soluble in water than in hexane. We can separate a mixture of iodine and salt in water by:

1. putting the mixture in a **separating funnel**

TERMS

Separating funnel: A piece of glassware used to separate two immiscible liquids or a solute which is more soluble in one liquid than another.

2. adding hexane to the separating funnel
3. mixing the solutions by shaking
4. allowing the layers to separate on standing. The iodine moves to the hexane layer and the salt remains in the water
5. running off the layer of salt in water. This leaves the iodine in the hexane layer

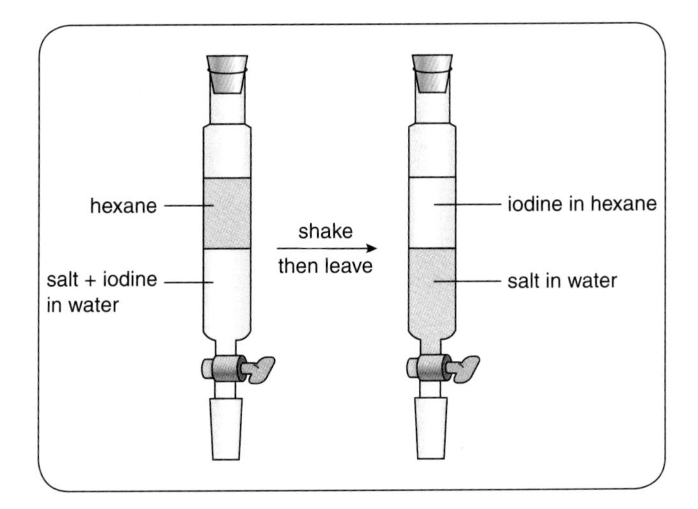

Figure 3.02

Sample answer

Cyclohexane is a volatile solvent. Potassium iodide is soluble in water but not in cyclohexane. Iodine dissolves in both cyclohexane and a solution of potassium iodide in water.

Describe in detail how you would obtain solid iodine from a solution of iodine and potassium iodide in water using cyclohexane. **[8]**

Add cyclohexane to the mixture [1]. Shake the mixture [1] in a separating funnel [1] then allow the layers to settle [1]. Some of the iodine goes into the cyclohexane layer [1] but potassium iodide does not [1]. Then remove the layer of cyclohexane [1] and allow the cyclohexane to evaporate [1] (This leaves the solid iodine).

3.04 Filtration

Filtration separates undissolved solids from a liquid. Molecules of liquid and dissolved substances can flow through the holes in the filter paper but larger particles of solid are too big to pass through. They remain on the filter paper. An example is separating sand from seawater.

- The **filtrate** is the solution passing through the filter paper.
- The **residue** is the solid remaining on the filter paper.
- Traces of solution remaining between particles of solid are removed by washing with a suitable solvent.

TERMS

Filtrate: In filtration, the liquid which goes through the filter paper.

Residue: In filtration, the solid that is trapped on the filter paper.

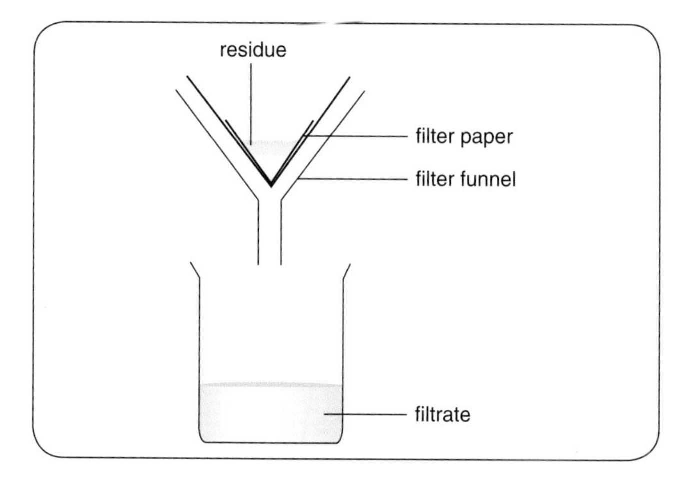

Figure 3.03

3.05 Crystallisation

Crystallisation separates a soluble solid from a solution, for example copper(II) sulfate from its aqueous solution:

1. Gently heat the solution in an evaporating basin to concentrate it.
2. Evaporate the solvent until the crystallisation point (a drop of the solution forms crystals when placed on a cold tile).
3. Leave the saturated solution to form crystals.
4. Filter off the crystals then dry them between filter papers.

3.06 Simple distillation

Simple distillation is used to separate a **volatile** liquid from a solution of a non-volatile solid, for example to separate copper(II) sulfate and water from an aqueous solution of copper(II) sulfate (Figure 3.04).

TERMS

Simple distillation: A method of separating a volatile from a non-volatile substance by evaporation and condensation.

Volatile: Easily changed to a vapour. Volatile substances have low boiling points.

Distillation involves boiling and condensation. It works because the components to be separated have different boiling points.

To separate water from copper(II) sulfate by simple distillation:

1. The solution of copper(II) sulfate in water is heated in a distillation flask.
2. The water boils first because it is volatile. The steam turns to liquid in the condenser.
3. The copper(II) sulfate remains in the distillation flask because it is not volatile – it has a much higher boiling point than water.

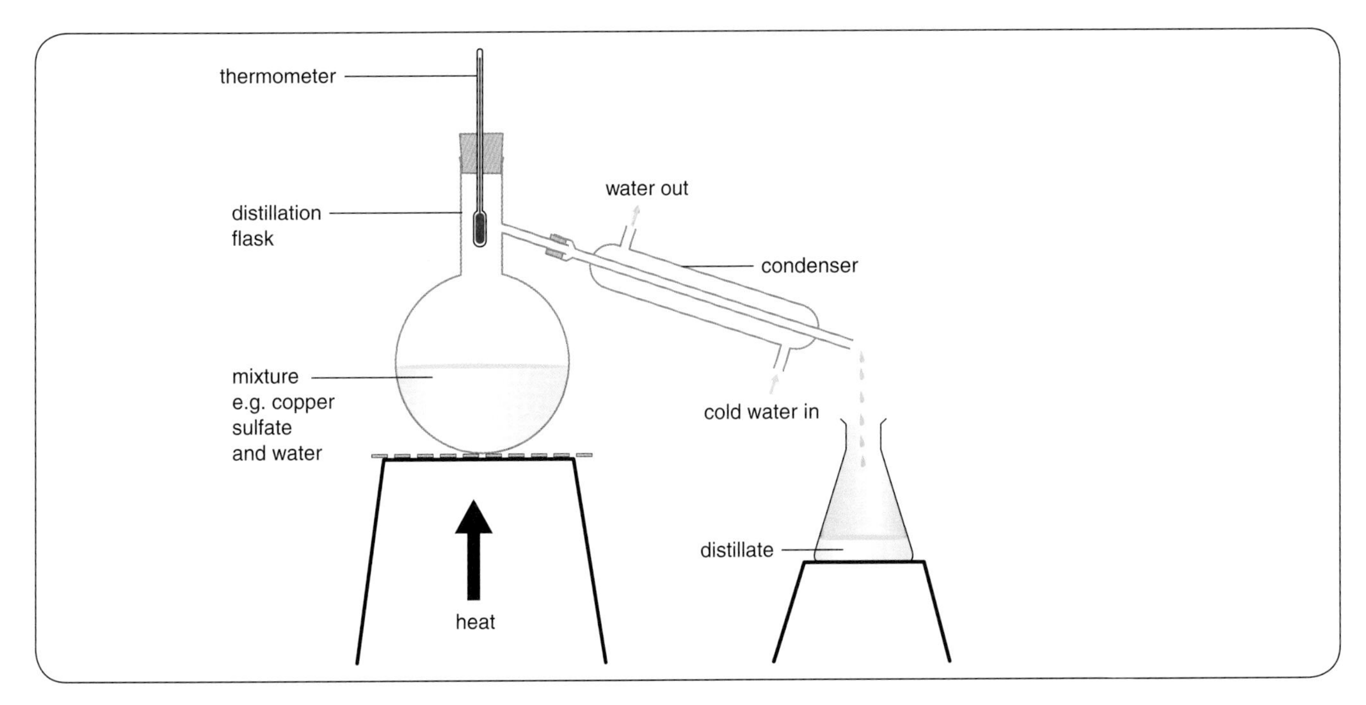

Figure 3.04 Apparatus for simple distillation

3.07 Fractional distillation

TERMS

Fractional distillation: The separation of liquids with different boiling points from a mixture of liquids by evaporation and condensation in a long column.

Fractional distillation is used to separate miscible liquids with different boiling points. It is used to separate petroleum fractions (see Unit 27) and to purify alcohol from a mixture of water and alcohol (Figure 3.05).

- In fractional distillation there is a temperature gradient in the column: higher at the bottom and lower at the top.
- The more volatile components in the mixture (lower boiling points) move faster up the column.
- The less volatile components in the mixture (higher boiling points) do not move as fast.
- The components of the mixture reach the condenser in turn, the most volatile first. They change from vapour to liquid in the condenser and the fractions containing particular components of the mixture are collected one at a time.

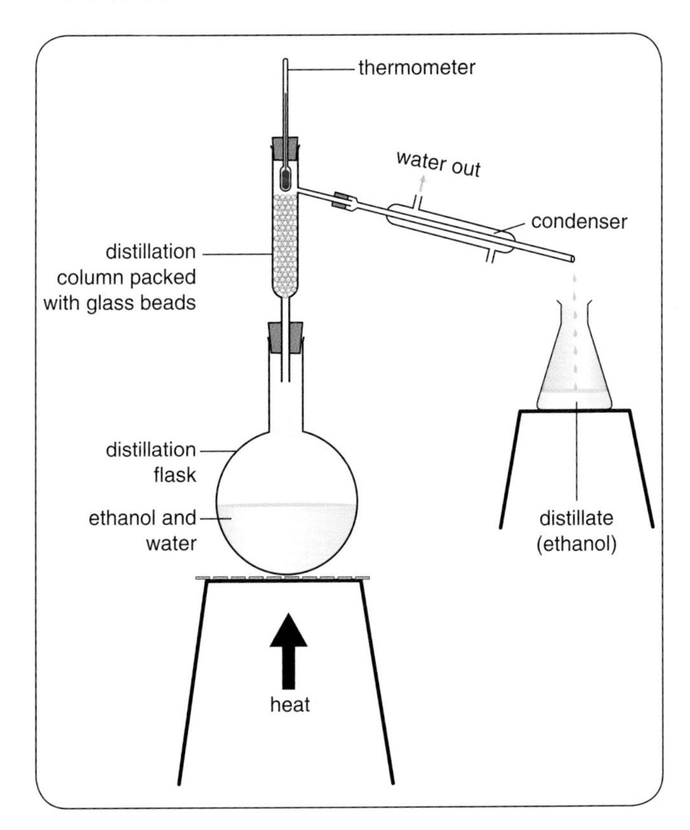

Figure 3.05

3.08 Which purification technique?

To chose the best method to purify a substance from a mixture you have to know the following properties of the different substances in the mixture:

- state at room temperature and pressure
- solubility in water or other solvent
- differences in boiling points

Mixture requiring separation	Separation method
insoluble solid and liquid	filtration
crystals of solid from a solution	crystallisation
two soluble solids with different solubility in water and organic solvent	using an organic solvent and separating funnel
a volatile liquid from a solution of a non-volatile solid	simple distillation
a mixture of liquids with different boiling points	fractional distillation

Table 3.02

Worked example 3.02

Describe how you would separate a mixture of sand and aqueous copper(II) sulfate to obtain purified sand, copper(II) sulfate and water. Give an explanation for each stage you use. **[7]**

In order to choose the correct method of separation you need to know the solubility of each component, for example sand is insoluble in water and copper(II) sulfate is soluble in water. So the first step is to separate the solid from the solution.

Filter off the sand **[1]**. Sand is solid but copper(II) sulfate is in solution **[1]**.

You then need to separate the component of the solution. Distillation is appropriate because water has a low boiling point and copper(II) sulfate has a high boiling point.

The filtrate is a solution of copper(II) sulfate **[1]**. This is distilled **[1]**. The copper(II) sulfate remains in the flask **[1]** because it has a high boiling point **[1]**. The water condenses in the condenser and is collected as a distillate **[1]**.

NOTE: simple distillation is used rather than crystallisation, because purified water was required.

Progress check

3.07 What is the best method to separate water from aqueous copper(II) sulfate? [1]

3.08 Hexane and octane are liquids with slightly different boiling points. Suggest the best method of separating these liquids. [1]

3.09 What is the simplest method of separating powdered chalk from water? [1]

3.10 What is the best method of separating two dyes with different solubility in water? [1]

Exam-style questions

Question 3.01

Liquid A has a boiling point of 74 °C. Liquid B has a boiling point of 49 °C.

a Describe with the aid of a diagram how you would obtain purified samples of liquid A and B from a mixture of liquids A and B. [6]

b Liquid B dissolves in both water and pentane. Pentane has a boiling point of 36 °C. Which is the most volatile, B or pentane? Give a reason for your answer. [1]

c The sample of liquid A is impure. What test would you do to show that it is impure and what result would you expect? [2]

d Liquid B reacts with sodium to form a crystalline solid, C. Describe how you could make pure dry crystals of C from an aqueous solution of C. [4]

Question 3.02

Chromatography is used to separate a mixture of carbohydrates which are soluble in an organic solvent. The result is shown in the diagram (Figure 3.06).

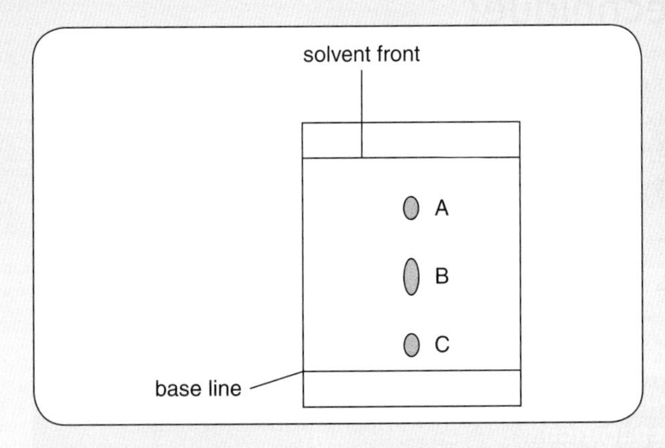

Figure 3.06

a Carbohydrates are colourless compounds. Suggest how the spots can be made visible. [2]

b A student suggested that there were three different carbohydrates in the mixture. What information in the diagram suggests this? [1]

c Why can we not be absolutely certain that there are just three carbohydrates in the mixture? [2]

d Calculate the R_f value of carbohydrate A. [2]

e Which carbohydrate was least soluble in the organic solvent? Give a reason for your answer. [1]

Revision checklist

You should be able to:

- Describe how paper chromatography is used to separate mixtures
- Deduce R_f values from chromatograms
- Understand the purpose of locating agents in making colourless spots visible
- Understand the importance of purity of substances
- Identify substances and assess their purity from melting point and boiling point information
- Describe and explain methods of purification by the use of a suitable solvent, filtration, crystallisation and distillation
- Suggest suitable purification techniques, given information about the substances involved

Atomic structure

Learning outcomes

By the end of this unit you should:

- State the relative charges and approximate relative masses of protons, neutrons and electrons
- Be able to define proton number and nucleon number
- Use proton number and the simple structure of atoms to explain the basis of the Periodic Table
- Be able to define isotopes
- Know why isotopes of the same element have the same chemical properties
- State the two types of isotopes as being radioactive and non-radioactive
- Know some medical and industrial uses of radioactive isotopes
- Be able to describe the electronic structure of atoms and the significance of the noble gas electronic structures

4.01 Atomic structure

TERMS

Atom: The smallest part of an element that can take part in a chemical change.

- **Atoms** contain sub-atomic particles called protons, neutrons and electrons.
- In the middle of each atom is a tiny nucleus containing protons and neutrons.
- Outside each atom, the electrons are arranged in electron shells (Figure 4.01).

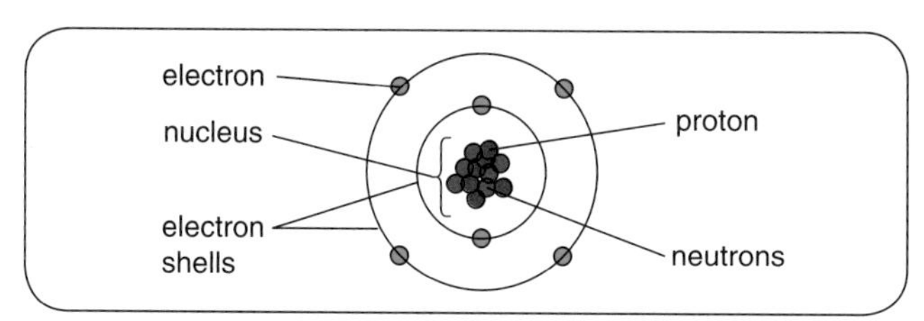

Figure 4.01

The table shows some information about the sub-atomic particles.

Sub-atomic particle	Symbol for the particle	Relative mass	Relative charge
proton	p	1	+1
neutron	n	1	no charge
electron	e^-	0.00054	-1

Table 4.01

4.02 Proton number and the Periodic Table

TERMS

Proton number (atomic number): The number of protons in the nucleus of an atom.

- The number of protons in the nucleus of an atom is called the **proton number (atomic number)**.
- Each atom of the same element has the same number of protons.
- The atoms of the elements are arranged in order of proton number. Hydrogen has one proton, helium has two, lithium three, beryllium four and so on.

- In a neutral atom, the number of electrons = the number of protons.
- The Periodic Table is an arrangement of elements in order of increasing proton number so that elements with the same number of electrons in their outer shell fall in the same vertical column (Group). So Group I elements have 1 electron in their outer shell, Group II elements have 2 electrons in their outer shell and so on.

Sample answer

Describe the structure of a helium atom. Use your Periodic Table to help you.

In your answer include the type, number and position of each sub-atomic particle present. **[5]**

Helium has a nucleus containing protons and neutrons **[1]**. There are two protons **[1]** and two neutrons **[1]**. The two electrons **[1]** spin round in a shell outside the nucleus **[1]**.

4.03 Isotopes

TERMS

Isotopes: Atoms of the same element which have the same proton number but a different nucleon number.

Nucleon number (mass number): The total number of protons and neutrons in the nucleus of an atom.

- Atoms of the same element can have different numbers of neutrons. These atoms are called **isotopes.**
- The total number of protons and neutrons in the nucleus of an atom is called the **nucleon number (mass number).**
- So isotopes are atoms with the same number of protons but different numbers of neutrons.

TIP

Remember that isotopes refer to atoms not elements. For example, in a molecule with more than one chlorine atom, there may be more than one isotope of chlorine present.

Isotopes can be written with their full name, for example carbon-14, uranium-235. The number after the name is the nucleon number. We usually describe isotopes using standard notation. This shows the chemical symbol, the nucleon number and the proton number.

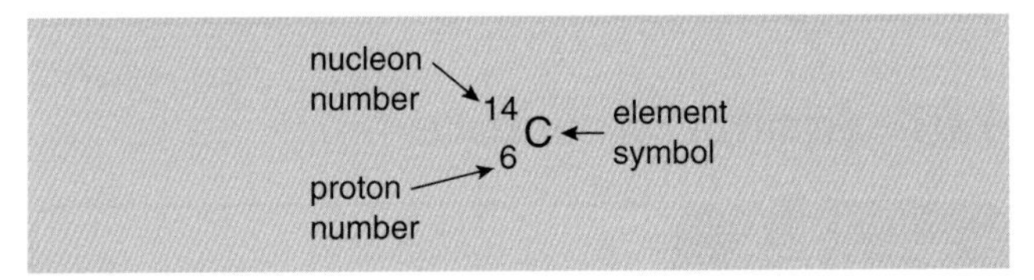

Three isotopes of hydrogen are shown in Figure 4.02 with their standard notation.

How many neutrons?

Number of neutrons = nucleon number − proton number

So using the notation for chromium with a nucleon number of 52:

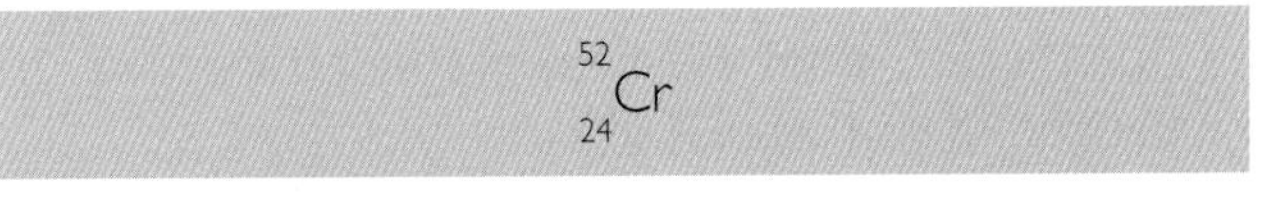

There are 24 protons and (52 − 24) = 28 neutrons.

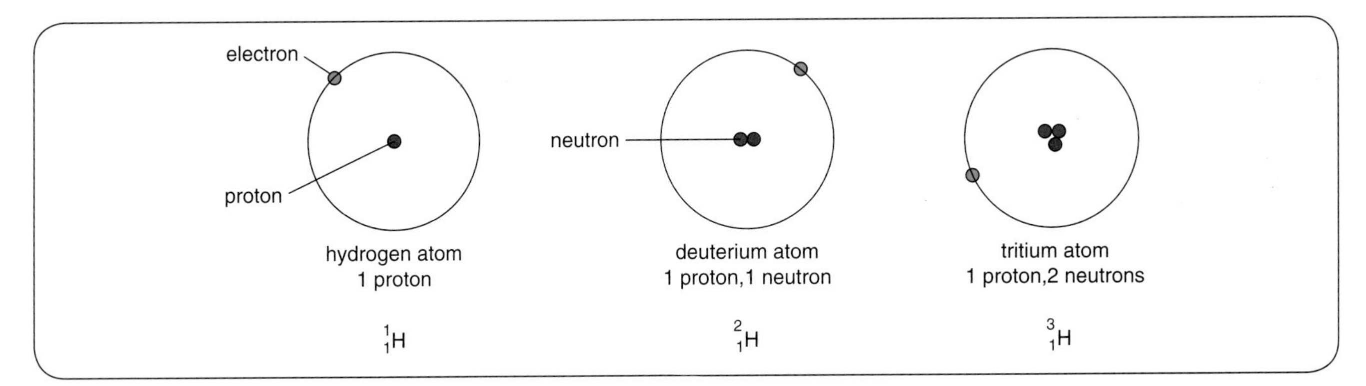

Figure 4.02

The nucleon number is usually shown on the Periodic Table as well as the proton number. The nucleon number is always the larger of the two figures.

Properties of isotopes

Isotopes of the same element have the same chemical properties because they have the same number of electrons in their outer shell. They may have slightly different physical properties.

Progress check

4.01 Give the relative charges and relative masses of a proton, a neutron and an electron. [6]

4.02 What are *isotopes*? [1]

4.03 Deduce the number of neutrons in an isotope of iron with a nucleon number of 54. [1]

4.04 Deduce the total number of neutrons in a molecule of sulfur $^{32}_{16}S$, which contains 8 sulfur atoms. Show all your working. [2]

4.04 Radioactivity and its uses

Some isotopes are non-radioactive. Others are radioactive. Radioactive isotopes have unstable nuclei. The nuclei break down (decay). As it decays, the nucleus gives out particles or rays. The measurement of the rate of emission of these particles or rays has found many uses.

- Medical uses: cancer treatment, treatment for an overactive thyroid gland, generation of electric current in heart pacemakers, sterilising medical equipment, location of tumours.
- Industrial uses: measuring and controlling the thickness of paper, measuring fluid flow and locating leaks in pipelines, measuring engine wear, energy generation in nuclear power stations.

The syllabus only requires you to know one medical and one industrial use of radioactive isotopes. Make sure that you know the difference between medical and industrial uses.

Worked example 4.01

Complete the table to show the number of sub-atomic particles in these atoms. **[4]**

How to get the answer:

Atom	Number of protons	Number of neutrons	Number of electrons
$^{127}_{53}I$	*this is given by the number at the lower left* = 53	*number of neutrons is top left number (nucleon number) minus number of protons (127 − 53)* = 74	*number of electrons in an atom = the number of protons* = 53
$^{176}_{71}Lu$	*this is given by the number at the lower left* = 71	*number of neutrons is top left number (nucleon number) minus number of protons (176 − 71)* = 105	*number of electrons in an atom = the number of protons* = 71

NOTE: 1 mark is given for the first column being correct, 1 mark for the last column being correct and 1 mark each for the number of neutrons.

Table 4.02

Sample answer

Tritium is an isotope of hydrogen. It is radioactive and has a nucleon number of 3.

a What is meant by the term radioactive? **[1]**

It gives off radiation or particles from unstable atoms [1].

b Describe the similarities and differences in the atomic structure of tritium and ordinary hydrogen. **[4]**

Both have one proton [1] and one electron [1]. Tritium has 2 neutrons [1] but hydrogen does not have any [1].

c Suggest one industrial use of a radioactive isotope. **[1]**

Measuring the thickness of paper [1].

4.05 Electron shells

TERMS

Electron shells: The regions at different distances from the nucleus where one or more electrons are found.

In a simple model of the atom the electrons are arranged in orbits around the nucleus. These orbits are called **electron shells** (Figure 4.03).

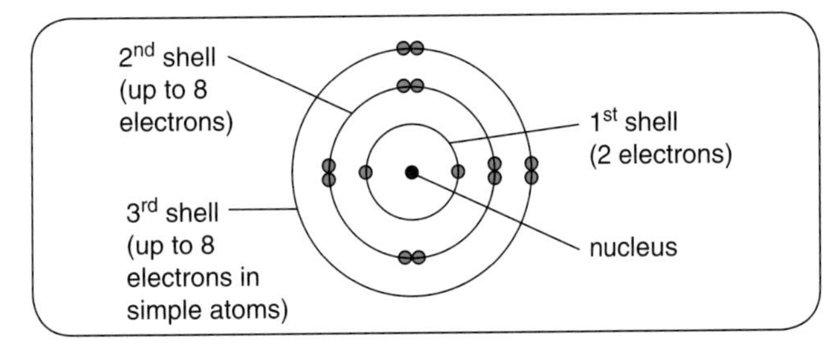

Figure 4.03

- The first shell, nearest the nucleus holds a maximum of two electrons.
- The second shell, further away from the nucleus, holds a maximum of eight electrons.
- The third shell is even further away. It starts filling up when the second shell has eight electrons.
- The fourth shell starts filling up when the third shell has eight electrons.
- Up to calcium (proton number = 20), the shells fill in order 1,2,3,4.

4.06 Deducing electronic structures

TERMS

Electronic arrangement (electronic structure): The number and arrangement of electrons in the electron shells of an atom.

The **electron arrangement** in an atom (also called the electron configuration or electronic structure) is deduced by adding electrons, one at a time, to the electron shells starting with the first shell.

We can write the number of electrons in each shell as a number separated by commas. For example:

- A hydrogen atom has 1 proton, so has 1 electron. This goes into the 1st shell. So the electron arrangement is 1.
- A lithium atom has 3 protons, so has 3 electrons. Two electrons go into the first shell. This shell is then full, so the 3rd electron goes into the 2nd shell. So the electron arrangement is 2,1.
- A sodium atom has 11 protons, so has 11 electrons. Two electrons go into the 1st shell and 8 electrons into the second. The second shell is full, so the 11th electron goes into the 3rd shell. So the electron arrangement is 2,8,1.
- Figure 4.04 shows the electronic structures of some atoms.

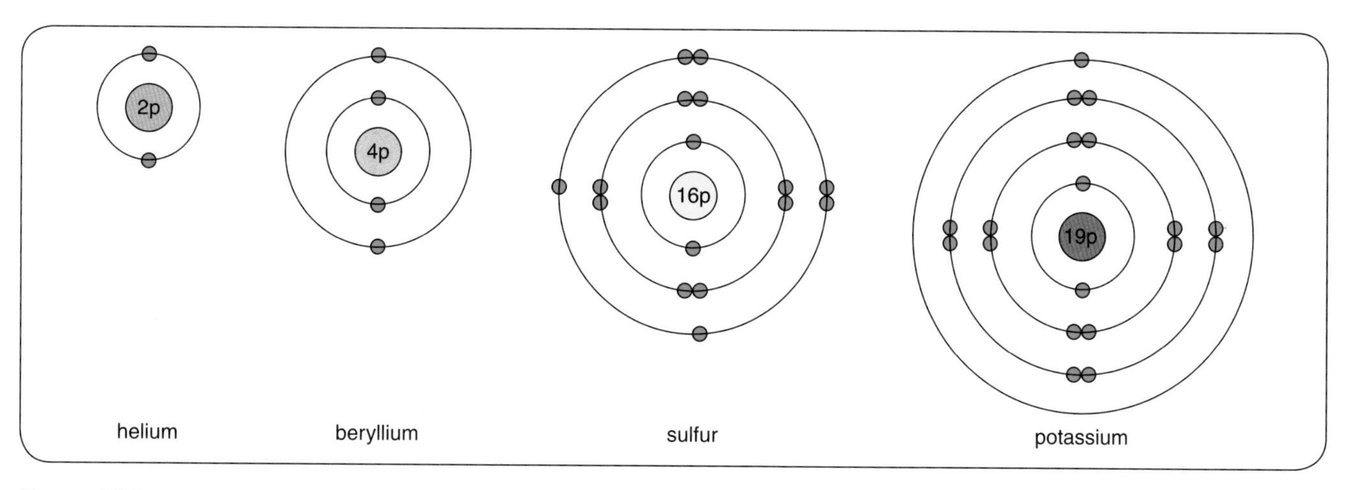

Figure 4.04

The electronic structures of the first 20 elements are:

Element	Number of electrons	Electronic structure	Element	Number of electrons	Electronic structure
hydrogen	1	1	sodium	11	2,8,1
helium	2	2	magnesium	12	2,8,2
lithium	3	2,1	aluminium	13	2,8,3
beryllium	4	2,2	silicon	14	2,8,4
boron	5	2,3	phosphorus	15	2,8,5
carbon	6	2,4	sulfur	16	2,8,6
nitrogen	7	2,5	chlorine	17	2,8,7
oxygen	8	2,6	argon	18	2,8,8
fluorine	9	2,7	potassium	19	2,8,8,1
neon	10	2,8	sodium	20	2,8,8,2

Table 4.03

You should be able to deduce the electronic structures of the first 20 elements from the information in the Periodic Table because in a neutral atom the number of protons = the number of electrons.

4.07 Electronic structures and the Periodic Table

- The number of outer shell electrons corresponds with the Group number, for example Group III elements have three outer shell electrons and Group VII elements have seven outer shell electrons. The electrons in the outer shell of an atom are sometimes called the valency electrons.
- The number of shells containing electrons gives the Period number.
- The number of electrons in the outer shell determines the chemical properties of an element. Group I metals are very reactive because they can easily lose their outer shell electron.
- The elements in Group I have similar chemical properties because they have the same number of outer shell electrons. The same goes for some other Groups, for example Group II and Group VII.
- The Group VIII elements (noble gases) have a 'full' outer shell of electrons, the maximum number of

Worked example 4.02

a Draw the electronic structures of a chlorine atom and a chloride ion, Cl^-. **[3]**

Use your Periodic Table to find out the number of electrons in a chlorine atom. This is the same as the proton number (17). Then fill the shells starting from the lowest until you get this number until you get to 17. A chloride ion has one more electron than a chlorine atom.

b Explain in terms of electronic structure why chloride ions and bromide ions are both stable and have similar chemical properties. **[5]**

There are two parts to this question: the stability and the chemical properties. Make sure that you answer both. The stability relates to the noble gas structure and the reactivity to the number of electrons in the outer shell.

Chlorine and bromine are in Group VII so the electronic structure of their ions is similar **[1]**. Both ions have eight electrons in their outer shell **[1]**. This is the noble gas structure **[1]** which is stable **[1]**. Since both ions have the same number of electrons in the outer shell they have similar chemical properties **[1]**.

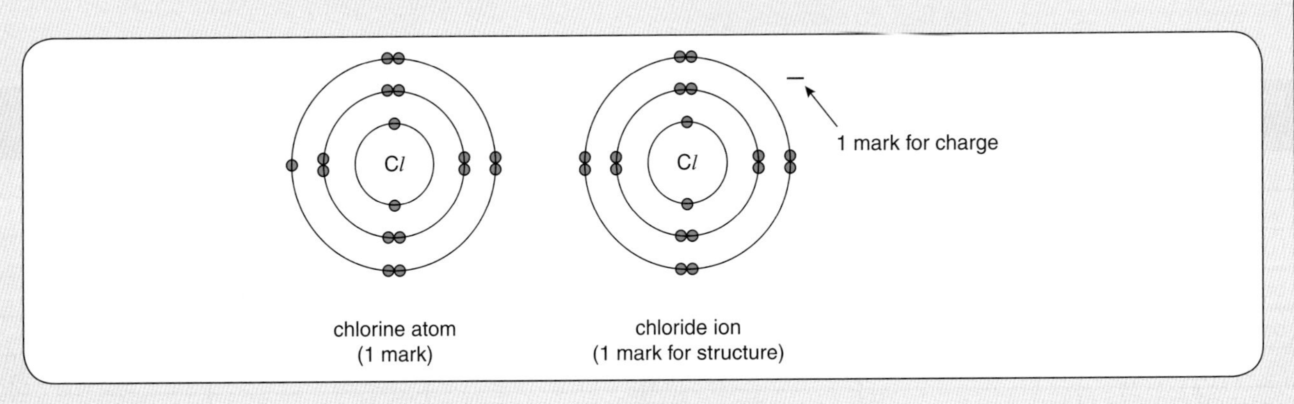

Figure 4.05

electrons the shell can hold (two for helium and eight for the others). This is called the **noble gas structure**.

- A full outer shell of electrons makes these atoms stable because it takes a lot of energy to remove an electron from a full shell.

TERMS

Noble gas structure: The electronic structure of ions or atoms with a complete outer shell of electrons.

Progress check

4.05 Write the electronic structure of

a a silicon atom

b a calcium atom

c a fluorine atom. [3]

4.06 How many electrons are in the second electron shell, when the third shell starts filling? [1]

4.07 a Write the electronic structures of an aluminium ion, Al^{3+} and an oxide ion, O^{2-}. [2]

b Explain why these ions are stable. [2]

Exam-style questions

Question 4.01

The electronic structures of the atoms of four elements are:

A 2,8,4 B 2,2 C 2,8 D 2,8,8,1

a Which one of these elements is in Group I of the Periodic Table? Explain your answer. [2]

b Which one of these elements has the lowest proton number? Explain your answer. [2]

c Which one of these elements is in Period 4 of the Periodic Table? Explain your answer. [2]

d Which element is a noble gas? Explain your answer. [2]

e Element A has three naturally-occurring isotopes

i What is meant by the term *isotope*? [1]

ii An isotope of A has a nucleon number of 30. State the number of electrons, protons and neutrons in this isotope. [3]

Question 4.02

Two isotopes of bromine are $^{79}_{35}Br$ and $^{81}_{35}Br$.

a Deduce the number of neutrons in each of these isotopes. [1]

b Bromine has the electronic structure 2,8,18,7.

i Explain how this structure shows that bromine is in Group VII of the Periodic Table. [1]

ii Explain how this structure shows that bromine has a proton number of 35. [2]

c Magnesium reacts with bromine to form magnesium bromide.

i Write the electronic structure for a bromide ion. [1]

ii Write the electronic structure for a magnesium ion, Mg^{2+}. [1]

Revision checklist

You should be able to:

- ■ State the relative charges and approximate relative masses of protons, neutrons and electrons
- ■ Define nucleon number (mass number) as the total number of protons and neutrons in the nucleus of an atom
- ■ Explain the basis of the Periodic Table in terms of the number of protons
- ■ Explain that isotopes are atoms of the same element with different numbers of neutrons
- ■ Understand that isotopes have the same properties because they have the same number of electrons in their outer shell
- ■ State the two types of isotopes as being radioactive and non-radioactive
- ■ State one medical and one industrial use of radioactive isotopes
- ■ Describe the electronic structure of atoms
- ■ Understand the importance of the noble gas electronic structure

Elements, compounds and mixtures

Learning outcomes

By the end of this unit you should:

- Know the difference between physical and chemical changes
- Be able to describe the differences between elements, compounds and mixtures
- Be able to describe the differences between metals and non-metals
- Be able to describe an alloy as a mixture of a metal with other elements

5.01 Physical and chemical changes

TERMS

Physical change: A change in a physical property, for example melting, boiling.

Chemical change: How elements or compounds react with other substances.

- Physical properties do not generally depend on the amount of substance present. They do not involve a chemical reaction. Examples include density, melting point, electrical conductivity, hardness.
- A **physical change** involves a change in a physical property, for example melting and boiling are physical changes
- Chemical properties and **chemical changes** describe how elements or compounds react with other substances, for example magnesium reacts with oxygen to form magnesium oxide.

5.02 Elements

TERMS

Element: A substance containing only one type of atom.

- A chemical **element** contains only one type of atom in its structure.
- An element cannot be broken down into anything simpler by chemical reactions.

The structures of the elements in Figure 5.01 look different but each one of them only has one type of atom.

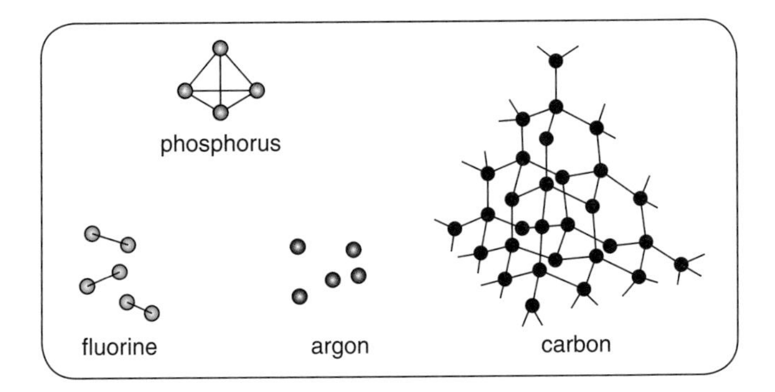

Figure 5.01

5.03 Compounds

TERMS

Compound: A substance made up of two or more different atoms (or ions) bonded together.

- When a substance contains two or more different types of atoms or ions which are bonded together it is called a **compound**.
- A compound has a fixed amount of each element in it. For example, a carbon dioxide molecule always has one carbon atom and two oxygen atoms.

When defining a compound, there are two things to remember: (1) there are two or more different atoms (2) the atoms are bonded (joined) together.

- Compounds usually have different properties from the elements from which they are made. The table shows how the properties of the elements sodium and chlorine differ from those of the compound sodium chloride.

Property	Sodium	Chlorine	Sodium chloride
state	solid	gas	solid
colour	silvery	green	white
boiling point / °C	883	−35	801
reaction with water	reacts to form an alkaline solution	reacts to form an acidic solution.	does not react – just dissolves

Table 5.01

5.04 Mixtures

Pure water is only made up of one substance (component). Pure sodium chloride (salt) has only one component in it.

- A **mixture** consists of two or more elements or compounds which are not chemically bonded together. For example, air is a mixture because it contains several components including oxygen, nitrogen and carbon dioxide.
- The components (parts) of a mixture can usually be separated by physical means, e.g. chromatography, distillation, crystallisation.
- A mixture can have varying numbers of each type of atom because you can change the ratio of the components. (In a compound there is always a fixed ratio of the different types of atom.)

TERMS

Mixture: An impure substance which contains two or more different components.

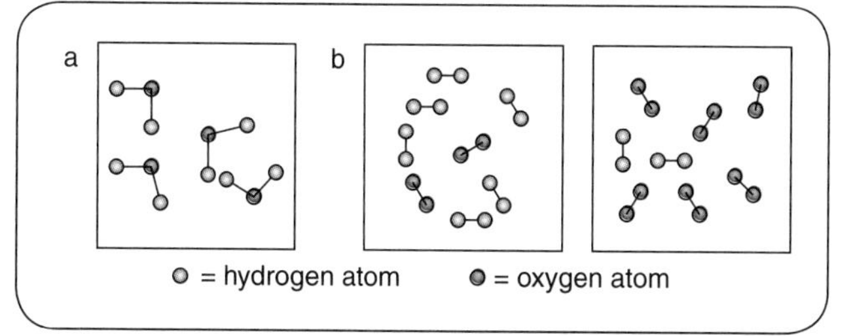

Figure 5.02 The difference between **a** a compound of oxygen and hydrogen (water) and **b** two mixtures of oxygen and hydrogen

A mixture can have varying numbers of each type of atom because you can change the ratio of the substances added together.

Sample answer

Suggest the most suitable method of obtaining purified samples of both substances in the following mixtures. In each case describe and explain why you chose the particular method.

a A solution of sodium chloride in water. **[4]**

Water is more volatile than sodium chloride [1] so simple distillation is used [1]. Water is the distillate [1] and sodium chloride remains in the flask [1].

b A mixture of ethanol (alcohol) and water. **[3]**

The two liquids have different boiling points [1], so fractional distillation is used [1]. The liquid with the lower boiling point comes off first from the distillation column [1].

c A mixture of sulfur (insoluble in water) and solid sodium chloride (soluble in water). **[6]**

We first need to dissolve the sodium chloride in the mixture by adding water and stirring [1] to dissolve the sodium chloride [1]. Filtration [1] is then used to filter off solid sulfur, which stays on the filter paper [1]. The solution of sodium chloride is heated to evaporate the water [1]. Crystals of sodium chloride remain in the evaporating basin [1].

Progress check

5.01 Which of the following are physical changes?

a the reaction of nitric acid with calcium carbonate,

b the sublimation of carbon dioxide,

c water vapour condensing,

d the formation of magnesium oxide when magnesium is heated in air. [2]

5.02 What is meant by the term *element*? [1]

5.03 Give two differences between a mixture and a compound. [4]

5.04 A solution of salt in water is a mixture. Explain why and describe an experiment to demonstrate your answer. [4]

5.05 Comparing compounds and mixtures

Compound (pure substance)	Mixture (impure substance)
composition has a fixed ratio of atoms	can have any composition
cannot be separated by physical means	can be separated by physical means
physical properties are different from the elements from which they are made	physical properties, e.g. colour, density, are the average of the substances in the mixture
compounds are formed by a chemical change	when mixtures are formed there is no chemical change
heat is absorbed or released when a compound is formed	heat is not usually absorbed or released when a mixture is formed (except when substances dissolve to form solutions)

Table 5.02

Worked example 5.01

Iron is a silvery metal and sulfur is a yellow non-metal. The compound iron sulfide is formed when iron and sulfur are heated together.

a Copy and complete the table to show the results of the tests. **[2]**

b Suggest the appearance of a mixture of iron and sulfur. **[1]**

c What information in the table suggests that iron sulfide is a compound and not a mixture? **[2]**

a

Iron and sulfur	Iron sulfide
The iron can be separated from the sulfur using a magnet.	*You need to recognise that the components in a compound cannot be separated because they are bonded together.* The iron cannot be separated from the sulfur. **(1 mark)**
The sulfur can be dissolved in an organic solvent leaving the iron as a solid.	Iron sulfide does not dissolve in organic solvents
You need to recognise that there is no energy change when solids are mixed. No heat is given off. **(1 mark)**	When the compound is formed, heat is given off.

Table 5.03

b *The appearance of a mixture is in-between that of the individual colours and in mixtures of solids, grains may be seen.* Yellow with silvery specks / silvery with yellow specks / grey **[1]**

c *Any piece of information from the table can be used but you must make clear that you are referring to both the compound and the mixture because both are given in the question, for example:*

The sulfur in the mixture dissolves in organic solvent **[1]** but the iron sulfide does not **[1]**.

5.06 Metals and non-metals

Metals can be distinguished from non-metals by differences in their physical properties.

Physical property	Metals	Non-metals
electrical and heat conduction	conducts	do not conduct (exception – carbon as graphite).
lustre (shininess)	lustrous	dull surface (exceptions – iodine and graphite).
malleability (can be beaten into different shapes)	malleable	not malleable; brittle when hit
ductility (can be drawn into wires)	ductile	not ductile; break easily when a pulling force is applied
sonorous (make a ringing sound when hit)	sonorous (there are exceptions)	not sonorous; make a dull sound when hit

Table 5.04

TERMS

Ductile: Can be drawn into wires by a pulling force.

Malleable: Can be beaten into different shapes.

Lustrous: Shiny like a mirror.

Other properties which are less useful for distinguishing metals from non-metals are the following:

- Density: many metals have a high density (exceptions are the Group I metals). Most non-metals have low densities.
- Melting and boiling points: many metals have high melting and boiling points (exceptions include the Group I metals and some others, e.g. mercury, gallium). Most non-metals have low melting points (exceptions are carbon, silicon, boron).
- Hardness: many metals are hard (exceptions include the Group I metals). Most non-metals are soft (exceptions diamond (carbon), boron).

TIP

When answering questions about the physical properties of metals, it is best to select the properties that are common to all of them. For example, it would be wrong to suggest that sodium is hard and has a high melting point.

We can also use chemical properties to distinguish metals from non-metals:

- Many metallic oxides are basic. Many non-metallic oxides are acidic.
- Many metals reacts with acids to produce hydrogen. Most non-metals do not react with acids.
- When they react, metals form positive ions by losing electrons. Non-metals form negative ions by gaining electrons. (Hydrogen is an exception because it can form positive ions.)

Alloys

TERMS

Alloy: A mixture of a metal with another element or elements.

- Most alloys are mixtures of metals, e.g. brass is an alloy of copper with zinc.
- Some alloys are mixtures of metals with non-metals, e.g. mild steel is a mixture of iron with a small amount of carbon.

Progress check

5.05 Give two physical properties of most non-metals which are not shown by most metals. [2]

5.06 Give the name of two metals which are soft. [2]

5.07 Give the name of a non-metal which conducts electricity. [1]

5.08 Explain why an alloy of iron is not a pure element. [1]

Exam-style questions

Question 5.01

Sodium is a soft, shiny metal with a fairly low melting point. Bromine is a red liquid with a low boiling point. Sodium reacts with bromine to form sodium bromide, a white, crystalline solid with a high melting point.

a i Describe three properties shown by all metals. [3]

ii Give two properties of sodium that are not typical of most metals. [2]

b Give two reasons why bromine is a non-metal. [2]

c Suggest why sodium bromide is a compound of sodium and bromine and not a mixture of sodium and bromine. [4]

Question 5.02

Sulfur is a yellow non-metal that does not react with acids. Iron is a silvery, magnetic metal that reacts with hydrochloric acid to produce an odourless gas. When heated together, iron and sulfur react to produce iron sulfide. Iron sulfide is non-magnetic when pure. It is a black solid which reacts with hydrochloric acid to produce a bad-smelling gas.

a Iron sulfide is a compound. What is the meaning of the term *compound*? [1]

b Sulfur is a non-metal. Describe three properties which are typical of most non-metals. [3]

c Describe two differences between a mixture of iron and sulfur and a compound of iron and sulfur. [4]

Revision checklist

You should be able to:

- Understand the difference between physical and chemical properties
- Define the terms *element* and *compound*
- Distinguish compounds from mixtures in terms of their composition and properties
- Distinguish metals from non-metals in terms of conductivity, malleability, ductility and lustre
- Describe an alloy, such as brass, as a mixture of a metal with other elements

Bonding and structure

Learning outcomes

By the end of this unit you should:

- Describe the formation of ions by electron loss or gain
- Describe the formation of ionic bonds between elements from Groups I and VII
- Describe the formation of ionic bonds between metallic and non-metallic elements
- Know that ionic compounds form lattice structures containing positive and negative ions
- Describe the formation of single covalent bonds in H_2, Cl_2, H_2O, CH_4, NH_3 and HCl as the sharing of pairs of electrons leading to the noble gas configuration
- Be able to describe the electron arrangement in more complex covalent molecules
- Know how ionic and simple covalent compounds differ in volatility, solubility and electrical conductivity
- Be able to explain why ionic and simple molecular compounds differ in their melting and boiling points by referring to the forces between their particles
- Be able to describe the giant covalent structures of diamond and graphite and relate their structures to their uses
- Be able to describe the structure of silicon(IV) oxide (silicon dioxide) and relate the similarity of its properties to those of diamond
- Be able to describe metallic bonding and explain why metals are malleable and conduct electricity

6.01 The formation of ions

TERMS

Ion: An electrically charged particle formed from an atom or group of atoms by loss or gain of electrons.

- Positive **ions** are formed by the loss of one or more electrons. For example:

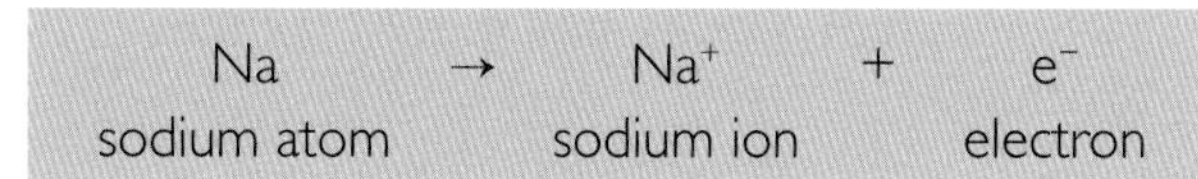

Na	→	Na^+	+	e^-
sodium atom		sodium ion		electron

- Negative ions are formed by the gain of one or more electrons. For example:

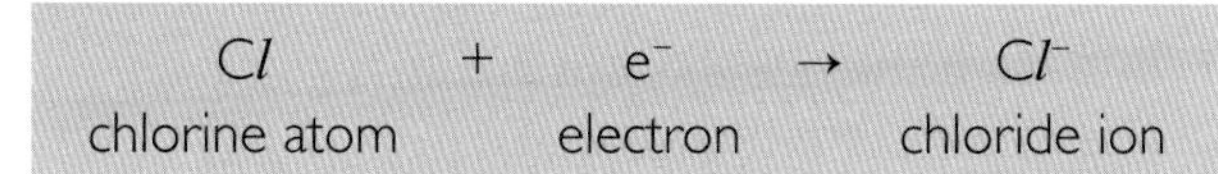

Cl	+	e^-	→	Cl^-
chlorine atom		electron		chloride ion

- The charge on the ion depends on the number of electrons lost or gained. For most metal ions the number of positive charges is the same as the group number. So aluminium in Group III forms 3+ ions:

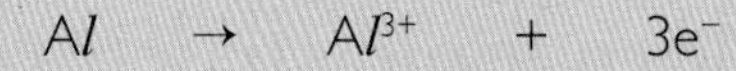

$$Al \rightarrow Al^{3+} + 3e^-$$

- For non-metal ions, the negative charge is eight minus the group number. So oxygen in Group VI forms 8 − 6 = 2^- ions (oxide ions):

$$O + 2e^- \rightarrow O^{2-}$$

- Ions are formed by the transfer of electrons from one atom to another. We can show this by a dot-and-cross diagram (Figure 6.01).
- The ions formed in electron transfer have the electronic structure of the nearest noble gas. This is a stable electronic structure (see Figure 6.01).
- **Ionic bonds** are formed in a similar way between other elements from Group I and VII. Figure 6.02 shows the formation of lithium fluoride by transfer of an electron from a lithium atom to a fluorine atom. Only the outer shell electrons are shown.

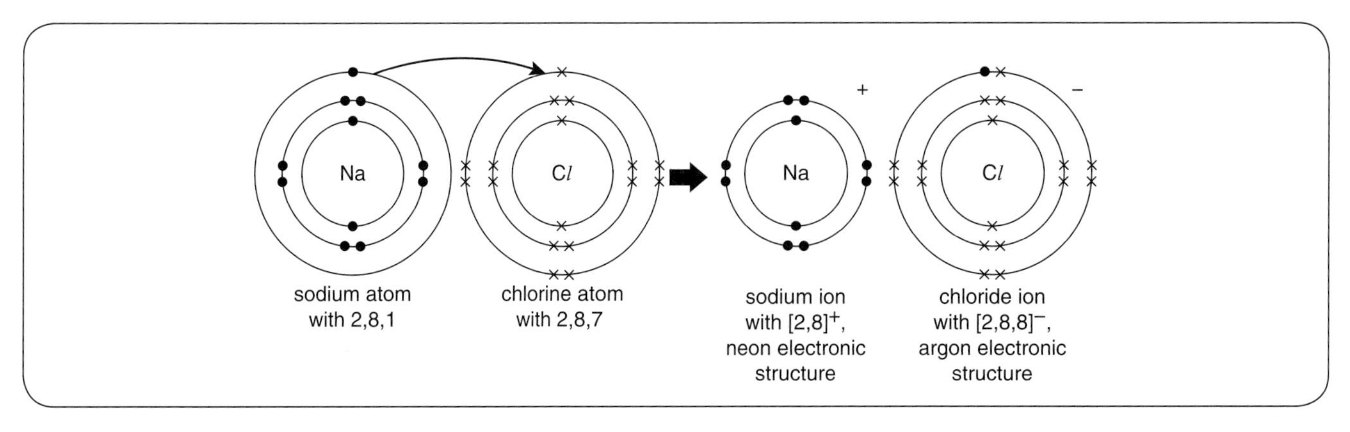

Figure 6.01

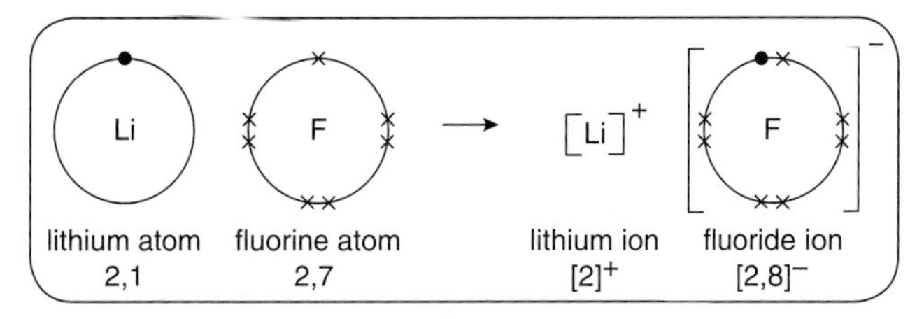

Figure 6.02

TERMS

Ionic bond: An ionic bond is a strong bond formed by the electrostatic attraction between positive and negative ions in an ionic structure.

6.02 More ionic structures

- Metals form positive ions and non-metals form negative ions.
- Ionic bonds are formed between reactive metals and reactive non-metals.

Example 1: Magnesium oxide

The two electrons in the outer shell of the magnesium atom are transferred to the oxygen atom.

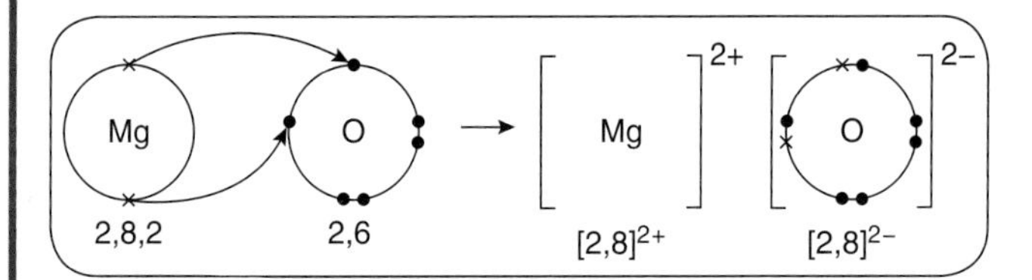

Figure 6.03

Example 2: Calcium chloride

A calcium atom loses its two outer electrons. A chlorine atom has only space in its outer shell for one electron. So, two chlorine atoms are needed to react with one calcium atom. Each chlorine atom gains one electron.

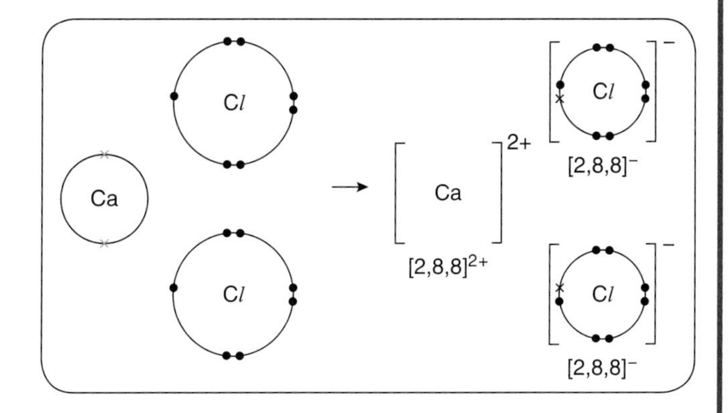

Figure 6.04

6.03 Ionic lattices

TERMS

Lattice: A continuous regular arrangement of particles which repeats itself throughout the structure.

- An ionic **lattice** has a lattice of positive and negative ions.
- In sodium chloride the ions are arranged in an alternating pattern (Figure 6.05).
- The ionic lattice is held together by strong electrostatic attractions.

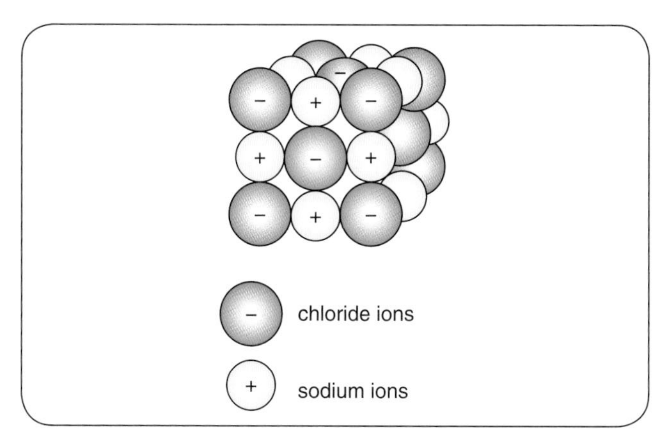

Figure 6.05

6.04 Covalent bonding

TERMS

Covalent bond: A bond formed by sharing a pair of electrons between two atoms.

Molecule: A particle having two or more atoms joined by covalent bonds.

A **covalent bond** between two hydrogen atoms in a hydrogen **molecule** is shown in Figure 6.06.

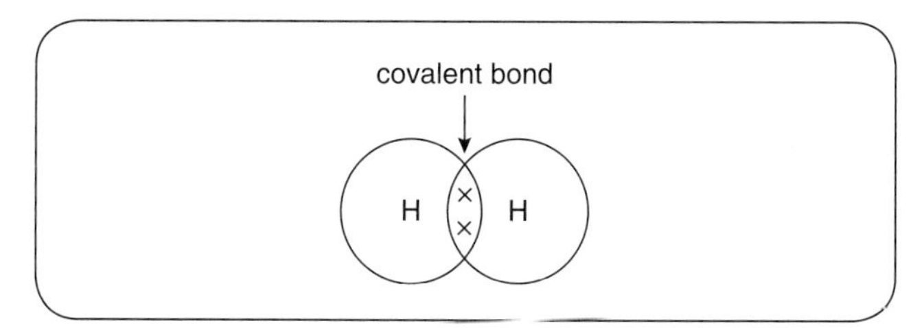

Figure 6.06

- A single covalent bond is shown by a line between the two atoms, e.g. H–H.
- The shared electrons in the covalent bond usually come from the outer shell of the atoms which combine.
- When some non-metal atoms combine, not all the electrons in the outer shell form covalent bonds. The pairs of electrons not used in covalent bonding are called **lone pairs**.

TERMS

Lone pair: A pair of electrons in the outer shell of an atom in a molecule which does not form a covalent bond.

- Some atoms are able to share two pairs of electrons to form a double bond. We show this by a double line. For example, O=O.

6.05 Drawing dot-and-cross diagrams for simple molecules

To draw a dot-and-cross diagram for a molecule:

- We usually show only the outer shell electrons.
- We use a dot for electrons from one of the atoms and a cross for the electrons of the other.
- The electrons are arranged so that the number of outer shell electrons in each atom corresponds to the nearest noble gas electronic structure.
- It is often helpful to show electrons in pairs. In the outer shell of a noble gas there are four pairs of electrons (or one pair for helium). Figure 6.07 shows how to pair the electrons in some simple molecules.
- Notice that in ammonia, NH_3, one pair of outer shell electrons on the nitrogen atom does not form a covalent bond.

Progress check

6.01 a Describe the formation of bromide ions from bromine atoms and potassium ions from potassium atoms. [2]

b Copy and complete the equation for the formation of lithium ions from lithium.

$$\ldots\ldots \rightarrow Li^{+} + \ldots\ldots$$ [2]

6.02 Draw a dot-and-cross diagram for:

a hydrogen chloride [2]

b water. [2]

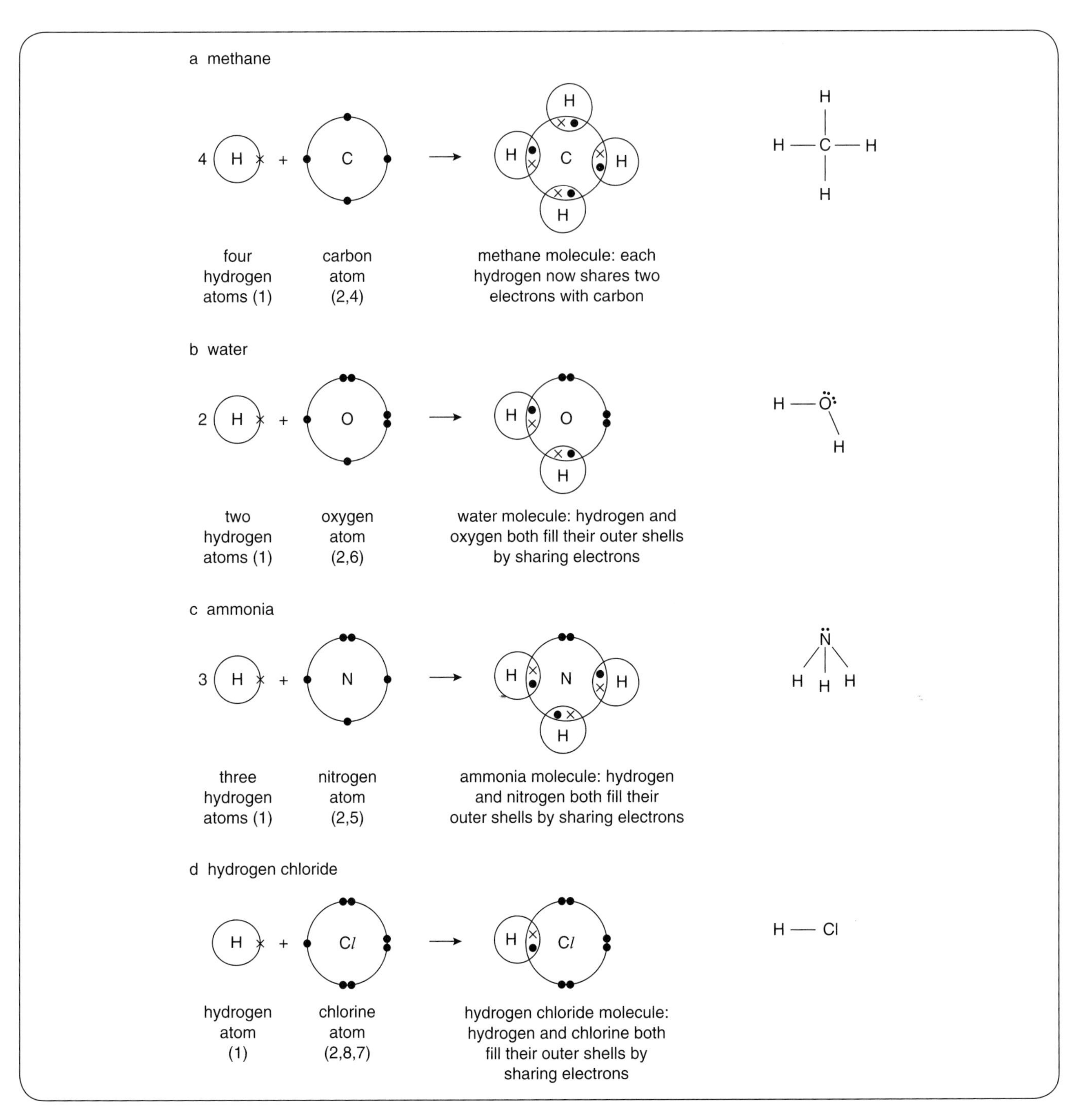

Figure 6.07

6.06 More complex covalent molecules

In an oxygen molecule, each oxygen atom (2,6) gains two electrons to complete its outer shell. It can only do this by sharing two pairs of electrons. A double bond is formed (Figure 6.08). When three pairs of electrons are shared, as in the nitrogen molecule, a triple bond is formed.

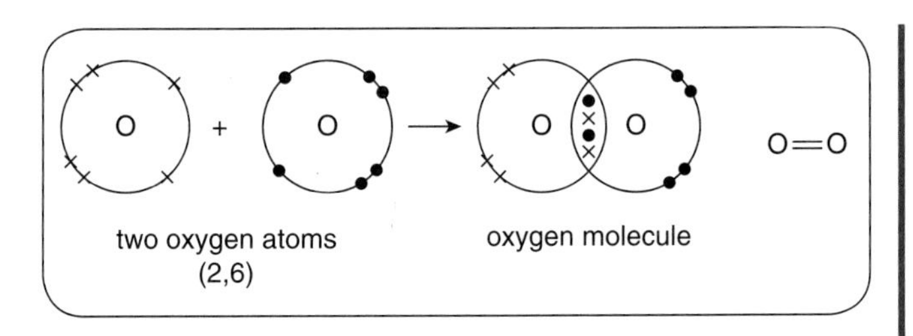

Figure 6.08

Figure 6.09 shows the dot-and-cross diagrams for some more complex molecules.

Note that if there are more than three types of atom in the molecule, you can use other symbols such as open circles or squares to show the origin of the electrons.

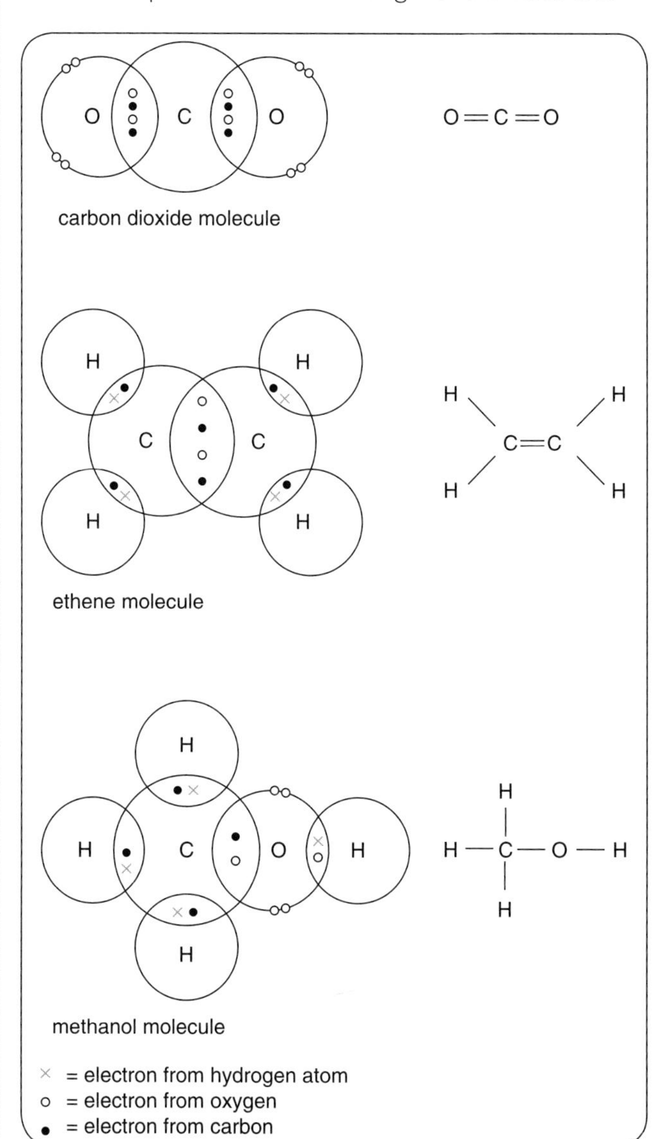

Figure 6.09

Worked example 6.01

Draw a dot-and-cross diagram to show the arrangement of the outer electrons in phosphorus trichloride, PCl_3. **[2]**

How to get the answer:

Step 1. Draw the electronic structure of a phosphorus atom and three chlorine atoms showing only the outer shell electrons (see Figure 6.10 ***a****). Refer back to Unit 4 if you are unsure how to do this.*

Step 2. Arrange the chlorine atoms so that they are around the central phosphorus atom (see Figure 6.10 ***b****).*

Step 3. Pair up the electrons, one from each chlorine atom and one from the phosphorus so that each atom has a noble gas structure. In this case it is four pairs of electrons (eight electrons) around each atom (see Figure 6.10 ***c****).*

Step 4. Make sure that there are eight electrons around the phosphorus atom by including the lone pair (see Figure 6.10 ***c****).*

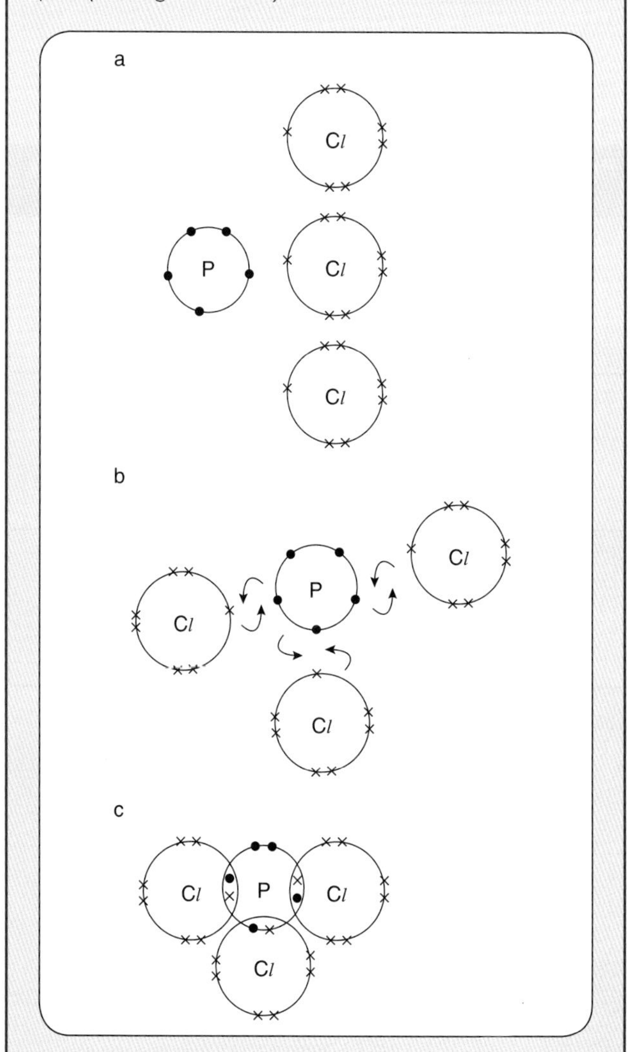

Figure 6.10

There is one mark for the correct pairing of each chlorine to the phosphorus and one mark for the lone pair on the phosphorus.

6.07 Differences between ionic and simple covalent compounds

- The particles present in ionic compounds are positive and negative ions.
- Examples of ionic compounds are salts, e.g. sodium sulfate, ammonium chloride and some metal oxides, and hydroxides, e.g. calcium oxide, sodium hydroxide.
- The particles present in simple covalent compounds are molecules.
- Examples of simple covalent molecules are octane and carbon dioxide.
- We can distinguish between ionic and simple molecular compounds by considering their volatility, solubility and electrical conductivity.

Property	Ionic compound	Simple covalent compound
volatility	Not volatile. They have high boiling points. They also have high melting points.	Many are volatile. Most have low boiling points. They also have low melting points.
solubility	Most are soluble in water but insoluble in organic solvents such as cyclohexane.	Most are insoluble in water but soluble in organic solvents. There are some exceptions.
electrical conductivity	The solids do not conduct. When molten or dissolved in water they do conduct.	Do not conduct in the solid or liquid state or in aqueous solution. There are several exceptions.

Table 6.01

- Some simple covalent compounds, e.g. ethanol, are soluble in water and not in organic solvents.
- Some simple covalent compounds, e.g. hydrogen chloride, react with water and conduct electricity in aqueous solution. This is because they form ions.

Remember that non-metallic elements having molecules, for example iodine and sulfur are simple covalent structures. These elements are volatile, insoluble in water and do not conduct electricity.

Worked example 6.02

Sodium chloride is an ionic compound.

a Describe how ions are formed from sodium and chlorine atoms. **[2]**

b Describe the electrical conductivity of solid and molten sodium chloride as well as the electrical conductivity of sodium metal and chlorine gas. **[4]**

How to get the answer:

a *Sodium is a metal. Metal ions have a positive charge. The charge on the sodium is positive because the uncharged sodium atom has lost an electron. So you just need to write:* Sodium ion is formed by loss of an electron **[1]**.

Chlorine is a non-metal. Non-metal ions have a negative charge. The charge on the chloride ion is positive because the uncharged chlorine atom has gained an electron. So you just need to write: A chloride ion is formed by the gain of an electron **[1]**.

b *Sodium chloride is an ionic solid. For a substance to conduct electricity it needs moving ions or electrons flowing through the whole structure. The ions in solid sodium chloride cannot move because they are in fixed positions. So 1 mark is awarded for* 'solid sodium chloride does not conduct electricity' **[1]**.

When ionic compounds are molten (liquid) the ions are free to move. So 1 mark would be awarded for 'molten sodium chloride conducts electricity' **[1]**.

Sodium is a metal. All metals conduct electricity. So 1 mark is awarded for 'sodium conducts electricity' **[1]**.

Chlorine is a non-metal and is a simple molecular compound. Non-metals (except graphite) do not conduct electricity. So 1 mark would be awarded for 'chlorine does not conduct electricity' **[1]**.

Progress check

6.03 A compound which is soluble in water has a melting point of 801 °C. What type of structure does this substance have? [1]

6.04 Compound X has a simple molecular structure. Suggest the properties of compound X in terms of volatility, electrical conductivity and solubility in water. [3]

6.08 Explaining differences in properties

- Ionic compounds have high melting and boiling points because of the strong attractive forces between the ions. It takes a lot of energy to overcome these forces.
- Simple covalent compounds have low melting and boiling points because the forces between their molecules (**intermolecular forces**) are weak.

TERMS

Intermolecular forces: The weak forces between molecules.

- Ionic compounds do not conduct electricity when in the solid state. This is because the ions cannot move. They do conduct in the molten (liquid) state or when dissolved in water because the ions separate out and can move.
- Simple covalent compounds do not conduct electricity when solid or molten. This is because they do not have ions or electrons to conduct. Molecules are not electrically charged.

TIP

Make sure that you use the phrase 'intermolecular forces' correctly. The forces between molecules are weak but the forces between the atoms within the molecule are strong covalent bonds.

Sample answer

a Sodium oxide, Na_2O is an ionic compound with a lattice structure.

Draw a dot-and-cross diagram for sodium oxide showing only the outer electron shells. Include the correct ionic charges and the correct number of ions. **[4]**

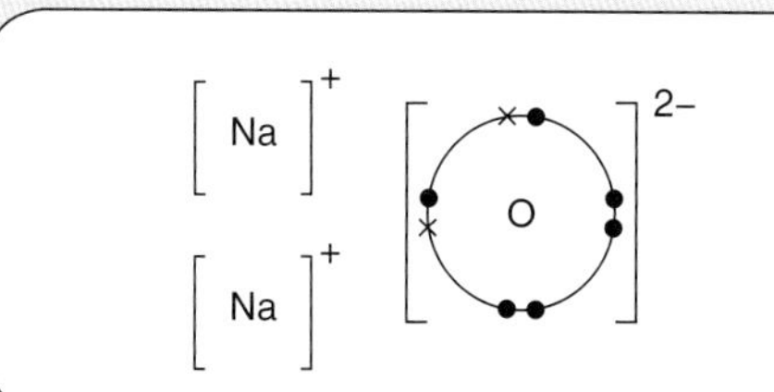

Fig 6.11

1 mark for showing 2 sodium ions and 1 oxide ion.

1 mark for the correct electronic configuration in each ion.

1 mark for each correct charge Na^+ **[1]** and O^{2-} **[1]**.

b i What is meant by the term ionic lattice? **[3]**

A regularly repeating arrangement **[1]** of positive and negative ions **[1]** which repeats itself throughout the structure **[1]**.

ii Explain why sodium oxide has a high melting point and does not conduct electricity at room temperature. **[4]**

Sodium oxide has a high melting point because it has strong forces between the ions **[1]** because the positive and negative ions attract each other electrostatically **[1]**. It needs a lot of energy to overcome these strong forces **[1]**.

It does not conduct because the ions are not free to move **[1]** in the solid.

6.09 Giant covalent structures

TERMS

Giant covalent structures (macromolecular structures): Structures with a lattice (network) of covalent bonds which repeats throughout the whole structure.

Diamond and graphite are both **giant covalent structures** made of carbon atoms (Figure 6.12).

In diamond, each carbon atom forms four covalent bonds with other carbon atoms. This network of strong bonds extends unbroken throughout the whole structure. In graphite, the carbon atoms are hexagonally arranged in layers.

TIP

Make sure that you do *not* use the term intermolecular forces when referring to diamond. If referring to graphite, only use the term intermolecular forces when writing about the forces *between* the layers.

Similar properties of diamond and graphite

The properties of diamond and graphite can be explained by their structure and bonding:

- They have high melting and boiling points. It is difficult to break down the network of strong covalent bonds. It needs a high temperature to overcome the huge numbers of strong bonds and melt the solid.
- They are insoluble in water and in organic solvents. The network of covalent bonds is too strong to allow solvent molecules to form strong enough bonds with the carbon atoms.

When answering questions about why giant structures have high melting points it is incorrect to write about strong forces between *molecules*. The best answers refer to strong bonds between the *atoms*.

How the properties of diamond and graphite differ

Hardness:

- Diamond is hard. The network of strong covalent bonds makes it difficult to scratch the surface of the crystal. Because it is so hard, diamond is used for the edges of tools such as glass cutters and drill bits.
- Graphite is soft. It is easy to scratch. The forces between the layers of graphite are weak. The layers can slide over each other when a force is applied. The layers flake away easily. So graphite is used as a lubricant and in the 'leads' of pencils.

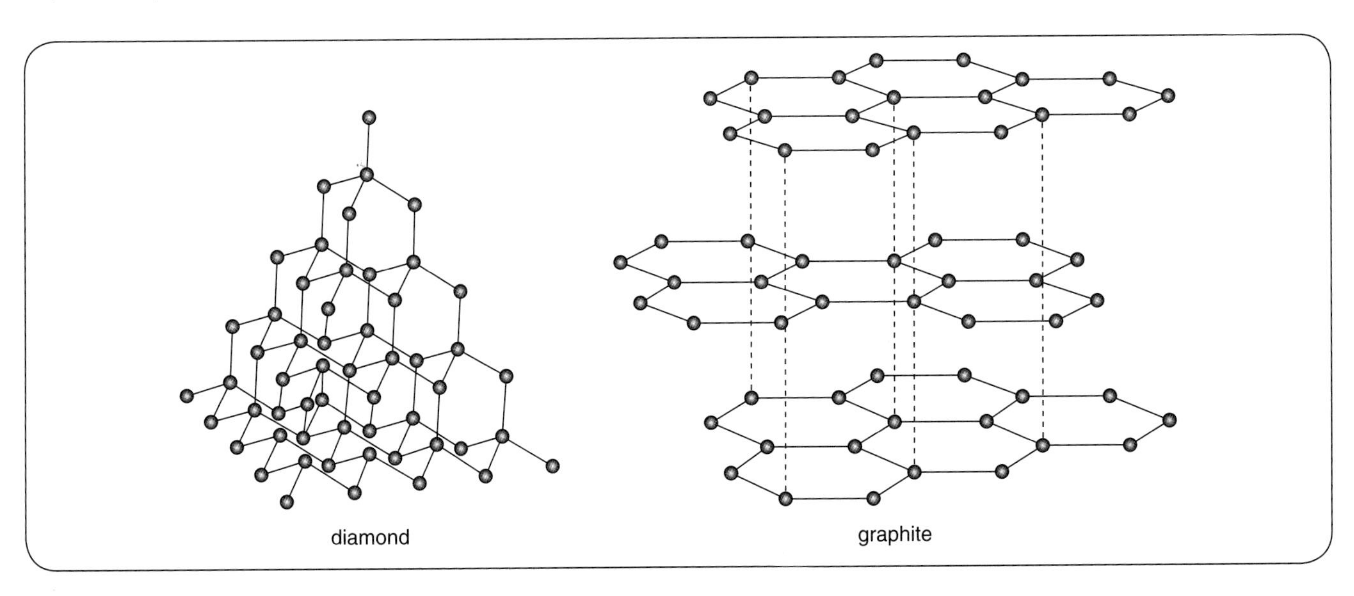

Figure 6.12

Electrical conductivity:

- Diamond does not conduct electricity. It is a covalent compound with no ions. None of the electrons is free to move along the structure because they are all involved in bonding.
- Graphite conducts electricity. Three of the electrons in each carbon atom are used to form covalent bonds. The fourth electron in each atom is free to move around and along the layers. These electrons are called **delocalised electrons**. They can also be called free or mobile electrons. When a voltage is applied, the delocalised electrons can move along the layers.

TERMS

Delocalised electrons: Electrons which are not associated with any particular atom and are able to move between atoms or ions.

Sample answer

Ammonia, NH_3, is a simple covalent molecule. Graphite is a macromolecule with a giant covalent structure. Compare the volatility and electrical conductivity of these two molecules and draw a dot-and-cross diagram for ammonia. **[7]**

Ammonia is volatile and has a low boiling point **[1]**. Graphite is a giant structure so has a high boiling point **[1]**. Ammonia does not conduct **[1]** but graphite does **[1]**.

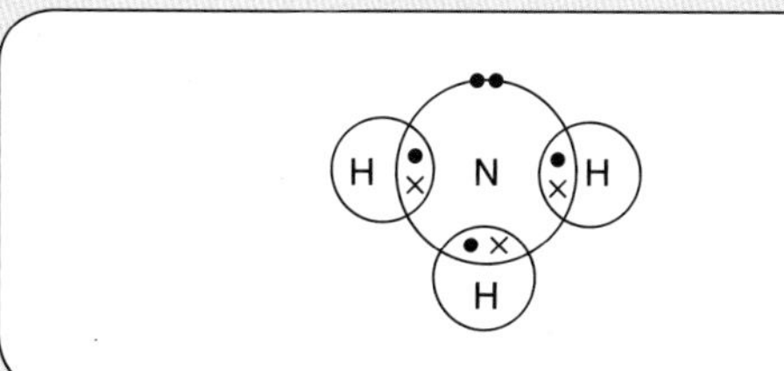

Figure 6.13

Bonding pairs of electrons between each of the three N and H atoms **[2]**.

If this not scored then a pair of bonding electrons between one of the N and H atoms **[1]**.

Lone pair of electrons on the N atom **[1]**.

Progress check

6.05 Compound T is a greyish-black solid which is insoluble in water. Compound T sublimes at 3697 °C. It conducts electricity when it is solid. Identify compound T and state its type of structure. [3]

6.06 Relate the structure of graphite to its use as a lubricant and diamond to its use in cutting tools. [4]

6.10 The structure and properties of silicon(IV) oxide

There are several forms of silicon(IV) oxide (commonly called silicon dioxide). The silicon dioxide found in quartz has a structure similar to diamond (Figure 6.14).

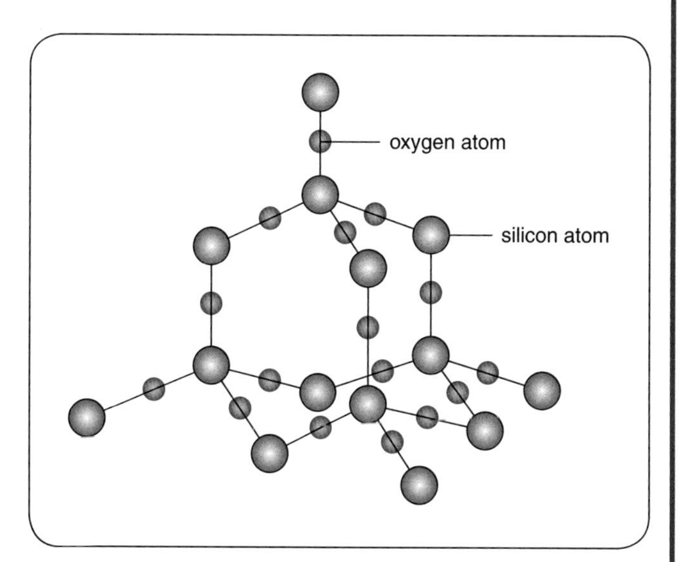

Figure 6.14

Each silicon atom is bonded to four oxygen atoms. Each oxygen atom is bonded to two silicon atoms, so the formula for silicon dioxide is SiO_2. The arrangement of the units in space is based on a tetrahedron (equal-sided pyramid with triangular base). Silicon dioxide has properties similar to diamond:

- It is very hard and has a high melting point because of the strong covalent bonding throughout the structure.

6.11 Metallic bonding

- The atoms in a metal are regularly arranged in a lattice of closely packed layers.
- Metal atoms in a lattice tend to lose their outer electrons and become positive ions.
- The electrons which are lost are free to move throughout the lattice. They form a 'sea' of delocalised electrons.
- A **metallic bond** is formed by the attractive forces between the delocalised electrons and the positive ions (Figure 6.15).

TERMS

Metallic bonding: A lattice of metal ions surrounded by a 'sea' of delocalised electrons.

TIP

When writing about the chemical properties of metals it is acceptable to refer to the ions as 'atoms'. But you must call them ions in a question about the structure of metals.

6.12 Explaining some properties of metals

Metals are malleable and ductile.

- When a force is applied to a metal, the layers slide over each other. When the pushing or pulling force is no longer present, new attractive forces can be formed between the delocalised electrons and the metal ions. So the properties of the metal stay the same.
- Metals conduct electricity when solid or molten. This is because the delocalised electrons can move between the ions in the metal lattice when a voltage is applied.

TIP

Make sure that you know the correct particles involved in electrical conduction. In metals and graphite it is delocalised electrons but in molten ionic compounds, it is the ions.

Progress check

6.07 Describe and explain the difference in melting points between magnesium oxide and sulfur dioxide. [6]

6.08 Describe metallic bonding and explain why metals conduct electricity. [5]

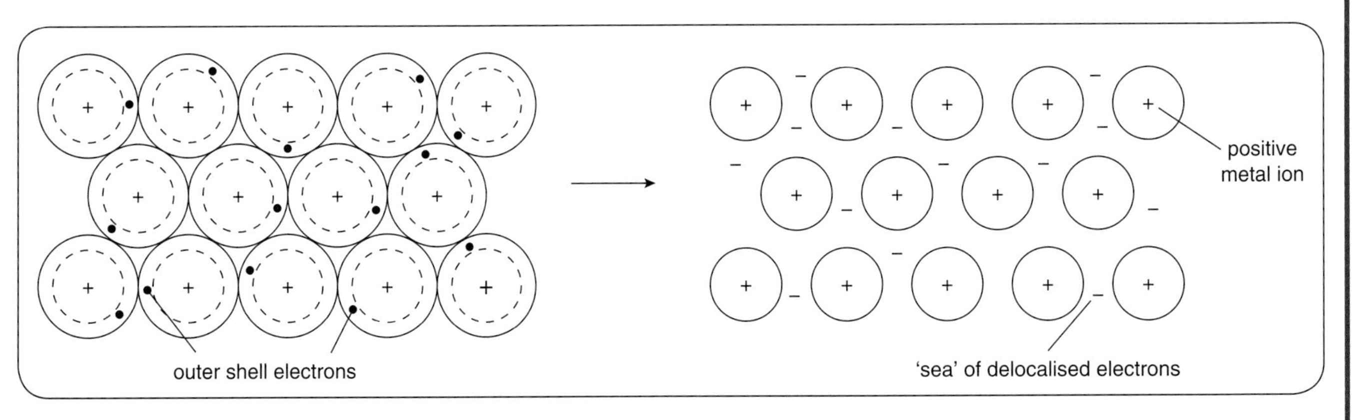

Figure 6.15

Exam-style questions

Question 6.01

Diamond and graphite are two forms of carbon.

a Describe two differences in the structure of diamond and graphite. [2]

b Explain why graphite conducts electricity but diamond does not. [2]

c Explain why diamond has a high melting point. [2]

d Methane, CH_4, is a covalent compound of carbon and hydrogen.

i Draw a dot-and-cross diagram for methane, CH_4. [1]

ii Octane is also a compound of hydrogen and carbon. Octane is a liquid at room temperature and pressure. It has a simple molecular structure. Describe its properties in terms of volatility, solubility and electrical conductivity. [3]

Question 6.02

The structures of carbon dioxide and silicon dioxide are shown in Figure 6.16.

a Explain why at room temperature and pressure, carbon dioxide is a gas but silicon dioxide is a solid with a very high melting point. [6]

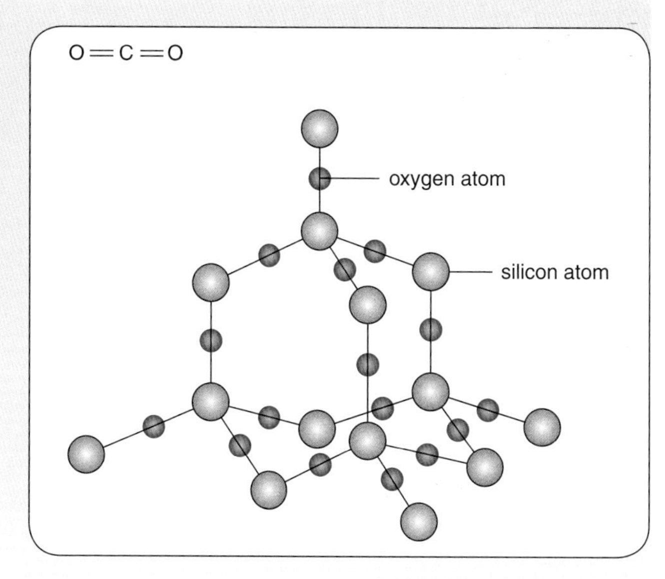

Figure 6.16

b Draw a dot-and-cross diagram to show the electronic structure of carbon dioxide. [2]

c Describe two ways in which the structure of silicon dioxide is similar to diamond. [2]

d Explain why silicon dioxide does not conduct electricity. [1]

e Silicon dioxide is found as an impurity in iron ore. Use ideas about metallic structure to explain:

i Why iron conducts electricity. [2]

ii Why iron is malleable. [4]

Revision checklist

You should be able to:

- Describe the formation of ions by electron loss or gain
- Describe the formation of ionic bonds between elements from Groups I and VII
- Describe the formation of ionic bonds between elements in Groups other than Groups I and VII
- Describe the lattice structure of ionic compounds
- Describe a covalent bond as a pair of electrons shared between two atoms
- Understand how sharing pairs of electrons leads to the noble gas electron structure of each atom
- Describe the electron arrangement in H_2, Cl_2, H_2O, CH_4, NH_3 and HCl

- Describe the electron arrangement in more complex covalent molecules such as N_2, C_2H_4, CH_3OH and CO_2
- Describe how volatility, solubility and electrical conductivity differ in ionic and covalent compounds
- Describe the giant covalent structures of graphite and diamond
- Relate the structures of diamond and graphite to their uses
- Describe the macromolecular structure of silicon(IV) oxide
- Describe the similarity in properties between diamond and silicon(IV) oxide, related to their structures
- Describe metallic bonding
- Explain the electrical conduction and malleability of metals in terms of bonding

Formulae and equations

Learning outcomes

By the end of this unit you should:

- Use the symbols of the elements and write the formulae of simple compounds
- Be able to deduce the formula of a simple compound from the relative number of atoms present or from a diagram of its structure
- Determine the formula of an ionic compound from the charges on the ions present
- Be able to construct word equations and simple balanced chemical equations
- Construct equations with state symbols, including ionic equations
- Deduce the balanced equation for a chemical reaction, given relevant information

7.01 Symbols and formulae

- Each element has its own symbol, for example C is carbon, Br is bromine, Na is sodium.
- Molecular formula: this gives the number of atoms of each particular element in one molecule of a compound or element, e.g. HCl, Br_2, H_2O.
- **Structural formula:** this shows how the atoms are bonded in a molecule. There are several ways of writing a structural formula:

```
      H   H   H
      |   |   |
  H - C - C - C - H
      |   |   |
      H   H   H
```

Full structural formulae (displayed formula)

$CH_3CH_2CH_3$

Simplified structural formula

- **Empirical formula:** this shows the simplest whole number ratio of atoms or ions in a compound. For example, the empirical formula of Al_2Cl_6 is $AlCl_3$.

TERMS

Structural formula: A formula showing how the atoms are bonded in a molecule.

Empirical formula: A formula showing the simplest whole number ratio of atoms or ions in a compound.

7.02 Naming compounds

Compounds with two different atoms:

- Metal and non-metal: the name of the metal comes first and the non-metal name changes to 'ide'. For example:

chlorine	+	sodium	→	sodium chlor<u>ide</u>
oxygen	+	magnesium	→	magnesium ox<u>ide</u>

- Two non-metals, one of which is hydrogen: the hydrogen comes first:

bromine	+	hydrogen	→	hydrogen brom<u>ide</u>

- Other combinations of non-metals: if the element has a lower Group number or is lower in a particular Group, the lower element comes first, e.g. nitrogen dioxide, sulfur dioxide.
- Some compounds are known by their common names, e.g. water, H_2O.

Compounds with particular groups of atoms:

- Sodium hydroxide, $NaOH$, contains the hydroxide ion, OH^-.
- Potassium nitrate, KNO_3, contains the nitrate ion, NO_3^-.
- Magnesium sulfate, $MgSO_4$, contains the sulfate ion, SO_4^{2-}.
- Calcium carbonate, $CaCO_3$, contains the carbonate ion, CO_3^{2-}.

7.03 Deducing formulae

We can use the idea of combining power with hydrogen (valency) of different atoms to work out the formula.

Group	I	II	III	IV	V	VI	VII	VIII
Combining power	1	2	3	4	3	2	1	0

Table 7.01

Some other combining powers are:

- hydrogen = 1
- ions = the charge on the ion, e.g.

nitrate, NO_3^- = 1; sulfate, SO_4^{2-} = 2; Fe^{3+} = 3

Worked example 7.01

What is the formula of aluminium sulfide?

How to get the answer:

Step 1. Put the combining power beneath each symbol

Al S
3 2

Step 2. Swap the numbers

Al_2S_3

Worked example 7.02

What is the formula of sodium oxide?

Step 1. Na O
1 2

Step 2. Na_2O *(We do not put the figure 1 if there is only one atom.)*

Worked example 7.03

What is the formula of magnesium nitrate?

Step 1. 2(Mg) 1(NO_3) *(the charge on the ions)*

Step 2. $Mg(NO_3)_2$ *(Note the brackets to keep the NO_3 together.)*

Deducing formulae from diagrams

Simple molecules: simply count up the number of atoms (Figure 7.01).

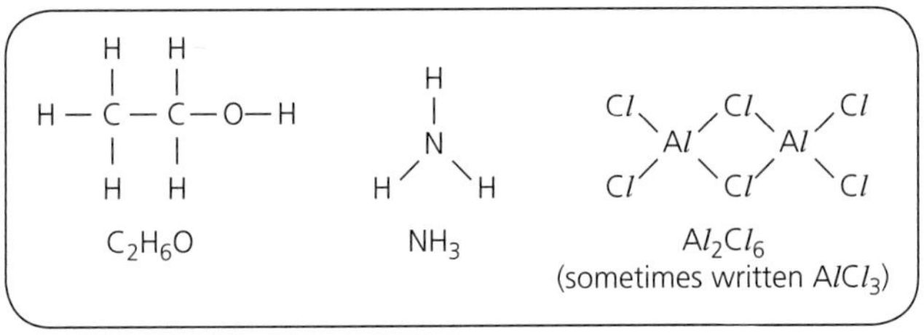

C_2H_6O NH_3 Al_2Cl_6 (sometimes written $AlCl_3$)

Figure 7.01

Ionic structures containing simple ions: count up each ion and then cancel to the lowest ratio (Figure 7.02).

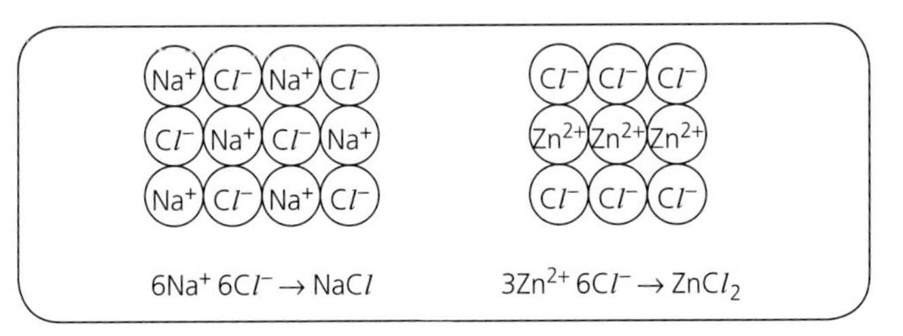

Figure 7.02

When writing the formula for an ionic compound by counting the ions, the formula is always its simplest ratio of the ions.

Deducing formulae of ionic compounds

You can write the formula of an ionic compound if you know the charge on the ions.

Some examples are:

iron(II) Fe^{2+}, copper(II) Cu^{2+}, silver(I) Ag^+, chromium(III) Cr^{3+}

Some ions contain more than one type of atom:

NH_4^+ ammonium	OH^- hydroxide	NO_3^- nitrate	SO_4^{2-} sulfate
CO_3^{2-} carbonate	HCO_3^- hydrogen carbonate		

Worked example 7.04

What is the formula of calcium bromide?

How to get the answer:

Step 1. Write the formulae of the ions: Ca^{2+} and Br^-

Step 2. Balance the charges: We need two Br^- to balance one Ca^{2+}

Step 3. Write the formula with the metal ion first: $CaBr_2$

Worked example 7.05

What is the formula of aluminium sulfate?

Step 1. Formulae of the ions: Al^{3+} and SO_4^{2-}

Step 2. Balance the charges: $Al^{3+} \times 2 = +6$ and $SO_4^{2-} \times 3 = -6$

Step 3. Write the formula: $Al_2(SO_4)_3$

Progress check

7.01 Write formulae for
- a aluminium oxide [1]
- b hydrogen sulfide [1]
- c magnesium nitrate [1]
- d silicon(IV) chloride. [1]

7.02 Write formulae for:
- a potassium sulfate [1]
- b magnesium carbonate [1]
- c aluminium nitrate. [1]

7.03 Deduce the formulae of the following compounds: [3]

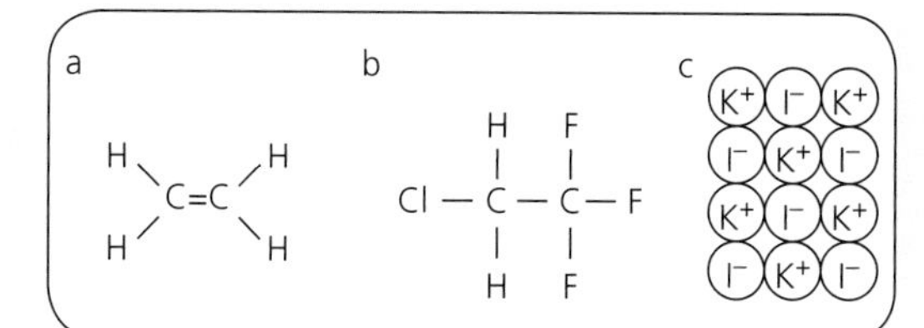

Figure 7.03

7.04 Here is a list of ions.

NH_4^+ SO_4^{2-} Fe^{2+} Ce^{3+} NO_3^-

Use this list to deduce the formulae of the following compounds.
- a ammonium sulfate
- b cerium nitrate
- c iron(II) nitrate
- d ammonium nitrate.

7.04 Word equations

- A word equation shows the reactants and products with their full chemical names.
- The arrow shows that the reactants are being converted to products.
- Particular conditions required, e.g. heat, catalyst, can be shown above the arrow.

7.05 Chemical equations (symbol equations)

A full chemical equation (often called a symbol equation) is a shorthand way of describing a chemical reaction. When writing these equations:

- The number of each type of atom in the reactants and products must be the same.
- You must not alter any of the symbols or formulae.
- The numbers used in balancing the equation must only be written in front of the formulae or symbols.
- Metals are written as their symbols e.g. Fe, Hg, Na, K. There are no subscripts.
- The following non-metals must always be written as molecules:

H_2, N_2, O_2, F_2, Cl_2, Br_2, I_2

- Other non-metals are usually written as their symbols, e.g. C, Si, P, S.
- A subscripted number multiplies the atom in front of it or all the atoms in brackets.

In $Fe_2(SO_4)_3$, there are 2 Fe atoms and 3 SO_4 units. In 3 SO_4 units there are 3 S atoms and $3 \times 4 = 12$ O atoms.

- A large number in front of a formula multiplies all the way through. So $2Ca(NO_3)_2$ has 2 Ca atoms, $2 \times 2 = 4$ N atoms and $2 \times 2 \times 3 = 12$ O atoms.

Worked example 7.06

Hydrogen reacting with oxygen to form water:

Step 1. Write the formulae for the reactants and products with an arrow between:

$$H_2 + O_2 \rightarrow H_2O$$

Step 2. Count the number of atoms of each element. Using dots or crosses may help.

In this case a dot represents a hydrogen atom and a cross an oxygen atom.

$$\begin{array}{ccccc} H_2 & + & O_2 & \rightarrow & H_2O \\ \bullet\bullet & & \times\times & & \bullet\bullet\times \end{array}$$

There are 2 oxygen atoms on the left but only 1 on the right.

Step 3. Balance the atoms by putting a number in front of one of the reactants or products. In this case the oxygen needs balancing. Then count again.

$$\begin{array}{ccccc} H_2 & + & O_2 & \rightarrow & 2H_2O \\ \bullet\bullet & & \times\times & & \bullet\bullet\bullet\bullet\times\times \end{array}$$

Step 4. Balance the other atoms. In this case it is hydrogen:

$$\begin{array}{ccccc} 2H_2 & + & O_2 & \rightarrow & 2H_2O \\ \bullet\bullet\bullet\bullet & & \times\times & & \bullet\bullet\bullet\bullet\times\times \end{array}$$

The equation is now balanced.

TIP

When balancing equations:

- You must not change any of the formulae.
- Numbers used to balance go in front of the formula.

Progress check

7.05 Write word equations for these reactions:

a $Cu(NO_3)_2 + Mg \rightarrow Cu + Mg(NO_3)_2$ [1]

b $CaO + H_2SO_4 \rightarrow CaSO_4 + H_2O$ [1]

c $Cl_2 + 2KBr \rightarrow Br_2 + 2KCl$ [1]

d $2NH_3 + H_2SO_4 \rightarrow (NH_4)_2SO_4$. [1]

7.06 Balance these equations:

a $Mg + O_2 \rightarrow MgO$ [1]

b $P + O_2 \rightarrow P_2O_3$ [1]

c $Na + H_2O \rightarrow NaOH + H_2$ [1]

d $Na + O_2 \rightarrow Na_2O$ [1]

e $SO_2 + O_2 \rightarrow SO_3$. [1]

Worked example 7.07

(More difficult example)

Barium hydroxide reacting with nitric acid to form barium nitrate and water:

Step 1. Write the formulae for the reactants and products:

$$Ba(OH)_2 + HNO_3 \rightarrow Ba(NO_3)_2 + H_2O$$

Step 2. Count the number of atoms or groups of atoms. Groups of atoms such as OH, SO_4, CO_3, NO_3 should remain together.

$$\begin{array}{ccccccc} Ba(OH)_2 & + & HNO_3 & \rightarrow & Ba(NO_3)_2 & + & H_2O \\ 1Ba\ 2OH & & 1H\ 1NO_3 & & 1Ba\ 2NO_3 & & 2H\ 1O \end{array}$$

Step 3. Balance the nitrate, NO_3.

$$\begin{array}{ccccccc} Ba(OH)_2 & + & 2HNO_3 & \rightarrow & Ba(NO_3)_2 & + & H_2O \\ 1Ba\ 2OH & & 2H\ 2NO_3 & & 1Ba\ 2NO_3 & & 2H\ 1O \end{array}$$

Step 4. Balance the hydrogen and oxygen. Note that you cannot change the formula for water.

$$Ba(OH)_2 + 2HNO_3 \rightarrow Ba(NO_3)_2 + 2H_2O$$

Sample answer

Write a balanced equation for the reaction of propane gas, C_3H_8, with oxygen to form carbon dioxide and water. **[2]**

Correct formulae: $C_3H_8 + O_2 \rightarrow CO_2 + H_2O$ **[1]**

Correct balance: $C_3H_8 + 5O_2 \rightarrow 3CO_2 + 4H_2O$ **[1]**

TIP

In equations involving combustion reactions (see sample answer) balance the carbon first, then the hydrogen. Balance the oxygen last.

7.06 State symbols

- The state symbols are (s) for solid, (l) for liquid, (g) for gas and (aq) for an aqueous solution.
- State symbols can be written after the formula for each reactant and product, e.g. $Cl_2(g)$ is chlorine gas, $Br_2(l)$ is liquid bromine, $H_2O(g)$ is steam, $NaCl(aq)$ is aqueous sodium chloride.

- State symbols can be included in both molecular and ionic equations:

$$CaCO_3(s) + 2HCl(aq) \rightarrow CaCl_2(aq) + CO_2(g) + H_2O(l)$$

$$Fe^{2+}(aq) + 2OH^-(aq) \rightarrow Fe(OH)_2(s)$$

7.07 Ionic equations

TERMS

Ionic equation: An equation showing only those ions and molecules taking part in the reaction

Spectator ions: Ions which do not take part in a reaction.

In an aqueous solution of an ionic compound, the ions are separate from each other.

Before you can write an **ionic equation**, you have to recognise the types of substances which are ionic. These include:

- soluble salts, e.g. sodium chloride, potassium sulfate
- acids, e.g. hydrochloric acid, sulfuric acid
- Group I and some Group II hydroxides, e.g. sodium hydroxide, barium hydroxide

Worked examples

7.08 Write the ionic equation for the reaction of aqueous chlorine with aqueous potassium iodide.

Step 1. Write down the balanced equation.

$$2KI(aq) + Cl_2(aq) \rightarrow 2KCl(aq) + I_2(aq)$$

Step 2. Identify the substances which are ionic and write down the ions separately.

$$2K^+(aq) + 2I^-(aq) + Cl_2(aq) \rightarrow 2K^+(aq) + 2Cl^-(aq) + I_2(aq)$$

*Step 3. Rewrite the equation, deleting the ions which are the same on each side of the equation (**spectator ions**). In this case the K^+ ions.*

$$2I^-(aq) + Cl_2(aq) \rightarrow 2Cl^-(aq) + I_2(aq)$$

7.09 If two solutions are mixed and a precipitate (solid) is formed. For example:

$$FeCl_2(aq) + 2NaOH(aq) \rightarrow Fe(OH)_2(s) + 2NaCl(aq)$$

Step 1. Write the formula of the precipitate as the product:

$$\rightarrow Fe(OH)_2(s)$$

Step 2. Write down the ions that go to make up the precipitate as the reactants:

$$Fe^{2+}(aq) + OH^-(aq) \rightarrow Fe(OH)_2(s)$$

Step 3. Balance the equation.

$$Fe^{2+}(aq) + 2OH^-(aq) \rightarrow Fe(OH)_2(s)$$

TIP

- When writing an ionic equation, you need to identify the products which are not ions. These are usually precipitated solids or simple molecules like water, bromine or carbon dioxide.
- When writing ionic equations involving acids, the acid is represented by a hydrogen ion, H^+.

Progress check

7.07 Write balanced equations, including state symbols, for these reactions:

a calcium + hydrochloric acid → calcium chloride + hydrogen [3]

b potassium + bromine vapour → potassium bromide [3]

c aluminium + oxygen → aluminium oxide [3]

d sulfuric acid + aqueous ammonia → ammonium sulfate [3]

e magnesium carbonate + hydrochloric acid → magnesium chloride + water + carbon dioxide. [3]

7.08 Write ionic equations for these reactions:

a $Mg(s) + H_2SO_4(aq) \rightarrow MgSO_4(aq) + H_2(g)$ [2]

b $Cl_2(aq) + 2KBr(aq) \rightarrow Br_2(aq) + 2KCl(aq)$ [2]

c $FeCl_3(aq) + 3NaOH(aq) \rightarrow Fe(OH)_3(s) + 3NaCl(aq)$. [2]

Exam-style questions

Question 7.01

Magnesium carbonate reacts with hydrochloric acid to form calcium chloride, carbon dioxide and water.

a Copy and complete the symbol equation for this reaction.

$$MgCO_3 + \ldots HCl \rightarrow MgCl_2 + \ldots\ldots\ldots + H_2O$$ [2]

b Magnesium reacts with steam.

$$Mg(s) + H_2O(g) \rightarrow MgO(s) + H_2(g)$$

Write a word equation for this reaction. [2]

c Write the formulae for magnesium sulfate and magnesium sulfide. [2]

Question 7.02

At high temperature and pressure, in the presence of a catalyst, nitrogen reacts with hydrogen to form ammonia, NH_3.

a Write a word equation for this reaction. Include all the information in the question. [2]

b Complete the balanced equation for this reaction.

$$N_2 + \ldots\ldots\ldots \rightarrow 2NH_3$$ [2]

c Aqueous ammonia reacts with sulfuric acid to form a salt.

i Copy and complete the equation for this reaction.

$$\ldots NH_3 + H_2SO_4 \rightarrow (NH_4)_2SO_4$$ [1]

ii State the name of the salt formed. [1]

Question 7.03

This question is about potassium and potassium bromide.

a Potassium reacts with water to form potassium hydroxide and hydrogen.

Write a balanced equation for this reaction to include state symbols. [3]

b An aqueous solution of magnesium bromide reacts with aqueous silver nitrate, $AgNO_3$, to form a precipitate of silver bromide and a solution of magnesium nitrate.

i Write a full symbol equation for this reaction. [2]

ii Convert this equation into an ionic equation to include state symbols. [2]

iii Which ions do not take part in this reaction? [1]

Question 7.04

Iron(III) oxide, Fe_2O_3, and aluminium are heated together. Iron and aluminium oxide, are formed.

a Write a balanced equation for this reaction. [2]

b The equation shows the reaction of iron with sulfuric acid.

$$Fe(s) + H_2SO_4(aq) \rightarrow FeSO_4(aq) + H_2(g)$$

Explain the meaning of i (g) and ii (aq) in this equation. [2]

c When iron reacts with hydrochloric acid the soluble salt, $FeCl_2$, is formed as well as hydrogen gas. Write an ionic equation for this reaction. [2]

d Iron(III) salts contain the Fe^{3+} ion. Write the formula for iron(III) sulfate. [1]

Revision checklist

You should be able to:

- Use the symbols of the elements and write the formulae of simple compounds
- Deduce the formula of an ionic or molecular compound from a diagram of its structure
- Deduce the formula of a compound from the charges on its ions
- Construct word equations and simple balanced chemical equations
- Construct equations with state symbols, including ionic equations
- Write chemical equations with state symbols

Masses and moles

Learning outcomes

By the end of this unit you should:

- Define relative atomic mass, A_r
- Define relative molecular mass, M_r and relative formula mass
- Define the mole and the Avogadro constant
- Be able to do calculations involving reacting masses involving simple proportion
- Be able to calculate stoichiometric reacting masses
- Be able to do calculations involving limiting reactants
- Be able to do calculations involving the molar gas volume
- Be able to calculate empirical and molecular formulae from data provided
- Be able to calculate percentage yield, percentage composition and percentage purity

8.01 Relative masses

When doing calculations, we have to compare masses of atoms accurately. To do this we use relative masses compared with a standard atom, carbon-12. For **relative atomic mass**, A_r, we use the average mass of atoms because many naturally-occurring elements are mixtures of isotopes. That is why the A_r of chlorine is 35.5. The relative atomic masses are shown in your Periodic Table.

Relative molecular mass, M_r is found by adding together the masses of all the atoms in a molecule. So the relative molecular mass of sulfur dioxide (SO_2) is found using the A_r values S = 32 and O = 16:

$$32 + (2 \times 16) = 64$$

TERMS

Relative atomic mass, A_r: The average mass of naturally occurring atoms of an element on a scale where a carbon-12 atom has a mass of exactly 12 units.

Relative molecular mass, M_r: The sum of the relative atomic masses of all the atoms in a molecule.

- For ionic compounds we use the term relative formula mass. Relative formula mass is the relative mass of one formula unit of a compound on a scale where an atom of the carbon-12 isotope has a mass of exactly 12 units.

We calculate relative formula masses in exactly the same way as relative molecular masses. For example, the relative formula mass of magnesium hydroxide $Mg(OH)_2$ is found by using the A_r values Mg = 24, O = 16, H = 1:

$$24 + (2 \times (16 + 1)) = 58$$

- Note that relative masses do not have any units.

TIP

When doing calculations involving relative atomic masses, you should use the data given on your Periodic Table.

8.02 The mole and the Avogadro constant

When doing chemical calculations both the number of reacting atoms and their mass are important. But even 1 000 000 molecules of water is too small to be weighed. So we have to scale up by a large number to get an amount we can weigh accurately.

- The amount of substance that contains 6.02×10^{23} atoms, molecules or ions is called a **mole**.
- The number of specific particles (atoms, molecules, electrons or ions) in a mole (6.02×10^{23}) is called the **Avogadro constant**.

TERMS

Mole: The amount of substance which has the same number of specific particles (atoms, molecules or ions) as there are atoms in exactly 12 g of the isotope carbon-12.

Avogadro constant: The number of defined particles (atoms, molecules, electrons or ions) in one mole of those particles.

Moles and mass

A mole (abbreviation mol) is the relative mass (atomic, molecular or formula mass) in grams. So a mole of sulfur atoms has a mass of 32 g and a mole of chlorine molecules, Cl_2 ($A_r = 35.5$) has a mass of $2 \times 35.5 = 71$ g.

We often refer to the mass of one mole of a substance as its molar mass.

Molar mass has the unit g/mol. So the molar mass of sodium is 23 g/mol.

TIP

When using moles, it is important to make clear the type of particle present. For example, 160 g of bromine, Br_2, is 1 mol of bromine molecules but it is 2 moles of bromine atoms.

The number of moles of substance is found by dividing the mass of substance in grams by either the relative atomic mass for metals or the relative molecular or formula mass for compounds or elements such as Br_2.

$$\text{number of moles (mol)} = \frac{\text{mass of substance in grams (g)}}{\text{molar mass (g/mol)}}$$

Worked example 8.01

Deduce the number of moles of calcium hydroxide, $Ca(OH)_2$, in 18.5 g of calcium hydroxide. A_r values: Ca = 40, O = 16, H = 1.

Formula mass of $Ca(OH)_2 = 40 + (2 \times (16 + 1)) = 74$

$$\text{mol} = \frac{\text{mass}}{\text{molar mass}} = \frac{18.5}{74} = 0.25 \text{ mol } Ca(OH)_2$$

Worked example 8.02

What is the mass of 0.6 mol of sodium sulfate, Na_2SO_4?

A_r values: Na = 23, S = 32, O = 16.

Formula mass of Na_2SO_4 =
$(2 \times 23) + 32 + (4 \times 16) = 142$
Rearranging the equation to make mass the subject:

mass = number of moles × molar mass =
$0.6 \times 142 = 85.2$ g Na_2SO_4

Progress check

8.01 Calculate the relative formula mass of the following:

a P_2O_3
b C_3H_8
c $CaCO_3$
d $Mg(NO_3)_2$
e $Al_2(SO_4)_3$. [5]

8.02 Calculate the number of moles in the following:

a 12.8 g of sulfur atoms
b 117 g sodium chloride, NaCl
c 5 g of calcium carbonate, $CaCO_3$
d 7.1 g of chlorine molecules, Cl_2
e 56.7 g of zinc nitrate, $Zn(NO_3)_2$. [5]

8.03 Calculate the mass of:

a 0.75 mol of NaOH
b 0.35 mol MgO
c 0.2 mol of K_2SO_4
d 12 mol $CuSO_4$
e 0.4 mol of $(NH_4)_2SO_4$. [5]

8.03 Calculating reacting masses

Using simple proportion

Worked example 8.03

A student reacts magnesium with excess aqueous copper sulfate:

$$Mg + CuSO_4 \rightarrow MgSO_4 + Cu$$

When 1.2 g of magnesium are completely reacted, 3.2 g of copper are formed.

Calculate the mass of magnesium needed to produce 48 g of copper.

3.2 g Cu is produced from 1.2 g Mg

So 48 g copper requires $\frac{48}{3.2} \times 1.2 = 18$ g Mg

Using molar masses or moles

If we want to find the mass of a particular product formed in a reaction we need to know:

- the relative number of moles of reactants and products in the equation (the **stoichiometry**)
- the mass of a particular reactant
- the molar mass of this reactant

There are two ways of doing this. Choose which method suits you best.

TERMS

Stoichiometry: The (mole) ratios of the reactants and products shown in the balanced equation.

Worked example 8.04

Method 1

Calculate the maximum mass of iron formed when 770 g of iron(III) oxide, Fe_2O_3, is reduced by excess carbon monoxide, CO. A_r values: Fe = 56, O = 16.0, C = 12.0.

Step 1. Write the balanced equation for the reaction. You will usually be given this.

$$Fe_2O_3 + 3CO \rightarrow 2Fe + 3CO_2$$

Step 2. Calculate the relevant formula masses. In this case, for Fe_2O_3 and Fe.

Fe = 56 $\quad Fe_2O_3 = (2 \times 56) + (3 \times 16) = 160$

Step 3. Multiply each formula mass in grams by the relevant number of moles in the balanced equation:

1 mole of $Fe_2O_3 \rightarrow$ 2 moles of Fe, so:
160 g of $Fe_2O_3 \rightarrow 2 \times 56$ g = 112 g Fe

Step 4. Use simple proportion to calculate the mass of iron produced:

$$\frac{112}{160} \times 770 = 539 \text{ g Fe}$$

Method 2

Step 1. Write the balanced equation.

Step 2. Calculate the number of moles of Fe_2O_3

$= \frac{770}{160} = 4.8125$ mol

Step 3. Use the stoichiometry of the equation:

1 mol $Fe_2O_3 \rightarrow$ 2 mol Fe
so 4.81 mol $Fe_2O_3 \rightarrow$ 9.625 mol Fe

Step 4. Calculate the mass of iron (mass = mol × molar mass):

mass of iron = 9.625 × 56 = 539 Fe

Sample answer

Calculate the mass of carbon that reacts with excess red lead oxide, Pb_3O_4, to form 124.2 g of lead, Pb. A_r values C = 12, Pb = 207.

$Pb_3O_4 + 4C \rightarrow 3Pb + 4CO$ **[4]**

4 mol C forms 3 mol Pb **[1]**

So 4 × 12 g → 3 × 207 g Pb **[1]**

48 g C → 621 g Pb **[1]**

So 124.2 g lead → $48 \times \frac{124.2}{621} = 9.6$ g C **[1]**

Which is the limiting reactant?

The **limiting reactant** is the reactant which is not in excess. It is used up first. When it is used up, the reaction stops.

TERMS

Limiting reactant: The reactant that is not in excess.

When deducing which reactant is limiting, we need to take account of the ratios of each reactant appearing in the balanced equation.

Worked example 8.05

A student reacts 9.2 g of sodium with 8.0 g of sulfur to form sodium sulfide, Na_2S. Which reactant is in excess? A_r values Na = 23, S = 32.

Step 1. Calculate the number of moles of each reactant:

$$\text{mol Na} = \frac{9.2}{23} = 0.40\,\text{mol}$$

$$\text{mol S} = \frac{8.0}{32} = 0.25\,\text{mol}$$

Step 2. Write the equation and determine the ratio of moles which react.

$2Na$	+	S	→	Na_2S
2 mol		1 mol		

Step 3. Compare the moles of each reactant:

To react completely with 0.4 mol of Na, it needs $\frac{1}{2} \times 0.4\,\text{mol S} = 0.20\,\text{mol}$

So S is in excess by $0.25 - 0.20 = 0.05$ mol

And Na is the limiting reactant.

When doing mole calculations, always take account of the number of moles of relevant reactants and products in the balanced equation.

8.04 Moles and gas volumes

- At room temperature and pressure 1 mole of any gas occupies $24\,dm^3$.
- The value $24\,dm^3$ is the molar gas volume.
- Since there are equal numbers of molecules in the same volume of any gas under the same conditions of temperature and pressure, we can use the molar gas volume to work out the number of moles reacting, e.g.

$H_2(g)$	+	$Cl_2(g)$	→	$2HCl(g)$
1 mol		1 mol		2 mol
1 volume		1 volume		2 volumes
$24\,dm^3$		$24\,dm^3$		$48\,dm^3$

Worked example 8.06

Calculate the volume of 3.08 g of carbon dioxide at room temperature and pressure. A_r values C = 12, O = 16.

Step 1. Calculate moles of CO_2:

$$\frac{3.08}{12 + (2 \times 16)} = 0.070\,\text{mol}$$

Step 2. Calculate the volume using: volume = mol × molar gas volume in dm^3.

$$= 0.070 \times 24 = 1.68\,dm^3$$

Worked example 8.07

Calculate the mass of propane, C_3H_8, in $480\,cm^3$ of propane gas at room temperature and pressure. A_r values C = 12, H = 1.

Step 1. Change volume in cm^3 to volume in dm^3:

$$480\,cm^3 = \frac{480}{1000} = 0.480\,dm^3$$

Step 2. Calculate the number of moles using:

$$\frac{\text{volume }(dm^3)}{24} = \frac{0.480}{24} = 0.020\,\text{mol}$$

Step 3. Calculate mass using:
mass (g) = moles × molar mass

molar mass of propane = $(3 \times 12) + (8 \times 1) = 44$

mass of propane = $0.020 \times 44 = 0.88$ g propane

Sample answer

Calculate the volume of oxygen at room temperature and pressure needed to burn 26.4 g of propane. A_r values C = 12, H = 1.

$$C_3H_8 + 5O_2 \rightarrow 3CO_2 + 4H_2O \text{ [5]}$$

1 mol propane reacts with 5 mol oxygen **[1]**

1 × 44 g of propane reacts with 5 × 32 g oxygen

44 g of propane react with 160 g oxygen **[1]**

26.4 g propane react with $160 \times \frac{26.4}{44} = 96$ g oxygen **[1]**

Moles $O_2 = \frac{96}{32} = 3$ mol **[1]**

Volume = 3 × 24 = 72 dm^3 O_2 **[1]**

Progress check

8.04 Calculate the maximum mass of sodium peroxide formed when 6.9 g of sodium is burned in excess oxygen. A_r values Na = 23, O = 16.

$$2Na + O_2 \rightarrow Na_2O_2 \text{ [2]}$$

8.05 Tin(IV)oxide is reduced to tin by carbon. Calculate the mass of carbon that reacts exactly with 28 g of tin(IV) oxide. A_r values C = 12, O = 16, Sn = 119.

$$SnO_2 + 2C \rightarrow Sn + 2CO \text{ [2]}$$

8.06 Calculate the volume of oxygen needed to react exactly with 60 g of methane, CH_4. A_r values H = 1, C = 12.

$$CH_4 + 2O_2 \rightarrow CO_2 + 2H_2O \text{ [3]}$$

8.05 Calculating empirical and molecular formulae

- A **molecular formula** shows the number and type of atom of each element in a molecule. For example, H_2O_2 for hydrogen peroxide.
- An empirical formula shows the simplest ratio of atoms or ions in a compound. For example, HO for hydrogen peroxide.
- The molecular formula is a whole number multiple of the empirical formula. For example, the empirical formula of butene is CH_2 and its molecular formula is C_4H_8.

TERMS

Molecular formula: The formula showing the number and type of atom of each element in a molecule.

Worked example 8.08

Calculating the empirical formula

A compound of carbon and hydrogen contains 85.7% of carbon and 14.3% hydrogen by mass. Deduce the empirical formula of this compound. A_r values H = 1, C = 12.

Step 1. Write down the percentages by mass and divide each by the atomic mass.

C	H
$\frac{85.7}{12} = 7.142$	$\frac{14.3}{1} = 14.3$

Step 2. Divide each by the lowest figure.

$$\frac{7.142}{7.142} = 1 \qquad \frac{14.3}{7.142} = 2$$

Step 3. Write the empirical formula: CH_2

Worked example 8.09

Molecular formula from the empirical formula

A compound has the empirical formula $C_2H_3Br_2$ and a relative molecular mass of 374.

Calculate the molecular formula of this compound. A_r values H = 1, C = 12, Br = 80.

Step 1. Find the empirical formula mass:

$$(2 \times 12) + (3 \times 1) + (2 \times 80) = 187$$

Step 2. Divide the relative molecular mass by the empirical formula mass.

$$\frac{374}{187} = 2$$

Step 3. Multiply the number of all the atoms in the empirical formula by the number in step 2.

$2 \times C_2H_3Br_2$ so the molecular formula is $C_4H_6Br_4$

8.06 Calculating percentage composition

The percentage composition is found by calculating the percentage by mass of each particular element in a compound.

Worked example 8.10

Calculate the percentage by mass of nitrogen in ammonium sulfate, $(NH_4)_2SO_4$. A_r values H = 1, N = 14, O = 16, S = 32.

Step 1. Calculate the molar mass of the compound:

Molar mass of $(NH_4)_2SO_4$
$= 2 \times [14 + (4 \times 1)] + 32 + (4 \times 16)$
$= 132$

Step 2. Add the atomic masses of the element required (nitrogen):

Moles of nitrogen atoms in 1 mole of $(NH_4)_2SO_4 = 2$

The sum of these atomic masses is $14 + 14 = 28$

Step 3. Calculate percentage:

$$\% \text{ by mass of N} = \frac{28}{132} \times 100 = 21.2\%$$

8.07 Percentage yield

$$\text{percentage yield} = \frac{\text{actual yield}}{\text{theoretical yield}} \times 100$$

- The actual yield is obtained by experiment.
- The theoretical yield assumes that there is 100% conversion of reactants to products.

Worked example 8.11

A student reacts 9 g of aluminium powder with excess chlorine. The mass of aluminium chloride produced is 26.7 g. Calculate the percentage yield of aluminium chloride. A_r values O = 16, Al = 27.

$$2Al + 3Cl_2 \rightarrow 2AlCl_3$$

Step 1. Calculate the relevant formula masses:

1 mol $Al \rightarrow$ 1 mol $AlCl_3$
27 g $Al \rightarrow 27 + (3 \times 35.5) = 133.5$ g $AlCl_3$

Step 2. Calculate the theoretical yield:

$$9\text{ g} \rightarrow \frac{9}{27} \times 133.5 = 44.5\text{ g } AlCl_3$$

Step 3. Calculate the percentage yield:

$$\frac{\text{actual yield}}{\text{theoretical yield}} \times 100 = \frac{26.7}{44.5} \times 100 = 60\%$$

8.08 Percentage purity

$$\text{percentage purity} = \frac{\text{mass of pure substance}}{\text{mass of impure substance}} \times 100$$

Worked example 8.12

When 6.0 g of impure calcium carbonate were heated, 2.2 g of carbon dioxide were formed. Calculate the percentage purity of the calcium carbonate.

A_r values C = 12, O = 16, Ca = 40.

$$CaCO_3 \rightarrow CaO + CO_2$$

Step 1. Calculate the relevant formula masses:

1 mol $CaCO_3 \rightarrow$ 1 mol CO_2
$40 + 12 + (3 \times 16)$ $12 + (2 \times 16)$
$100 \rightarrow 44$

Step 2. Calculate the theoretical mass of calcium carbonate used if pure.

From 2.2 g CO_2 we expect: $\frac{2.2}{44} \times 100 = 5$ g

Step 3. Calculate the percentage purity:

$$\frac{\text{mass of pure substance}}{\text{mass of impure substance}} \times 100 = \frac{5}{6} \times 100$$
$$= 83.3\%$$

Progress check

8.07 A compound of tin and chlorine has the composition by mass 45.6% tin and 54.4% chlorine. Calculate the empirical formula of this chloride.

A_r values Cl = 35.5, Sn = 119. [3]

8.08 Calcium carbonate reacts with hydrochloric acid.

$$CaCO_3 + 2HCl \rightarrow CaCl_2 + CO_2 + H_2O$$

a Calculate the number of moles of hydrochloric acid required to just react with 7 g of calcium carbonate. [2]

b Calculate the volume of carbon dioxide produced when 0.25 mol of hydrochloric acid reacts with excess calcium carbonate. [2]

8.09 A student added 6.4 g of iron to 0.10 mol of sulfuric acid.

$$Fe + H_2SO_4 \rightarrow FeSO_4 + H_2$$

a Show by calculation that sulfuric acid is the limiting reagent. [2]

b Calculate the volume of hydrogen produced. [2]

Exam-style questions

Question 8.01

Compound A contains 1.92 g of carbon, 0.32 g of hydrogen and 5.78 g of chlorine.

a Calculate the empirical formula of A. [2]

b The relative molecular mass of A is 99. Write the molecular formula of A. [2]

c Another compound of hydrogen, carbon and chlorine was made by reacting 13 g of ethyne, C_2H_2 with 69 g of chlorine.

$$C_2H_2 + 2Cl_2 \rightarrow C_2H_2Cl_4$$

Show by calculation which of ethyne or chlorine is in excess. [2]

Question 8.02

When 20.4 g of impure sodium hydrogen carbonate was heated, 2000 cm^3 of carbon dioxide was formed at room temperature and pressure.

a Calculate the percentage purity of the sodium hydrogen carbonate.

$$2NaHCO_3 \rightarrow Na_2CO_3 + CO_2 + H_2O$$ [5]

b A pure sample of sodium hydrogen carbonate was heated. Calculate the mass of sodium carbonate formed from 26.1 g of sodium hydrogen carbonate if the yield is 72%. [4]

Question 8.03

0.80 g of calcium is added to an aqueous solution containing 3.0 g of ethanoic acid. Which is the limiting reagent and what mass of the other reagent is in excess?

$$Ca + 2CH_3COOH \rightarrow (CH_3COO)_2Ca + H_2$$ [6]

Revision checklist

You should be able to:

- Define relative molecular mass, M_r and relative formula mass
- Define the mole in terms of number of specified particles compared with the carbon-12 isotope
- Define the Avogadro constant
- Do simple calculations involving reacting masses using simple proportion
- Do calculations involving reacting masses using masses, moles and the stoichiometric equation
- Do calculations involving the molar gas volume
- Understand the concept of limiting reactants
- Calculate empirical formulae and molecular formulae
- Calculate percentage yield and percentage purity

Solution concentration

Learning outcomes

By the end of this unit you should:

- Know the terms solute, solvent and solution
- Know that solution concentration is expressed in g/dm^3 or mol/dm^3
- Be able to use the relationship

  concentration (in mol/dm^3) $= \frac{\text{amount of substance (in mol)}}{\text{volume (in dm}^3\text{)}}$
- Be able to know how to do a titration
- Be able to calculate stoichiometric reacting masses from information about concentrations of solutions
- Be able to understand and use the results from a titration to calculate solution concentration

9.01 Solute, solvent and solution

- A solute is a substance that dissolves in a solvent.
- A solvent is a substance which dissolves a solute.
- A solution is formed by dissolving a solute in a solvent.

TIP

When making up solutions accurately, dissolve the solute in a small amount of solvent and then make it up to the volume required by adding more solvent. You do not add the solid to the volume of solvent required.

9.02 Calculating solution concentration

Concentration

$$(\text{mol/dm}^3) = \frac{\text{number of moles of solute (mol)}}{\text{volume of solution (dm}^3\text{)}}$$

TERMS

Concentration (of solution): The amount of solute dissolved in a defined volume of solution (usually mol/dm^3 or g/dm^3).

- To get the volume in dm^3, we divide volume in cm^3 by 1000.
- Moles of solute (mol) = concentration (mol/dm^3) × volume (dm^3).
- We can also express solution concentration as g/dm^3

$$\text{concentration (g/dm}^3\text{)} = \frac{\text{mass of solute (g)}}{\text{volume of solution (dm}^3\text{)}}$$

Worked example 9.01

Calculate the concentration in mol/dm^3 of a solution of sodium hydroxide containing 0.30 g NaOH in 50 cm^3 solution. A_r values H = 1, O = 16, Na = 23.

Step 1. Convert grams to moles: $\frac{0.30}{23+16+1}$ = 0.0075 mol

Step 2. Change cm^3 to dm^3: 50 cm^3 = 0.05 dm^3

Step 3. Calculate concentration: $\frac{0.0075}{0.05}$ = 0.15 mol/dm^3

Worked example 9.02

Calculate the mass of calcium chloride (M = 111) in 150 cm^3 of a 0.400 mol/dm^3 solution of calcium chloride.

Step 1. Change cm^3 to dm^3: 150 cm^3 = 0.150 dm^3

Step 2. Calculate number of moles: 0.400 × 0.150 = 0.060 mol

Step 3. Convert moles to grams: 0.060 × 111 = 6.6 g

Progress check

9.01 Calculate the concentration in mol/dm^3 of the following solutions:

a 4 g NaOH in 250 cm^3 solution. [1]

b 22.2 g $CaCl_2$ in 360 cm^3 solution. [1]

c 6 g ethanoic acid, CH_3COOH, in 125 cm^3 solution. [1]

9.02 Calculate the mass of solute needed to make these solutions.

a 2 dm^3 of aqueous sodium chloride, $NaCl(aq)$, of concentration 0.50 mol/dm^3. [1]

b 50 cm^3 of aqueous copper(II) sulfate, $CuSO_4(aq)$, of concentration 0.10 mol/dm^3. [1]

c 100 cm^3 of aqueous magnesium nitrate, $Mg(NO_3)_2(aq)$, of concentration 1.50 mol/dm^3. [1]

9.03 Calculate the volume of the following solutions:

a A solution containing 2 mol of sulfuric acid, H_2SO_4, of concentration 0.5 mol/dm^3. [1]

b A solution containing 0.05 mol of sodium hydroxide, NaOH, of concentration 0.8 mol/dm^3. [1]

9.03 Titrations

- If we want to determine the amount of substance present in a solution of acid or alkali of unknown concentration we carry out a **titration**.

TERMS

Titration: A method for finding out the amount of substance (in moles) in a solution of unknown concentration (usually by using a burette and indicator).

- Figure 9.01 shows the apparatus used.
- A volumetric pipette is used to put the alkali into the flask.

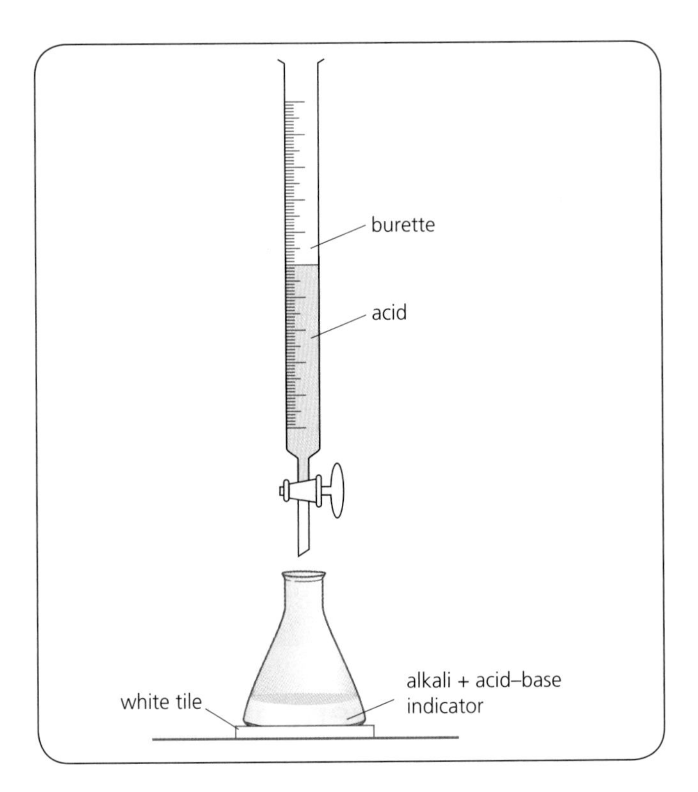

Figure 9.01

- We add acid from the burette to the alkali in the flask until the alkali has been neutralised (see Unit 15 for more information about neutralisation).
- We use a few drops of acid–base **indicator** added to the flask to find out when the acid has just reacted completely with the alkali. The point where this happens is called the end point of the titration.
- At the end point, the colour of the indicator changes suddenly.
- The experiment is repeated until you have at least two successive titration results (titres) that are not more than 0.1 cm^3 apart.

TERMS

Indicator: A substance that changes colour over a narrow range of pH to show the end point of a titration.

When doing a titration, you should only average the titration readings that are within 0.1 cm^3 of each other. You should not include the rough titration.

9.04 Calculating solution concentration from titration results

In calculations to find out solution concentration from titration results, we need to know the following:

- The volume of the solution of unknown concentration in the flask.
- The concentration of the solution in the burette.
- The volume of the solution in the burette found by experiment.
- The balanced equation for the reaction.

TIP

When carrying out calculations involving concentrations remember the units of volume. You may need to convert cm^3 to dm^3 or cm^3 to dm^3 at an appropriate point in the calculation.

Worked example 9.03

$25.0\,cm^3$ of potassium hydroxide solution is exactly neutralised by $6.20\,cm^3$ of sulfuric acid of concentration $0.200\,mol/dm^3$. Calculate the concentration, in mol/dm^3 of the potassium hydroxide solution.

$$2KOH(aq) + H_2SO_4(aq) \rightarrow K_2SO_4(aq) + 2H_2O(l)$$

Step 1. Calculate moles of reagent for which both concentration and volume are known.

In this case it is the acid.

$$\text{moles (mol)} = \text{concentration (mol/dm}^3) \times \text{volume (dm}^3)$$

$$\text{moles acid} = 0.200 \times \frac{6.20}{1000} = 0.00124\,\text{mol}$$

Step 2. Use the mole ratio in the balanced equation to calculate moles of KOH.

2 mol KOH reacts with 1 mol H_2SO_4

So moles KOH = $2 \times 0.00124 = 0.00248$ mol KOH

Step 3. Calculate the concentration of KOH using:

$$\text{concentration (mol/dm}^3) = \frac{\text{number of moles of solute (mol)}}{\text{volume of solution (dm}^3)}$$

volume of solution (dm^3)

$$\text{concentration of KOH} = 0.00248 \times \frac{1000}{25.0}$$
$$= 0.0992\,\text{mol/dm}^3$$

Sample answer

0.111 g of calcium hydroxide is completely dissolved in water. Calculate the volume, in cm^3, of $0.120\,mol/dm^3$ hydrochloric acid required to just neutralise the calcium hydroxide.

$Ca(OH)_2(aq) + 2HCl(aq) \rightarrow CaCl_2(aq) + 2H_2O(l)$ **[3]**

Moles of calcium hydroxide: $\frac{0.111}{74.0} = 0.00150$ mol $Ca(OH)_2$ **[1]**

Using the mole ratio in the balanced equation to calculate moles of acid.

1 mol $Ca(OH)_2$ reacts with 2 mol HCl

So moles $HCl = 2 \times 0.00150 = 0.00300$ mol HCl **[1]**

Volume of hydrochloric acid:

$$\text{volume (dm}^3) = \frac{\text{number of moles of solute (mol)}}{\text{concentration of solution (mol/dm}^3)}$$

$$\text{volume of } HCl = \frac{0.00300}{0.120} = 0.0250\,\text{dm}^3$$

$= 25.0\,cm^3$ **[1]**

Progress check

9.04 Sodium chloride dissolves in water to form aqueous sodium chloride.

Name

a the solution

b the solvent

c the solute. [2]

9.05 In a titration, what is meant by the term *end point*? [2]

9.06 Some titration results are shown:

titration number	rough	1st	2nd	3rd	4th	5th
titre/cm³	26.0	25.1	25.5	25.6	25.5	25.0

Table 9.01

Which results should be used to find the most accurate titre? Explain your answer. [3]

9.07 A sample of $25.0\,cm^3$ of aqueous potassium hydroxide of concentration $0.120\,mol/dm^3$ is neutralised by exactly $42.5\,cm^3$ of hydrochloric acid.

$$KOH + HCl \rightarrow KCl + H_2O$$

a Calculate the moles of potassium hydroxide. [1]

b Deduce the number of moles of hydrochloric acid. [1]

c Calculate the concentration of hydrochloric acid in mol/dm^3. [1]

Exam-style questions

Question 9.01

$20\,cm^3$ of a solution of sodium carbonate of concentration $0.20\,mol/dm^3$ is neutralised by $32.5\,cm^3$ of hydrochloric acid.

$$Na_2CO_3 + 2HCl \rightarrow 2NaCl + CO_2 + H_2O$$

a Calculate the number of moles of sodium carbonate. [1]

b How many moles of hydrochloric acid reacted with the number of moles of sodium carbonate you calculated in part a? [1]

c Calculate the concentration of the hydrochloric acid. [1]

Question 9.02

$25\,cm^3$ of a solution of barium hydroxide of concentration contains 1.71 g of barium hydroxide. This solution is neutralised by $22.5\,cm^3$ of hydrochloric acid.

$$Ba(OH)_2 + 2HCl \rightarrow BaCl_2 + 2H_2O$$

a Calculate the concentration of the solution of barium hydroxide in g/dm^3 and mol/dm^3. [2]

b How many moles of hydrochloric acid reacted with the number of moles of barium hydroxide you calculated in part a? [1]

c Calculate the concentration of the hydrochloric acid. [1]

Question 9.03

$25\,cm^3$ of a solution of potassium hydroxide solution of concentration $1.0\,mol/dm^3$ is neutralised by $22.4\,cm^3$ of sulfuric acid.

$$2KOH + H_2SO4 \rightarrow K_2SO_4 + 2H_2O$$

a Calculate the concentration of the solution of potassium hydroxide in mol/dm^3. [1]

b How many moles of sulfuric acid reacted with the number of moles of potassium hydroxide you calculated in part a? [1]

c Calculate the concentration of the sulfuric acid. [1]

Revision checklist

You should be able to:

- Use the terms solute, solvent and solution
- Recognise the units of solution concentration
- Know how to carry out a titration
- Calculate solution concentration, number of moles or volume using the relationship between these quantities and relevant information
- Calculate solution concentration from titration results

Electrochemistry

Learning outcomes

By the end of this unit you should:

- Describe the reasons for the use of copper and steel-cored aluminium in cables
- Be able to describe why plastics and ceramics are insulators
- Be able to define electrolysis
- Describe the electrode products and the observations made during the electrolysis of lead(II) bromide, concentrated hydrochloric acid, concentrated aqueous sodium chloride and dilute sulfuric acid between inert electrodes
- State the general principle that metals or hydrogen are formed at the negative electrode (cathode), and that non-metals (other than hydrogen) are formed at the positive electrode (anode)
- Be able to predict the products of electrolysis of a molten binary compound
- Be able to describe the reactions at the electrodes during electrolysis
- Relate the products of electrolysis to the electrolyte and electrodes used, exemplified by the specific examples in the Core together with aqueouscopper(II) sulfate using carbon electrodes and using copper electrodes
- Predict the products of electrolysis of a specified halide in dilute or concentrated aqueous solution
- Be able to describe the transfer of charge during electrolysis
- Be able to construct ionic half-equations for the reaction taking place during electrolysis
- Be able to outline electroplating and give some of its uses
- Know how aluminium is extracted from aluminium oxide
- Describe, in outline, the manufacture of chlorine, hydrogen and sodium hydroxide from concentrated aqueous sodium chloride
- Be able to describe the production of energy from electrochemical cells

10.01 Conductors and insulators

Electrical conductors allow electrical charge to pass through them easily.

They can be:

- solids, e.g. metals or graphite (conduction by free moving electrons).
- liquids, e.g. molten lead bromide or molten metals (conduction by moving ions)
- solutions, e.g. a solution of sodium chloride in water or aqueous solutions of acids (conduction by moving ions)

Copper is used for electrical wiring because it is an excellent conductor.

Aluminium cables with a steel core are used in high voltage power lines. Aluminium is a good electrical conductor but not as good as copper. But it is less dense and cheaper than copper. The steel core to the cable gives strength to prevent the cable breaking under its own weight.

- Insulators resist the flow of electricity. They are poor conductors of electricity.
- Most insulators are solids, e.g. plastics or ceramics.
- Plastics are used as insulators around electrical wiring and for the handles of some tools.
- Ceramics are used in high voltage power lines to prevent contact with the metal of the electricity pylons.

10.02 Electrolysis

When an electric current is passed through a molten ionic compound, the compound decomposes (breaks down). This process is called **electrolysis**. Ionic solutions also undergo electrolysis.

TERMS

Electrolysis: The breakdown of an ionic compound (molten or in aqueous solution) by the passage of electricity.

The vessel in which electrolysis takes place is called an electrolysis cell. Figure 10.01 shows the main parts of an electrolysis cell. The direction of electron flow in the external circuit is shown by the arrows.

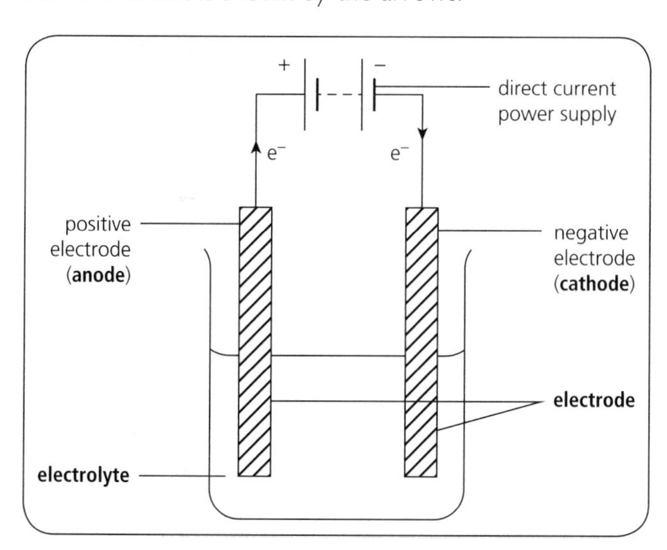

Figure 10.01

TERMS

Electrode: A rod of metal or graphite which leads an electric current to or from an electrolyte.

Electrolyte: An ionic compound which conducts electricity when molten or dissolved in water.

Anode: The positive electrode.

Cathode: The negative electrode.

10.03 The ions present in electrolytes

- Molten salts: the particles present are ions. Electrolysis can only take place when the salt is molten so that the ions separate from each other and are free to move:

$$PbBr_2(s) \rightarrow Pb^{2+}(l) + 2Br^-(l)$$

- Aqueous solutions of salts: when salts dissolve in water, they separate into their ions which are free to move. Positive ions (**cations**) and negative ions (**anions**) are formed:

$$NaCl(s) + (aq) \xrightarrow{heat} Na^+(aq) + Cl^-(aq)$$

- Solutions of acids: hydrogen ions and negative ions are formed:

$$HCl(g) \rightarrow H^+(aq) + Cl^-(aq)$$

TERMS

Anion: A negative ion.

Cation: A positive ion.

TIP

Cations move to the **C**athode and **A**nions move to the **A**node during electrolysis.

10.04 The products of electrolysis

When using inert (unreactive) graphite or platinum electrodes:

- Metals or hydrogen are formed at the negative electrode (cathode). This is because the ions moving towards the negative electrode are positively charged (opposite charges attract).
- Non-metals other than hydrogen are formed at the positive electrode (anode). This is because the ions moving towards the positive electrode are negatively charged.

- For a molten compound of a metal with a non-metal (a binary compound), the cathode product is always the metal and the anode product is always the non-metal.
- The table shows the electrode products formed during the electrolysis of particular electrolytes.

Electrolyte	Cathode product	Anode product
molten lead(II) bromide	lead	bromine
molten sodium iodide	sodium	iodine
molten zinc chloride	zinc	chlorine
concentrated aqueous sodium chloride	hydrogen	chlorine
concentrated hydrochloric acid	hydrogen	chlorine
dilute sulfuric acid	hydrogen	oxygen

Table 10.01

Progress check

10.01 Why are steel-cored aluminium cables used for overhead power lines? [3]

10.02 What do these terms mean? Electrical insulator, electrolysis, anode, electrolyte. [5]

10.03 Predict the products at the anode and cathode when the following are electrolysed:

a dilute sulfuric acid [2]

b concentrated hydrochloric acid [2]

c molten lithium bromide [2]

d concentrated aqueous sodium chloride. [2]

10.05 Reactions at the electrodes

During electrolysis:

- Electrons move from the power supply to the cathode.
- Positive ions in the electrolyte move to the negative cathode.
- The positive ions accept electrons from the cathode. Metals or hydrogen are formed:

$$Zn^{2+} + 2e^- \rightarrow Zn$$
$$2H^+ + 2e^- \rightarrow H_2$$

- The reaction at the cathode is a reduction reaction because electrons are gained (see Unit 14).
- Electrons move from the anode to the power supply.
- Negative ions in the electrolyte move to the anode.
- If the anode is inert, the negative ions lose electrons to the anode. Non-metals or oxygen are formed:

$$2Br^-(aq) \rightarrow Br_2(aq) + 2e^-$$
$$4OH^-(aq) \rightarrow O_2(g) + 2H_2O(l) + 4e^-$$

- If the anode is a reactive electrode the metal atoms of the anode loose electrons and form positive ions. The positive ions go into solution and the anode becomes smaller:

$$Zn \rightarrow Zn^{2+} + 2e^-$$
$$Ag \rightarrow Ag^+ + e^-$$

- The reaction at the anode is an oxidation reaction because electrons are lost (see Unit 14).
- Equations showing electron loss or gain like this are called **ionic half equations**.

TERMS

Ionic half equation: An equation balanced by electrons which shows either oxidation or reduction.

TIP

You can also write the half-equations like this:

$$Zn^{2+} \rightarrow Zn - 2e^-$$

$$2Br^- - 2e^- \rightarrow Br_2$$

You should stick to one method or the other to avoid getting muddled.

10.06 What products are formed in aqueous solution?

- Aqueous electrolytes contain more than one type of anion or cation.
- This is because water ionises very slightly: $H_2O \rightleftharpoons H^+ + OH^-$

Aqueous sodium chloride contains the ions $Na^+(aq)$, $H^+(aq)$, $Cl^-(aq)$ and $OH^-(aq)$.

When the ions reach the electrodes the ions gain or lose electrons. The electrode product depends on the types of ion present in the electrolyte, their concentration and whether the electrode is inert or reactive.

- During electrolysis only one type of cation or anion is discharged (converted to atoms).
- Ions lower in the reactivity series (see Unit 21) are discharged in preference to the ones above them:

higher reactivity $\xrightarrow{Na^+\ Al^{3+}\ Zn^{2+}\ H^+\ Pb^{2+}\ Cu^{2+}\ Ag^+}$ lower reactivity

If Cu^{2+} and H^+ ions are present, Cu^{2+} ions are discharged at the cathode because they are lower in the reactivity series.

- We can also arrange anions in order of ease of discharge:

higher reactivity $\xrightarrow{SO_4^{2-}\ NO_3^-\ OH^-\ Cl^-\ Br^-\ I^-}$ lower reactivity

For example, in concentrated aqueous sodium chloride, Cl^- ions are discharged and not OH^- ions (from water).

- For anions, a concentrated ion tends to get discharged rather than a less concentrated ion. When a concentrated aqueous solution of sodium chloride is electrolysed, Cl^- ions are discharged and not OH^- ions. But if the sodium chloride solution is dilute, the OH^- ion is discharged. So oxygen is formed, and only a little chlorine.
- Graphite or platinum electrodes do not take part in the chemical reaction. A reactive anode, such as a copper anode dipping into aqueous copper sulfate, does take part in the reaction and loses electrons when the ions go into solution:

$$Cu \rightarrow Cu^{2+} + 2e^-$$

Worked example 10.01

Concentrated aqueous lead nitrate is electrolysed. Describe any observations you would make and explain, using relevant equations, the reactions occurring at the cathode. **[6]**

You should first identify the electrode products taking into account whether the ions are in aqueous solution or not, their concentration and their reactivity. Equations can then be written. Lastly do not forget the observations. For example, gases will bubble off and metals will be deposited.

Any six points of:

At the anode oxygen is given off **[1]**. This is because hydroxide ions come from ionisation of water **[1]**. Nitrate is above hydroxide in reactivity at the electrode **[1]** so oxygen is formed from hydroxide ions. The oxygen comes off as bubbles **[1]**.

At the cathode lead is formed **[1]**. The reaction is $Pb^{2+} + 2e^- \rightarrow Pb$ **[1]**. This is because lead ions are lower than hydrogen ions in reactivity at the electrode **[1]**. The lead is seen as a grey layer on the cathode **[1]**.

10.07 Some examples of electrode reactions

These reactions are with inert (graphite) electrodes.

Electrolyte	Cathode reaction	Anode reaction
molten $PbBr_2$	$Pb^{2+} + 2e^- \rightarrow Pb$	$2Br^- \rightarrow Br_2 + 2e^-$
concentrated aqueous $NaCl$	$2H^+ + 2e^- \rightarrow H_2$	$2Cl^- \rightarrow Cl_2 + 2e^-$
dilute aqueous $NaCl$	$2H^+ + 2e^- \rightarrow H_2$	$4OH^- \rightarrow O_2 + 2H_2O + 4e^-$
concentrated aqueous $CuSO_4$	$Cu^{2+} + 2e^- \rightarrow Cu$	$4OH^- \rightarrow O_2 + 2H_2O + 4e^-$
concentrated aqueous HCl	$2H^+ + 2e^- \rightarrow H_2$	$2Cl^- \rightarrow Cl_2 + 2e^-$
dilute aqueous H_2SO_4	$2H^+ + 2e^- \rightarrow H_2$	$4OH^- \rightarrow O_2 + 2H_2O + 4e^-$

Table 10.02

10.08 Extraction of aluminium

- Aluminium is extracted by the electrolysis of molten aluminium oxide.
- The aluminium oxide is dissolved in cryolite. This lowers the melting point of the electrolyte.
- The reaction at the cathode is: $Al^{3+} + 3e^- \rightarrow Al$
- The reaction at the anode is: $4OH^- \rightarrow O_2 + 2H_2O + 4e^-$

10.09 The electrolysis of brine

Brine is concentrated aqueous sodium chloride. The electrolysis of brine produces chlorine, hydrogen and sodium hydroxide. See Figure 10.02.

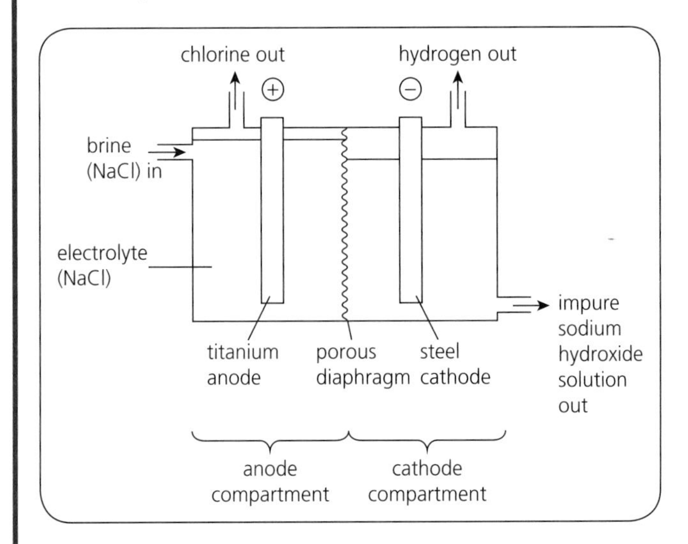

Figure 10.02

You do not have to remember the exact details of the diaphragm cell or the function of the diaphragm (porous partition).

- The electrolyte is concentrated sodium chloride.
- The ions present in the electrolyte are $Na^+(aq)$, $H^+(aq)$, $Cl^-(aq)$ and $OH^-(aq)$.

The electrode reactions are:

Anode: $2Cl^- \rightarrow Cl_2 + 2e^-$
Cathode: $2H^+ + 2e^- \rightarrow H_2$

As electrolysis proceeds the concentration of $H^+(aq)$ and $Cl^-(aq)$ ions in the electrolyte falls. This leaves the $Na^+(aq)$ and $OH^-(aq)$ ions in the electrolyte (aqueous sodium hydroxide).

Progress check

10.04 Explain why hydrogen and oxygen are produced and the acid becomes more concentrated when dilute sulfuric acid is electrolysed. [7]

10.05 Write half equations for the reactions at the anode and cathode when the following are electrolysed:

a concentrated hydrochloric acid [2]

b dilute aqueous sodium hydroxide [2]

c molten magnesium chloride [2]

d molten aluminium oxide. [2]

10.06 Explain where and why oxygen and hydrogen are formed when aqueous sodium sulfate is electrolysed. [6]

10.10 Electroplating

TERMS

Electroplating: The coating of the surface of one metal with a layer of another by an electrolytic reaction.

Electroplating is used to coat of the surface of one metal with a layer of a different metal. The metal which forms the covering (plating) is usually a less reactive metal. We electroplate articles because:

- it makes them more resistant to corrosion or scratching, e.g. chromium plating, nickel plating
- it improves their appearance, e.g. plating with silver or chromium makes the surface shiny

Figure 10.03 shows the apparatus used for electroplating an article with silver.

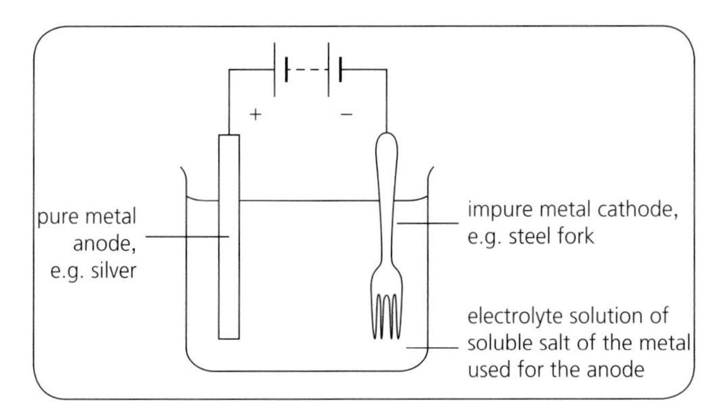

Figure 10.03

When we electroplate an article:

- The anode is made from the pure metal to be used for the plating, e.g. silver.
- The cathode is the object to be electroplated, e.g. a steel spoon.
- The electrolyte is an aqueous solution of a soluble salt of the pure metal at the anode, e.g. silver cyanide.

Worked example 10.02

a Solid nickel chloride does not conduct electricity. Explain why not and give two ways by which you could make it conduct. **[3]**

You have to remember that ions move in liquid or solution in electrolysis (not electrons).

In a solid, the ions cannot move **[1]**.

You can make it conduct by melting it **[1]** or dissolving it in water **[1]**.

b Describe how you would electroplate a steel spoon with nickel using a suitable electrolyte. **[4]**

Try to memorise the diagram for electroplating. Remember that the plating metal is the anode and that the electrolyte is a soluble salt of this. It is useful to remember that all nitrates are soluble.

A circuit connects the nickel via batteries to the spoon **[1]**. Nickel is made the anode **[1]** and steel is made the cathode **[1]**. The electrolyte is an aqueous solution of a nickel nitrate **[1]**.

In silver plating:

- Silver ions are formed at the anode from silver atoms:

$$Ag \rightarrow Ag^{+} + e^{-}$$

- The silver ions, being positive, are attracted to the cathode.
- At the cathode, silver ions accept electrons from the cathode and become silver atoms:

$$Ag^{+} + e^{-} \rightarrow Ag$$

- The silver atoms form the layer (silver plating) on the cathode.

10.11 Refining copper

Metals can be purified by electrolysis.

- The impure copper is made the anode. This is a reactive anode.
- A thin sheet of pure metal is the cathode.
- The electrolyte is an aqueous solution of a soluble salt of copper e.g. $CuSO_4(aq)$

Figure 10.04 shows the cell used for the purification of copper by electrolysis.

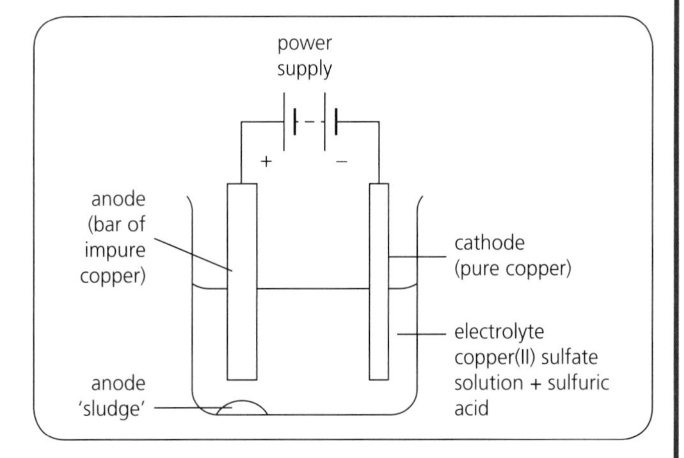

Figure 10.04

- Copper atoms at the anode lose electrons and form Cu^{2+} ions which go into solution:

$$Cu \rightarrow Cu^{2+} + 2e^{-}$$

- The anode becomes thinner and the impurities fall to the bottom of the cell as anode sludge.

- At the cathode Cu^{2+} ions gain electrons and form Cu atoms:

$$Cu^{2+} + 2e^- \rightarrow Cu$$

- The cathode becomes thicker because the pure metal is deposited on it.
- The colour of the solution remains the same depth of blue because the Cu^{2+} ions removed at the cathode are replaced by those going into solution at the anode.

Make sure that if you are asked for observations, you write about what you see, hear, smell or feel.

Sample answer

Describe the electrolysis of aqueous copper(II) sulfate with graphite electrodes and the copper electrodes. In each case, name the electrode products and describe what you would observe during the electrolysis. Include a half equation for the reaction at the cathode. **[10]**

With graphite electrodes, bubbles **[1]** of oxygen **[1]** come from the anode **[1]**. The cathode gets copper deposited on it **[1]**. The equation for this is $Cu^{2+} + 2e^- \rightarrow Cu$ **[1]**. The electrolyte gets lighter in colour **[1]**.

With copper electrodes the anode gets thinner **[1]** because it loses copper. The cathode gets thicker **[1]** because copper is deposited on it **[1]**. The equation for this is the same as for the graphite cathode. The electrolyte stays the same colour **[1]**.

Progress check

10.07 Describe how to electroplate a nickel object with copper. [4]

10.08 Describe two reasons for electroplating articles. [2]

10.12 Electrochemical cells

An **electrochemical cell** is a source of electrical energy. The simplest electrochemical cell consists of two metals of different reactivity dipping into an electrolyte and connected via an external circuit.

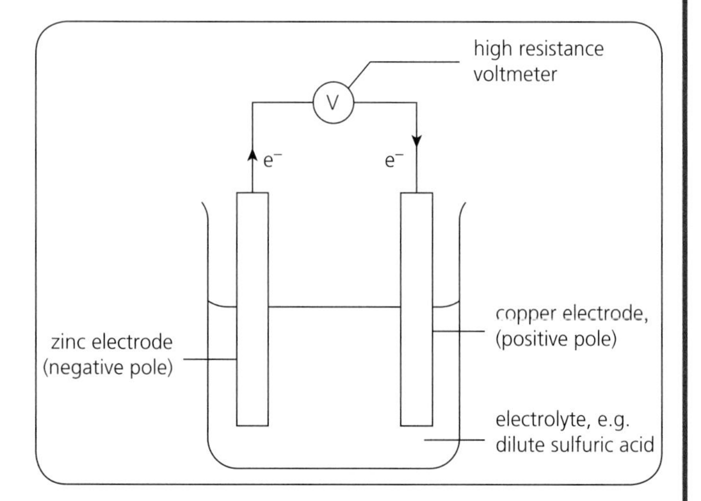

Figure 10.05 A simple electrochemical cell

TERMS

Electrochemical cell: A source of electrical energy where two metals of different reactivity dipping into an electrolyte are connected via an external circuit.

An electrochemical cell provides electrical energy whereas electrolysis requires energy from a power source.

When zinc and copper are placed in dilute sulfuric acid, a voltage is produced.

- Zinc is more reactive than copper, so it tends to lose electrons in preference to copper.

$$Zn(s) \rightarrow Zn^{2+}(aq) + 2e^-$$

- Zinc forms the negative electrode (pole) of the cell.
- At the copper electrode, hydrogen ions from the electrolyte gain electrons and form hydrogen gas:

$$2H^+(aq) + 2e^- \rightarrow H_2(g)$$

- The copper electrode forms the positive electrode (pole) of the cell.
- The difference in the ability of the electrodes to release electrons causes a voltage to be produced. The electrons travel round the external circuit from the negative pole to the positive pole.

Which gives the best voltage?

The table shows some voltages obtained with different combinations of metals in the cell.

Left-hand electrode	Right-hand electrode	Voltage /V	Order of reactivity of metals in right-hand electrode
copper	magnesium	+ 2.70	Most reactive
copper	tin	+ 0.48	↑
copper	copper	0	Least reactive
copper	silver	- 0.46	

Table 10.03

- The further apart the metals are in the reactivity series, the bigger is the voltage produced.
- We can use a table like this to calculate the voltages of other cells. For example, the voltage of a cell with magnesium and tin electrodes would be $(+2.70)-(+0.48) = 2.22V$.
- The voltage of the combination copper and silver is negative because copper is more reactive than silver and so copper is the negative electrode and silver is the positive electrode.

Progress check

10.09 The reactivity of five metals is shown below.

most reactive $\xrightarrow{\text{Mg Zn Fe Pb Cu}}$ least reactive

Which combination of metals will give the highest voltage when placed in an aqueous solution of sulfuric acid? Explain your answer. [2]

10.10 Give two differences between an electrochemical cell and an electrolysis cell. [2]

Exam-style questions

Question 10.01

A student electrolysed molten zinc chloride using graphite electrodes.

a Give two properties of graphite which make it suitable for use as an electrode during this electrolysis. [2]

b State the names of the products formed at the anode and cathode in this reaction. [2]

c Draw a labelled diagram to show this electrolysis. [3]

d Give the names of the electrolysis products formed at the anode and cathode when the electrolyte is concentrated aqueous sodium chloride. [2]

e What is meant by the term electrolyte? [2]

Question 10.02

A simple electrochemical cell is shown below:

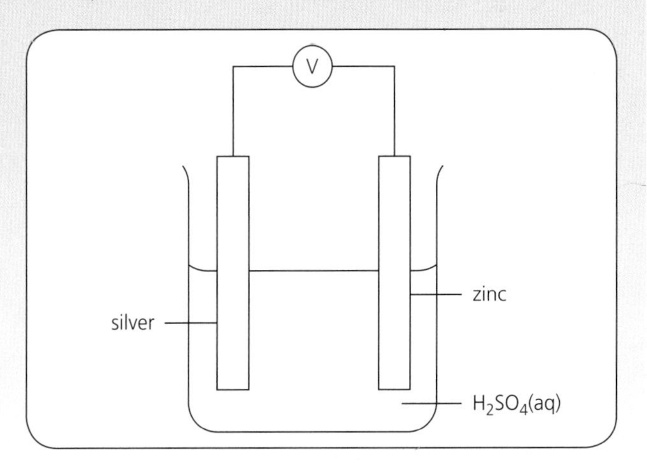

Figure 10.06

a Explain why zinc is the negative electrode and silver is the positive electrode. [2]

b Write half equations for the reactions at the positive electrode and the negative electrode [2]

c In which direction do the electrons move in the external circuit? Explain why? [2]

d Explain why the cell does not produce a voltage when both electrodes are zinc. [1]

e Magnesium is more reactive than zinc. Explain what happens when magnesium replaces silver as an electrode. [2]

Question 10.03

The electrolysis of an aqueous solution of sodium iodide produces iodine at the anode and hydrogen at the cathode.

a Write half equations for the reactions occurring at the cathode and anode. [4]

b Explain why sodium is not formed at the cathode. [2]

c Explain why iodine is formed at the anode rather than oxygen. [2]

d Explain why the solution becomes alkaline during electrolysis. [2]

e A small amount of oxygen is also formed at the anode during electrolysis. Explain why. [2]

f Explain why solid sodium iodide does not conduct electricity but an aqueous solution does conduct. [2]

Revision checklist

You should be able to:

- Describe the use of copper and describe and explain the use of steel-cored aluminium cables in the conduction of electricity
- Explain why plastics and ceramics are used as insulators
- Define the terms electrolysis, electrode, anode, cathode and electrolyte
- Describe the products of the electrolysis of a molten ionic compound containing two simple ions
- Describe the electrolysis of concentrated hydrochloric acid and dilute sulfuric acid
- State the products of electrolysis as metals or hydrogen at the cathode and non-metals at the anode
- Relate the products of electrolysis to the electrolyte and electrodes used
- Describe the electrolysis of aqueous copper(II) sulfate using copper electrodes (copper refining) or graphite electrodes
- Describe the reactions at the electrodes during electrolysis
- Describe the transfer of charge during electrolysis
- Construct ionic half-equations for the reactions at the electrodes
- Describe electroplating and some of its uses
- Describe the electrolysis of aluminium oxide
- Describe the manufacture of chlorine from concentrated aqueous sodium chloride
- Describe a simple electrochemical cell
- Relate the voltage of an electrochemical cell to the reactivity of the electrodes

Chemical energetics

Learning outcomes

By the end of this unit you should:

- Be able to describe the meaning of exothermic and endothermic reactions
- Be able to draw and label energy level diagrams for exothermic and endothermic reactions
- Be able to draw labelled energy level diagrams for exothermic and endothermic reactions using data provided
- Be able to describe bond breaking as an endothermic process and bond forming as an exothermic process
- Be able to calculate the energy of a reaction using bond energies
- Be able to describe the release of heat energy by burning fuels
- Be able to state the use of hydrogen as a fuel
- Know that uranium-235 is a source of energy
- Describe the reactions in a fuel cell

11.01 Exothermic and endothermic changes

Exothermic reactions release heat energy to the surroundings. The temperature of the surroundings increases. Examples of exothermic reactions include:

- the combustion of fuels
- the reactions of acids with metals
- the neutralisation reactions of acids with bases
- the production of energy from electrochemical cells

TERMS

Exothermic: A reaction or process which releases energy.

Endothermic: A reaction or process which absorbs energy.

Endothermic reactions absorb heat energy from the surroundings. The temperature of the surroundings decreases. Examples of endothermic reactions include:

- the thermal decomposition of carbonates
- electrolysis
- the first stages of photosynthesis (see Unit 12)

Progress check

11.01 What type of heat change occurs in the following reactions? Explain your answers.

a $2KNO_3 \xrightarrow{heat} 2KNO_2 + O_2$ [2]

b Burning paraffin. [3]

11.02 Copy and complete:

Exothermic reactions __________ heat to the __________. Reactions which are __________ absorb __________ from the surroundings. The __________ of fuels is exothermic. The thermal ______ of carbonates is endothermic. [6]

11.02 Energy level diagrams

- In an exothermic reaction, the reactants have more chemical energy than the products. So when the reaction occurs, heat energy is given out to the surroundings.
- In an endothermic reaction, the reactants have less chemical energy than the products. So when the reaction occurs, heat energy is absorbed (taken in) from the surroundings.

- We can show these energy changes by arrows (Figure 11.01):

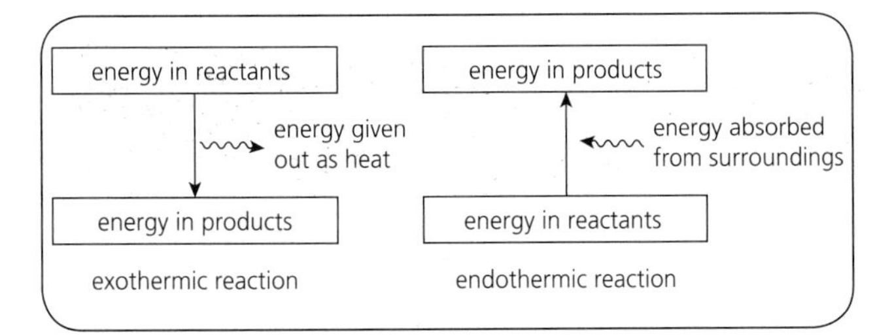

Figure 11.01

TERMS

Energy level diagram: A diagram showing the energy change of the reaction on the vertical axis and the reactants and products on the horizontal axis.

Figure 11.02 shows **energy level diagrams** for both an exothermic and an endothermic reaction.

- The heat energy content of the reactants and products is shown on the vertical axis.
- The reaction pathway is shown on the horizontal axis. The reaction pathway shows the reactants (on the left) to products (on the right).
- The arrow shows whether the reaction is exothermic or endothermic. A downward arrow describes an exothermic reaction. An upward arrow describes an endothermic arrow.

- We use the symbol ΔH (delta H) for the heat energy change.
- The units of heat energy change are kilojoules per mole (kJ/mol).

For an exothermic reaction:

- the energy of the reactants is higher than the energy of the products
- so energy is released to the surroundings and
- energy of products minus energy of reactants is a negative value
- so ΔH is negative, e.g:

  $CH_4(g) + 2O_2(g) \rightarrow CO_2(g) + H_2O(l)$ $\Delta H = -890$ kJ/ mol

For an endothermic reaction:

- the energy of the reactants is lower than the energy of the products
- so energy is absorbed from the surroundings
- energy of products minus energy of reactants is a positive value
- so ΔH is positive, e.g.

  $CaCO_3(s) \rightarrow CaO(s) + CO_2(g)$ $\Delta H = +572$ kJ/mol

Worked example 11.01

Draw an energy level diagram for the reaction of steam with carbon. The reaction is endothermic.

$H_2O(g) + C(s) \rightarrow H_2(g) + CO(g)$ **[4]**

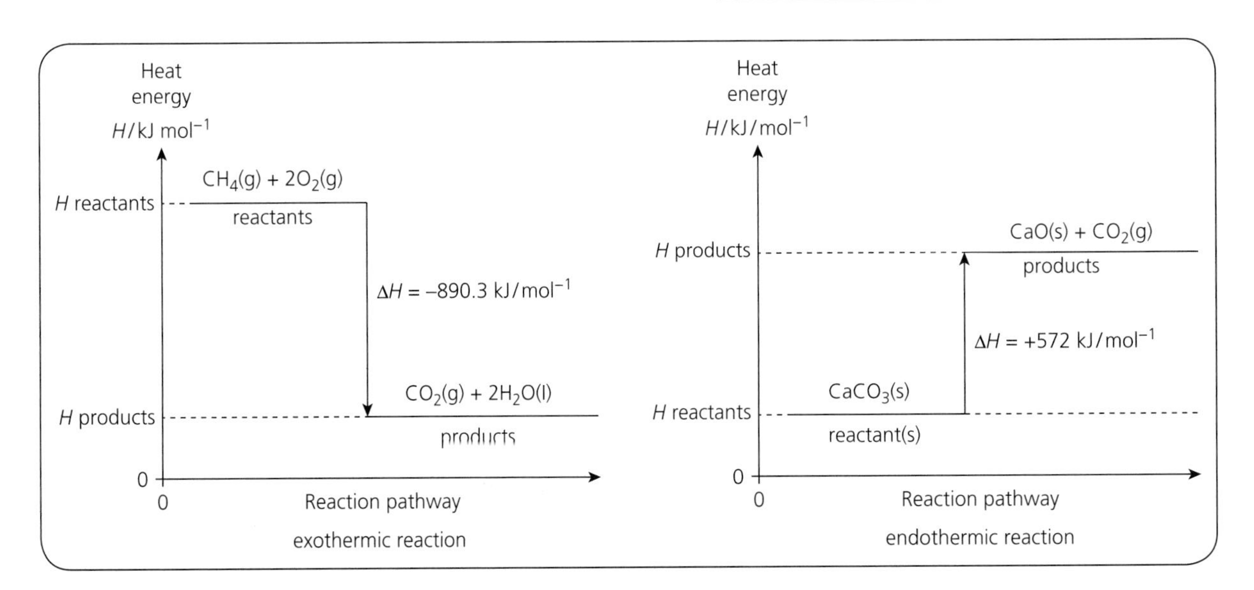

Figure 11.02

You should remember to label the axes and use the information in the question to draw the arrow in the correct direction. In this case the reaction is endothermic. You are putting energy in. So the arrow points upwards.

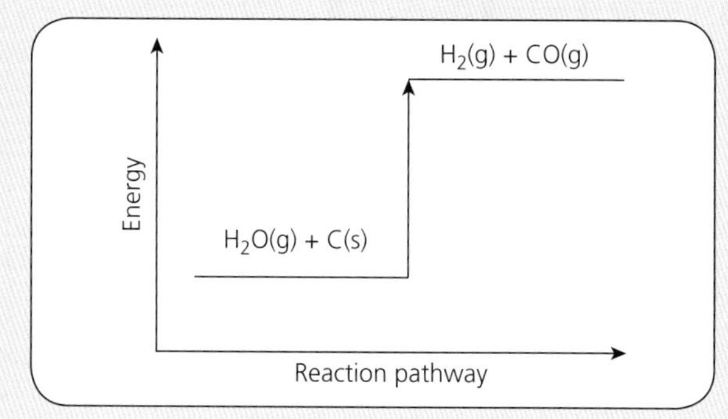

Fig 11.03

Axes correctly labelled **[1]**

Reactants on left and products on the right **[1]**

Product level above reactant level **[1]**

Arrow between reactants and products in correct direction (upwards) **[1]**

11.03 Making and breaking bonds

- Each type of bond needs a particular amount of energy to break it. For example, to break an O–H bond in water, it needs 464 kJ of energy per mole of bonds broken. This energy is called the **bond energy** (symbol *E*).

TERMS

Bond energy: The energy needed to break one mole of covalent bonds between two particular atoms.

- Energy is needed to break bonds. So bond breaking is endothermic.
- Energy is released when new bonds are formed. So bond making is exothermic.
- In an endothermic reaction, more energy is needed to break the bonds in the reactants than is given out making new bonds in the products.
- In an exothermic reaction, more energy is released on forming new bonds in the products than is needed to break the bonds in the reactants.

It is a common error to suggest that 'in an endothermic reaction more energy is used in breaking than making bonds'. This suggests that energy is used in making bonds which is incorrect.

Bond energy calculations

We can use bond energies to calculate how much heat is released or absorbed in a reaction. To do this, we need to know the bonds present in the reactants and products.

Worked example 11.02

Use the following bond energies to calculate the energy change in the reaction:

$$2H_2(g) + O_2(g) \rightarrow 2H_2O(g)$$

$E(\text{H}-\text{H}) = 436\,\text{kJ/mol}$, $E(\text{O=O}) = 498\,\text{kJ/mol}$, $E(\text{O–H}) = 464\,\text{kJ/mol}$.

Step 1: Identify the types of bonds in the reactants and products (you do not have to write these down unless you feel it helps you).

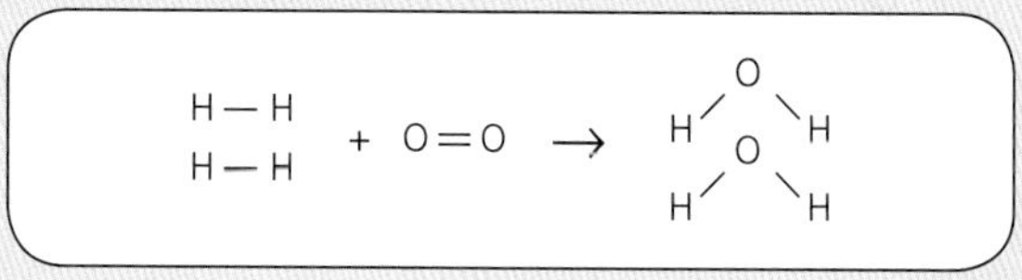

H—H
H—H + O=O → H–O–H, H–O–H

Fig 11.04

Step 2: Set out the bond energies as shown, taking into account the number of bonds broken and formed.

bonds broken in reactants (endothermic, ΔE positive)	bonds formed in products (exothermic, ΔE negative)
$2 \times E(\text{H–H}) = 2 \times 436$	$2 \times 2 \times E(\text{O–H}) = 4 \times 464$
$E(\text{O=O}) = 498$	
Total = +1370 kJ	Total = −1856 kJ

Step 3: Deduce the heat energy change:

The exothermic change is greater by 1856 − 1370 kJ = 486 kJ

So the heat energy change is −486 kJ.

Progress check

11.03 Oxygen can be broken down to ozone: $3O_2 \rightarrow 2O_3$. The bond energy *E*(O=O) is 805 kJ/mol. Calculate the energy change involved in breaking down three moles of O_2 into atoms. [1]

11.04 Explain why the reaction $CH_4(g) + 2O_2(g) \rightarrow CO_2(g) + H_2O(l)$ is exothermic in terms of bond breaking and bond making. [3]

11.04 Energy transfers

A fuel is a substance which releases energy when burned. When hydrocarbon fuels (see Unit 27) burn, carbon dioxide and water are produced. For example:

$C_3H_8(l) + 5O_2(g) \rightarrow 3CO_2(g) + 4H_2O(l)$ $\Delta H = -2219$ kJ/mol
propane oxygen carbon dioxide water

This type of reaction is called a combustion reaction.

When writing equations you do not include the words 'energy' or 'heat' on the left or right of the equation. This is because 'heat' is not a substance. The energy change is added as shown in the example above. You can, however, put 'heat' above the arrow.

We can measure the heat energy change when a substance undergoes combustion by burning a known mass of the substance. The heat energy released is used to raise the temperature of a known mass of water. To measure the heat energy released when a liquid fuel burns, we use a spirit burner and a metal can called a calorimeter (Figure 11.05).

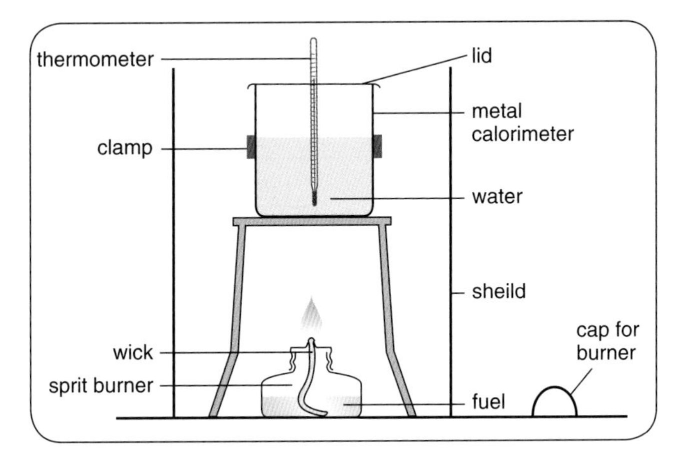

Figure 11.05

We can use this apparatus to compare the energy released per gram from different fuels:

- We measure the temperature rise of the water.
- We calculate the amount of fuel burned to give this temperature rise by weighing the spirit burner before and after the experiment.
- We keep the volume of water in the calorimeter the same and keep stirring this during the experiment.
- We compare the amount of fuel burned for the same temperature rise.

Sample answer

Suggest how you would carry out an experiment to compare the energy changes when different salts such as sodium chloride and copper sulfate are dissolved in water. **[8]**

I would take an insulated plastic beaker **[1]** to prevent heat losses and put a measured volume of water in it (50 cm^3) **[1]**. I would then take the temperature of the water with a thermometer **[1]** then add 2 grams **[1]** of copper sulfate and stir **[1]** until it dissolved. I would measure the highest temperature **[1]**. Then I would repeat the experiment with sodium chloride using the same volume of water **[1]** and the same mass of sodium chloride **[1]**.

Using hydrogen as a fuel

Hydrogen is used in rocket fuels and in fuel cells to power vehicles.

Advantages of hydrogen as a fuel:

- It releases more energy per kilogram than other fuels (apart from nuclear 'fuels').
- It is not polluting. The only combustion product is water.

Disadvantages of hydrogen as a fuel:

- It is expensive to produce and production may involve the use of energy sources which pollute the atmosphere e.g. burning fossil fuels
- It is difficult to store. It has to be stored in pressurised containers as a liquid.

Radioactive isotopes as a source of energy

Nuclear power stations use the heat given off by uranium-235 when it decays to heat up a gas. The hot gas heats up water and changes it into steam. The energy of the jet of steam produced is used to power turbines for electricity generation. The energy obtained from nuclear fuels is 'clean'. It does not produce carbon dioxide, sulfur dioxide or nitrogen oxides. Only a small amount of uranium-235 is needed to produce a large amount of energy but nuclear power is expensive and there is a possibility that accidents may cause the release of harmful radioactive materials.

Progress check

11.05 Name a radioactive isotope that is used as a source of energy. [1]

11.06 Give an advantage of using uranium-235 as a fuel rather than coal. [1]

11.05 Fuel cells

A **fuel cell** is an electrochemical cell in which a fuel gives up electrons at one electrode and oxygen gains electrons at the other electrode. Fuel cells are increasingly used instead of petrol or diesel to power buses and cars. One type of fuel cell is the hydrogen–oxygen fuel cell with an acidic electrolyte (Figure 11.06).

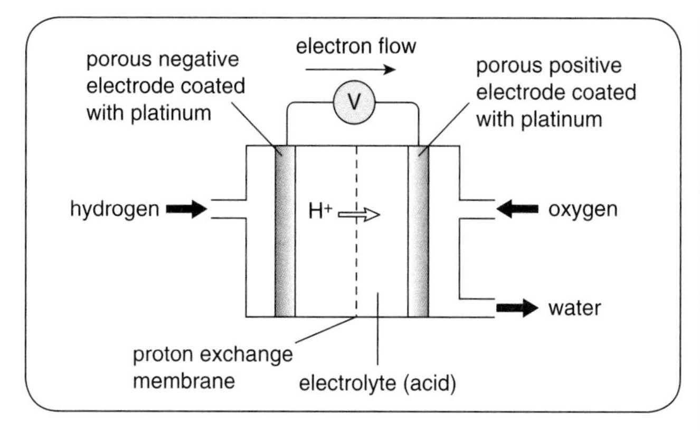

Figure 11.06

TERMS

Fuel cell: A type of electrochemical cell where hydrogen and oxygen undergo a reaction to produce electrical energy.

- Hydrogen and oxygen are bubbled through two porous electrodes where the half reactions take place.
- At the negative electrode, hydrogen loses electrons:

$$H_2(g) \rightarrow 2H^+(aq) + 2e^-$$

- At the positive electrode, oxygen gains electrons and combines with H^+ ions to form water.

$$4H^+(aq) + O_2(g) + 4e^- \rightarrow 2H_2O(l)$$

- The electrons move round the external circuit from the negative electrode to the positive electrode. As they do so, their energy can be used to do a useful job such as driving an electric motor.
- The overall reaction is $2H_2(g) + O_2(g) \rightarrow 2H_2O(l)$

You do not have to know the details of the construction of fuel cells or about how exactly they operate, but it is useful to be able to write the relevant half reactions from information provided.

Advantages of fuels cells over petrol engines:

- They are non-polluting since water is the only product made (no carbon dioxide or oxides of nitrogen are produced – see Unit 24).
- They produce more energy per kg of fuel burned than do petrol engines.

- The transmission of energy from the fuel cell to the motor is direct. There are no moving parts as in a petrol engine where energy is wasted as heat.

Limitations of fuel cells:

- The materials used to make the membrane and electrodes are expensive.
- High pressure tanks are needed to store sufficient amounts of hydrogen and oxygen.
- Fuel cells do not work very well at low temperatures.
- Hydrogen is expensive to produce and production may involve use of energy sources which pollute the atmosphere.

Progress check

11.07 Describe in simple terms how a fuel cell works. [3]

11.08 Give two advantages of a fuel cell compared with a petrol engine. [2]

Exam-style questions

Question 11.01

When magnesium reacts with hydrochloric acid, energy is released.

a State the name given to a chemical reaction which releases energy. [1]

b Copy and complete the energy level diagram for this reaction. [4]

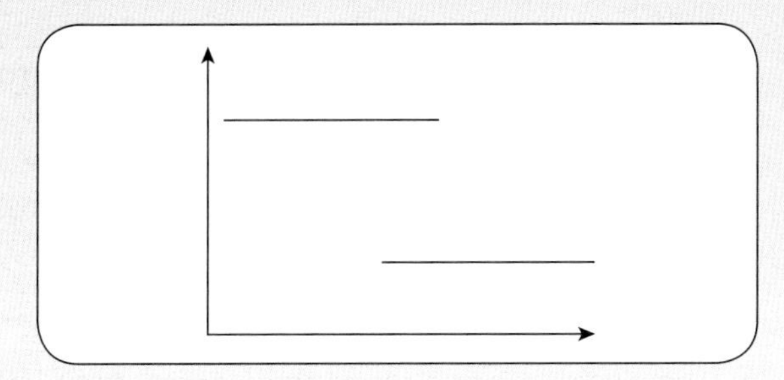

Figure 11.07

c Equal amounts of three compounds are dissolved in the same amount of water. The results are shown in the table.

Compound	Temperature of water at start / °C	Final temperature of solution / °C
sodium fluoride	23	27
magnesium chloride	19	25
lithium chloride	21	24

Table 11.01

i Which compound shows the greatest temperature change? [1]

ii When sodium fluoride dissolves in water, estimate the temperature change when:

- The amount of sodium fluoride is doubled keeping the amount of water the same. [1]
- The amount of water is doubled keeping the amount of sodium fluoride the same. Explain your answer. [2]

Question 11.02

When hydrogen reacts with oxygen, water is formed.

a Give two reasons why hydrogen is a better fuel for vehicles than petrol. [2]

b State one problem with the use of hydrogen as a fuel. [1]

c Draw an energy level diagram for the reaction of hydrogen with water. Label your diagram fully. [4]

d Hydrogen and oxygen react in a fuel cell containing an acid electrolyte.

i Write a half equation for the reaction involving hydrogen at the negative electrode. [1]

ii Explain why an electric current is produced in the external circuit. [3]

e Hydrogen reacts with chlorine to form hydrogen chloride:

$$H_2(g) + Cl_2(g) \rightarrow 2HCl$$

Use bond energy values to calculate the energy change in this reaction. [3]

$E(\text{H–H}) = 436\,\text{kJ/mol}$; $E(Cl\text{–}Cl) = 243\,\text{kJ/mol}$; $E(\text{H–}Cl) = 432\,\text{kJ/mol}$

Question 11.03

Nitrogen reacts with hydrogen to form ammonia:

$$N_2(g) + 3H_2(g) \rightleftharpoons 2NH_3(g)$$

The reaction is exothermic.

a Draw an energy level diagram for this reaction. [4]

b What is meant by the term *bond energy*? [1]

c Use bond energy values to calculate the energy change in this reaction. [3]

$E(N{\equiv}N) = 945$ kJ/mol; $E(H{-}H) = 436$ kJ/mol; $E(N{-}H) = 391$ kJ/mol.

d Hydrogen is used in fuel cells. In a fuel cell with an alkaline electrolyte, the electrode reaction at the cathode involves water reacting with oxygen to form hydroxide ions.

i Write the half equation for this reaction. [2]

ii Give two advantages and one disadvantage of a fuel cell. [3]

Revision checklist

You should be able to:

- Describe the meaning of exothermic and endothermic reactions
- Interpret energy level diagrams showing exothermic and endothermic reactions
- Draw and label energy level diagrams for exothermic and endothermic reactions
- Understand bond breaking and bond making in terms of the energy changes involved (endothermic and exothermic)
- Calculate the energy change of a reaction using bond energy values
- Describe the release of heat energy by burning fuels
- State the use of hydrogen as a fuel
- Describe the use of hydrogen in a hydrogen–oxygen fuel cell
- Describe the radioactive isotope uranium-235 as a source of energy

Rate of reaction

Learning outcomes

By the end of this unit you should:

- Know some practical methods for investigating the rate of reactions where a gas is produced
- Be able to devise and evaluate a suitable method for investigating the effect of a given variable on the rate of a reaction
- Be able to interpret data obtained from experiments concerned with rate of reaction
- Be able to describe and explain the effect of concentration, particle size, catalysts (including enzymes) and temperature on the rate of reactions
- Be able to link the effect of particle size to the danger of explosive combustion of fine powders
- Be able to explain the effects of concentration and temperature on reaction rate using the collision theory
- Be able to describe and explain the role of light in photochemical reactions
- Be able to describe the use of silver salts in photography
- Be able to describe photosynthesis

12.01 Reaction rate

Rate of reaction is the change in concentration of a reactant or product with time:

$$\text{rate} = \frac{\text{change in concentration of reactant or product}}{\text{time taken for this change}}$$

TERMS

Rate of reaction: The rate at which reactants are used up or the rate at which products are formed usually calculated as change in concentration in mol/dm^3 divided by time.

We do not usually measure concentration directly. So when carrying out experiments on rates of reaction, we measure something that is proportional to concentration, for example mass, volume of gas, colour intensity, electrical conductivity.

We can express rate as the change in volume of gas produced per second or change in mass of the reaction mixture per second or change in some other variable per unit time. The best unit to use to express rate is $mol/dm^3/s$.

12.02 Methods of following the course of a reaction

We can measure either:

- the decrease in mass (or some other property) of a particular reactant as the reaction progresses, or
- the increase in volume of gas (or some other property) of a particular product as the reaction progresses

Measuring gas volumes

Measuring the volume of gas produced at particular time intervals from the start of the reaction is a simple way of following the progress of the reaction. We use the apparatus shown in Figure 12.01.

We measure the volume of gas produced at particular time intervals, for example every 10 seconds.

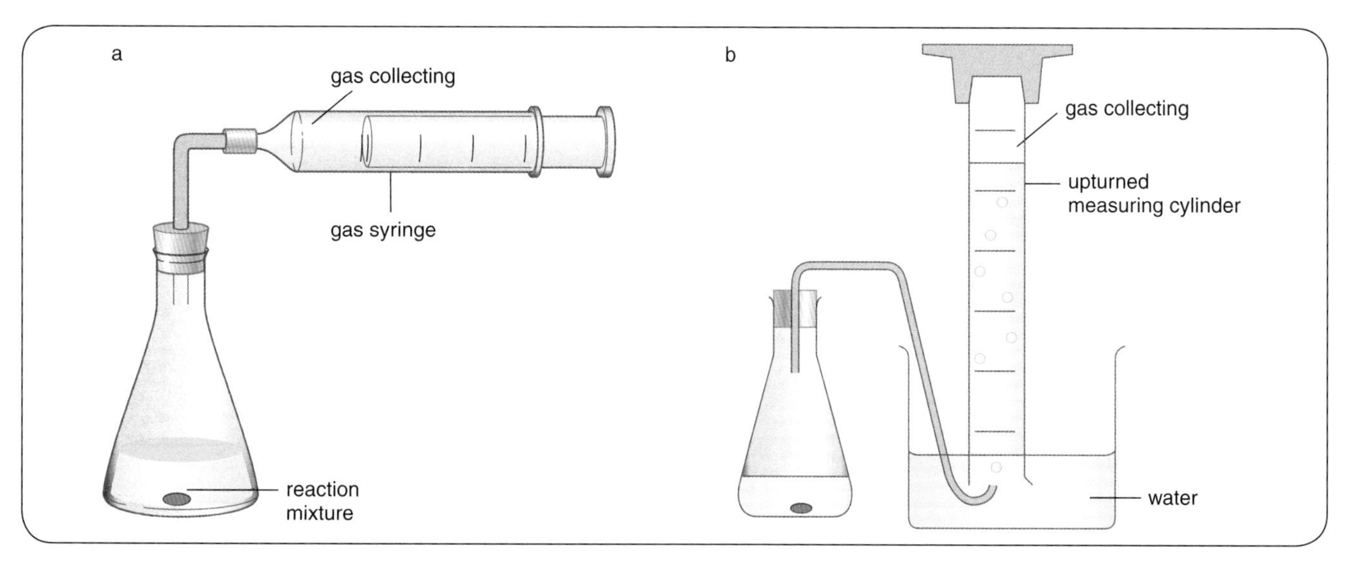

Figure 12.01

Progress check

12.01 What is meant by the term rate of reaction? [2]

12.02 Give two methods of measuring the rate of a reaction which gives off a gas. For each, describe the measurements to be taken. [4]

12.03 Describe how the rate of a reaction changes as the reaction proceeds. [2]

Other ways of following the course of a reaction

Figure 12.02 shows some other methods of following the course of a reaction. The method depends on the physical properties of the reactants and products.

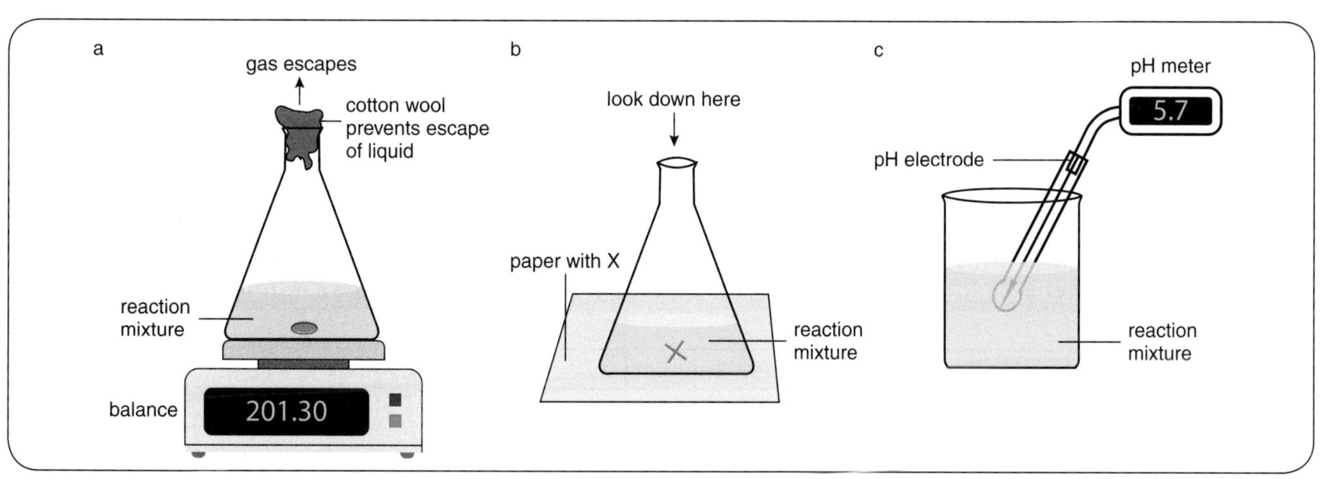

Figure 12.02

a Measuring the decrease in mass of the reaction mixture as a gas is released.

b Timing how long it takes for a precipitate formed during a reaction, to cover a cross so that it cannot be seen.

c Measuring a change in pH of the reaction mixture.

We can also take samples of the reaction mixture from time to time and analyse them by titration. For this, we have to stop the reaction by adding another chemical to the sample we have taken.

12.03 Interpreting data

Hydrogen gas is produced when magnesium reacts with excess hydrochloric acid:

$$Mg(s) + 2HCl(aq) \rightarrow MgCl_2(aq) + H_2(g)$$

Figure 12.3 shows how the volume of gas produced changes for this reaction.

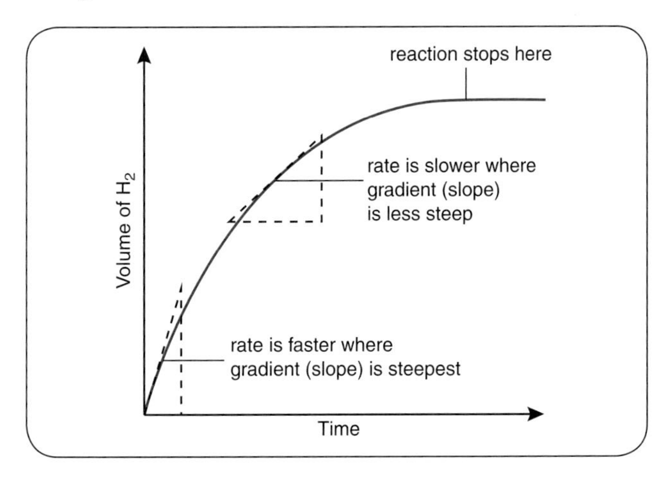

Figure 12.03

The graph shows us:

- how the rate changes with time. The gradient (slope) gets less as time goes on. This shows that the rate decreases as time increases
- when the reaction has stopped
- the volume of gas produced at a particular time

Figure 12.04 shows how the mass of the reaction mixture changes as the gas escapes.

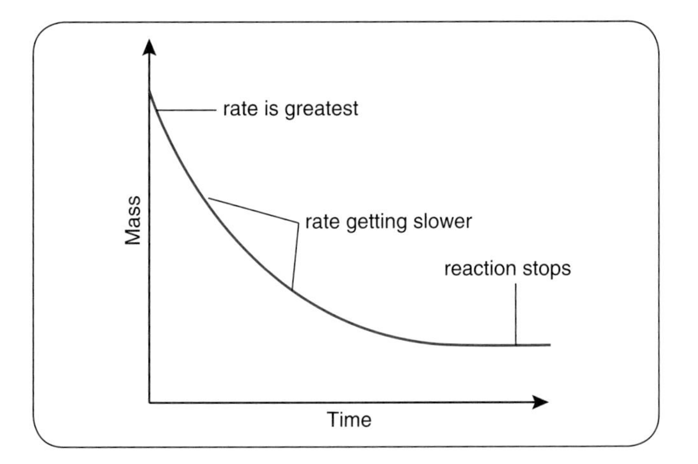

Figure 12.04

In some graphs the volume increase or mass decrease appears to be proportional to the time near the start of the reaction. You can calculate the initial rate of reaction by taking the initial gradient of the graph.

Worked example 12.01

The table compares the rate of reaction for the formation of iodine from iodide ions (I^-) and hydrogen peroxide (H_2O_2) in acid solution when different concentrations of reactants are used.

Describe how each of the reactants affects the rate of reaction. **[5]**

When answering this type of question you must choose a concentration in one of the columns so that the concentrations in the other two columns are kept the same.

By comparing A and B you are changing the concentration of H_2O_2 keeping concentration of iodide and acid constant.

So doubling concentration of H_2O_2 doubles rate **[2]**.

(But increase concentration of H_2O_2 increases rate gives **1 mark**.)

By comparing A and C you are changing the concentration of iodide but keeping concentration of peroxide and acid constant.

Doubling concentration of iodide doubles rate **[2]**.

(But increase concentration of iodide increases rate gives **1 mark**.)

By comparing A and D you are changing the concentration of acid but keeping concentration of peroxide and iodide constant.

Increase concentration of acid has no effect on rate **[1]**.

Experiment Number	Relative rate of reaction	Concentration of H_2O_2 in mol/dm^3	Concentration of I^- in mol/dm^3	Concentration of acid in mol/dm^3
A	1	0.2	0.2	0.2
B	2	0.4	0.2	0.2
C	2	0.2	0.4	0.2
D	1	0.2	0.2	0.4

Table 12.01

12.04 Factors affecting reaction rate

Concentration of reactant

Increasing the concentration of a reactant (keeping everything else constant) increases the rate of reaction.

Figure 12.05 shows how increasing the concentration of hydrochloric acid increases the rate of reaction between hydrochloric acid and calcium carbonate when the hydrochloric acid is in excess. The final volume of gas is the same because all the calcium carbonate was used up in each case:

$$CaCO_3(s) + 2HCl(aq) \rightarrow CaCl_2(aq) + H_2O(l) + CO_2(g)$$

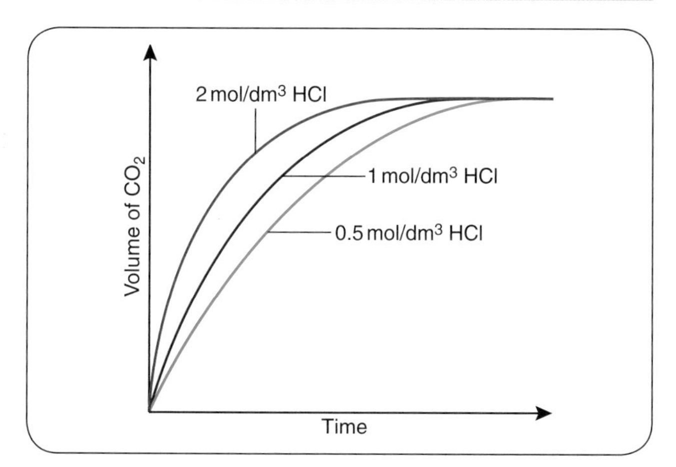

Figure 12.05

Surface area

Increasing the surface area of a solid (keeping everything else constant) increases the rate of reaction. So if we use the same mass of calcium carbonate, the rate increases as the particles get smaller. This is because the surface area is greater, the more the particles are 'cut up'.

large particles of $CaCO_3$	small particles of $CaCO_3$	powdered $CaCO_3$
——— faster rate of reaction ———>		

Industrial processes such as metal-working and milling grain cause fine powders to get into the air, for example particles of flour from flour mills, sawdust from sawmills, tiny particles of metal from metalworking or coal dust from coal mining. These fine powders are combustible. They burn rapidly in air and may even explode. This is because each particle has a very small volume and relatively large surface area. A lit match or spark from a machine can cause the fine powders suspended in air to explode. In coal mines methane gas mixed with air can also form an explosive mixture.

The effect of temperature

Increasing the temperature of the reaction mixture (keeping everything else constant) increases the rate of reaction. The rate approximately doubles for each 10 °C rise in temperature.

TIP

When comparing reaction rates at different concentrations or temperatures it is important that you use comparative words or phrases such as 'rate of reaction is fastER the highER the temperature' instead of 'the reaction is fast at high temperatures'.

Addition of a catalyst

TERMS

Catalyst: A substance which speeds up a chemical reaction without being used up or chemically changed after the reaction.

Enzyme: A biological (protein) catalyst.

- A **catalyst** is a substance which speeds up a chemical reaction and is not used up in the reaction.
- The mass of catalyst is the same at the end as at the start of the reaction.
- A small amount of catalyst can cause a big increase in reaction rate.
- The chemical composition of the catalyst at the end of the reaction is the same as that at the beginning.
- Catalysts can be solids, e.g. manganese dioxide, or substances in solution, e.g. hydrogen ions.
- **Enzymes** are biological catalysts. They are proteins which speed up the rate of most reactions taking place in living things. Enzymes work best over a

narrow temperature range. Most are inactivate below 0 °C and decrease in their activity above 40 °C. Most are permanently inactivated (denatured) above about 50 °C.

Sample answer

When zinc reacts with excess hydrochloric acid, hydrogen gas is produced.

a Describe how you could carry out an experiment to follow the progress of this reaction. **[5]**

b State the effect of the following on the rate of reaction: **i** decreasing concentration of acid, **ii** increasing the temperature, **iii** adding the same mass of zinc using larger pieces, **iv** adding a catalyst. **[4]**

a I would set up a flask with 20 cm^3 of hydrochloric acid **[1]** connected to a gas syringe by a tube **[1]**. Then I would add the zinc to the flask to start the reaction and quickly stopper the flask **[1]**. I would then measure the volume of hydrogen in the gas syringe **[1]** every 20 seconds **[1]**.

b **i** decreases rate **[1]**, **ii** increases rate **[1]**, **iii** decreases rate **[1]**, **iv** increases rate **[1]**.

Progress check

12.04 Explain why there is a danger of explosions occurring in buildings where grain is milled to make flour. [3]

12.05 How is reaction rate affected by

a decrease in temperature?

b increase in concentration?

c addition of a catalyst? [3]

12.06 Hydrogen gas is produced when iron reacts with hydrochloric acid. When powdered iron reacts with hydrochloric acid, 40 cm^3 of hydrogen are released in the first 30 seconds of the reaction. When lumps of iron are used (and all other conditions remain the same) only 10 cm^3 of gas are released in the first 30 seconds. Describe and explain this difference. [2]

12.05 Explaining changes in reaction rate

Reaction rate and change in concentration

- Increasing the concentration, decreases the distance between the reacting particles. There are more reactant particles per cm^3.
- The rate of collisions is increased. So there are more collisions per second between the particles.
- The kinetic theory suggests that when particles collide a proportion of them react. So, the more collisions per second, the faster is the reaction rate.

When explaining the effect of increasing concentration on reaction rate, remember:

- There are more particles *per unit volume* not just more particles.
- It is increased *rate* or increased *frequency* of collisions which is important, not just more collisions.

Reaction rate and change in surface area

The greater the surface area of a solid, the greater is the number of particles exposed. Breaking up calcium carbonate into smaller pieces exposes more surfaces, resulting in more particles being available to react. So, using the same mass of marble, smaller pieces react faster because:

- there are more reactant particles of solid exposed per cm^3 of solution
- the rate of collisions between the solid and reactant particles in solution is increased
- so the reaction rate is increased

Reaction rate and change in temperature

Particles must have a minimum amount of energy in order to react when they collide. This is called the **activation energy**. At low temperatures only a small number of the particles have sufficient activation energy. So the reaction is slow. The effect of temperature on reaction rate can be explained in the following way:

TERMS

Activation energy: The minimum energy needed for particles to react when they collide.

- An increase in temperature increases the kinetic energy of the particles. The particles move faster and have greater energy.
- The higher the temperature, the more energetic the collisions between the reactant particles.
- The higher the temperature, the greater the number of reactant particles have energy above or equal to the activation energy.
- The number of collisions leading to a reaction is increased. So the reaction rate increases.

TIP

An increase in temperature causes a very small increase in collision rate. The effect of the increase in the number of particles having sufficient activation energy with increasing temperature has far more effect. The activation energy explanation is better.

Progress check

12.07 Use the collision theory to explain:

a Why an increase in temperature increases reaction rate. [3]

b An increase in pressure in a reaction involving gaseous reactants and products, increases reaction rate. [3]

12.08 Suggest and explain a suitable method for following the progress of each of these reactions:

a $2HCl(aq) + MgO(s) \rightarrow MgCl_2(aq) + H_2O(l)$ [2]

b $2KNO_3(s) \rightarrow 2KNO_2(s) + O_2(g)$ [2]

12.09 Describe how and explain why magnesium oxide powder reacts more rapidly with hydrochloric acid than lumps of magnesium oxide. [3]

12.06 Photochemical reactions

TERMS

Photochemical reaction: A reaction which is catalysed by light or dependent on light for the reaction to occur.

Some reactions only occur in the presence of light or are catalysed by light. These are called **photochemical reactions.** The substitution of hydrogen atoms in methane by chlorine atoms in the presence of ultraviolet light (see Unit 27) is an example of a photochemical reaction:

$$CH_4(g) + Cl_2(g) \xrightarrow{\text{light}} CH_3Cl(g) + HCl(g)$$

The greater the intensity of the ultraviolet light, the faster is the reaction rate between the chlorine and methane.

Silver salts in photography

- The surface of film for black and white photography contain crystals of silver bromide.
- When light shines on the film, some of the silver bromide decomposes to silver:

$$2AgBr \rightarrow 2Ag + Br_2$$

- The silver bromide appears colourless at low concentrations but the silver appears grey-black.
- Parts of the film exposed to a stronger light appear dark grey or black and those exposed to weaker light appear lighter grey. Parts not exposed to light appear white.
- The reaction is a catalysed redox reaction (see Unit 14):

  Reduction: $Ag^+ + e^- \rightarrow Ag$ Oxidation: $2Br^- \rightarrow Br_2$
- A positive print is made by shining light through the negative onto a piece of photographic paper.

Photosynthesis

TERMS

Photosynthesis: The process of producing glucose and oxygen from carbon dioxide and water in plants in the presence of chlorophyll and light.

Photosynthesis is the process by which plants produce glucose and oxygen from carbon dioxide and water. The overall reaction is:

$$6CO_2 + 6H_2O \xrightarrow{\text{sunlight}} \underset{\text{glucose}}{C_6H_{12}O_6} + 6O_2$$

Sunlight is absorbed by pigments in the plant called chlorophylls. Chlorophylls act as catalysts in the first steps in the process. The greater the amount of sunlight absorbed, the faster is the rate of photosynthesis (as long as the carbon dioxide or chlorophyll are not limiting).

Progress check

12.10 Suggest two ways to speed up photosynthesis, other than by increasing the temperature. [2]

12.11 Explain why silver bromide goes dark grey in the light. [3]

Sample answer

A sheet of white paper was soaked in silver bromide solution and then allowed to dry. A metal letter 'L' and a piece of tissue paper were placed on top of the paper as shown.

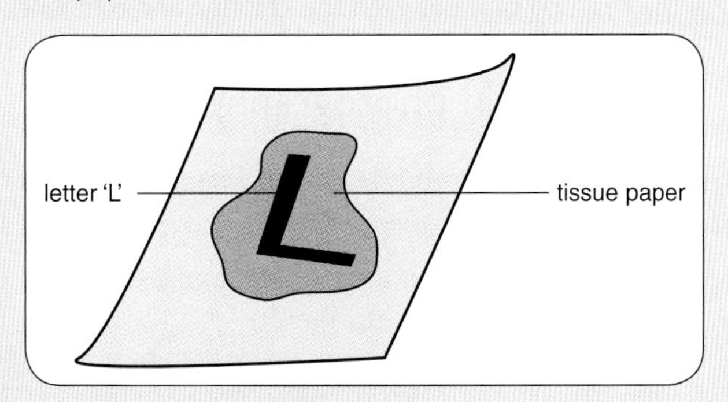

Figure 12.06

The paper was left in the light for a few minutes. Different parts of the paper turned different shades of grey. Describe and explain exactly what you would observe. Include an equation for the reduction of silver ions. **[7]**

The light causes a photochemical reaction **[1]** with the silver bromide. The light changes silver bromide to silver **[1]** which is grey. The equation for this reaction is $Ag^+ + e^- \rightarrow Ag$ **[1]**.

The L appears white **[1]** because the light cannot get through **[1]**. The area under the tissue appears grey **[1]** because only a small amount of light gets through **[1]**. The rest of the paper appears darker or black **[1]** because it has been exposed to more light **[1]**.

There are nine marking points. Three marks are for the first three marking points and any four of the other six marking points are credited.

Exam-style questions

Question 12.01

When excess calcium carbonate reacts with hydrochloric acid carbon dioxide is released. The graph shows how the volume of carbon dioxide changes when the experiment is carried out using two different concentrations of acid.

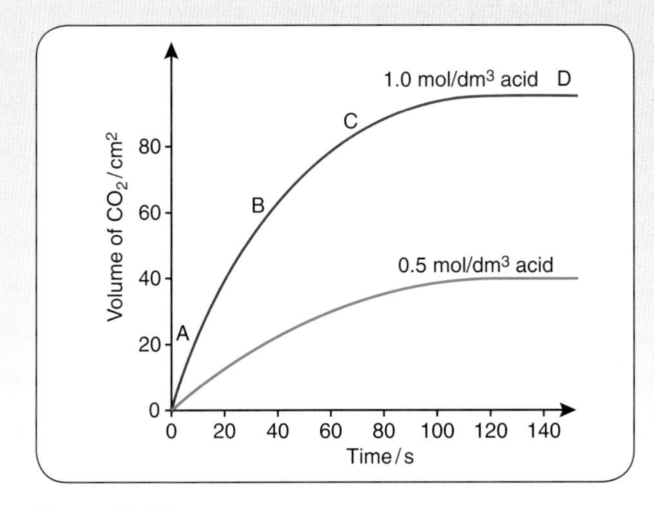

Figure 12.07

a Look at the line for the acid of concentration 1 mol/dm³.

i At what time did the reaction stop? [1]

ii What volume of carbon dioxide had been released 30 s after the start of the reaction? [1]

iii At which letter, A, B, C or D, is the rate of reaction fastest? Give a reason for your answer. [2]

b What information from the graph shows you that increasing the concentration of the acid, increases the rate of reaction? [1]

c How would the rate of reaction differ if smaller pieces of calcium carbonate were used? Explain your answer. Assume all other factors remain constant. [3]

Question 12.02

The table shows how changing the temperature changes the rate of reaction between aqueous sodium thiosulfate and hydrochloric acid. A precipitate of sulfur is formed. The time taken for the precipitate to obscure a cross at different temperatures was recorded using the same concentration of thiosulfate and acid each time.

Temperature / °C	Time to obscure cross / s	1/time
20	47	0.021
30	23	0.043
40	12	0.083
50	6	0.167

Table 12.02

a The third column is proportional to reaction rate. Explain why. [1]

b Use the information in the table to describe how the reaction rate varies with temperature. [2]

c Use ideas about particles to explain why reaction rate changes with temperature. [3]

d Describe and explain what will happen to the reaction rate when water is added to the reaction mixture. [3]

e Use ideas about colliding particles to explain why increasing the concentration of acid, increases the rate of this reaction. [2]

Revision checklist

You should be able to:

- ☐ Describe a method for investigating rate of reaction where a gas is given off
- ☐ Describe other methods for investigating rate of reaction
- ☐ Interpret data obtained from experiments concerned with rate of reaction
- ☐ Describe how particle size, concentration, temperature and catalysts affect rate
- ☐ Explain how particle size, concentration, temperature and catalysts affect rate
- ☐ Describe and explain the effect of light on silver bromide
- ☐ Describe photosynthesis in simple terms (equation, light, chlorophyll)

Reversible reactions

Learning outcomes

By the end of this unit you should:

- Understand that some chemical reactions can be reversed by changing the reaction conditions
- Be able to describe the reversible reactions of hydrated and anhydrous salts
- Be able to demonstrate knowledge and understanding of the concept of equilibrium
- Be able to predict the effect of changing the concentration, temperature and pressure on the position of equilibrium

13.01 Reversible reactions

Many chemical reactions go to completion. For example, when calcium carbonate reacts with excess hydrochloric acid, the reaction stops when all the calcium carbonate is used up. The products cannot be converted back to the reactants by a single reaction. The reaction is irreversible.

Some reactions can be reversed:

- When blue, **hydrated** copper(II) sulfate is heated, it decomposes into white, **anhydrous** copper(II) sulfate and water:

$$\underset{\text{hydrated copper(II) sulfate}}{CuSO_4.5H_2O(s)} \rightarrow \underset{\text{anhydrous copper(II) sulfate}}{CuSO_4(s)} + 5H_2O(l)$$

TERMS

Hydrated (salts): Salts containing water of crystallisation.

Anhydrous (salts): Salts without any water of crystallisation.

- When water is added to anhydrous copper(II) sulfate the reaction reverses and blue copper sulfate is formed:

$$CuSO_4(s) + 5H_2O(l) \rightarrow CuSO_4.5H_2O(s)$$

- We can show these two reactions in one equation:

$$CuSO_4.5H_2O(s) \rightleftharpoons CuSO_4(s) + 5H_2O(l)$$

TIP

In the formula of a hydrated salt, a dot separates the water from the rest of the formula. When calculating relative formula masses of hydrated salts the masses of the compound and the water are calculated separately then added together.

This is a **reversible reaction**. In reactions involving hydrated salts such as copper(II) sulfate, heating and adding water are not being carried out at the same time.

TERMS

Reversible reaction: A reaction in which the reactants combine to form products and the products can also react together to form the original reactants.

Worked example 13.01

Hydrated nickel chloride, $NiCl_2.6H_2O$, is green. Anhydrous nickel chloride is yellow.

Suggest how you could turn anhydrous nickel chloride green and write a symbol equation for this reaction. **[3]**

In forming hydrated salts from anhydrous salts, you simply put a dot after the formula of the salt and then add the correct number of water molecules.

Add (a controlled amount of) water to the anhydrous nickel chloride **[1]**.

$$NiCl_2 + 6H_2O \rightarrow NiCl_2.6H_2O$$

$NiCl_2$ on left and $NiCl_2.6H_2O$ on right **[1]**
$6H_2O$ on left **[1]**.

Progress check

13.01 What is meant by the terms a hydrated salt b reversible reaction? [2]

13.02 Write a symbol equation for the change which occurs on heating hydrated nickel sulfate $NiSO_4.7H_2O$. [2]

13.02 The concept of equilibrium

In **equilibrium reactions** the forward and the backward reactions are happening at the same time. The products are being converted to reactants at the same time as the reactants are being converted to products. So equilibrium reactions do not go to completion. We show an equilibrium reaction by the sign $\rightleftharpoons$.

TERMS

Equilibrium reaction: A reaction which does not go to completion and in which reactants and products are present in fixed concentration ratios at a particular temperature and pressure.

The characteristics of an equilibrium reaction are:

- It is dynamic: the molecules or ions of the substances on the left of the equation are being changed to those on the right at the same time as those on the right are being changed back to those on the left.
- The forward and the backward reactions occur at the same rate.
- The concentrations of reactants and products remain constant at equilibrium (at constant temperature and pressure). This is because the rate of the forward reaction equals the rate of the backward reaction.
- Equilibrium requires a closed system: none of the reactants or products can escape from the reaction vessel. Figure 13.01 shows the difference between an open and a closed system.

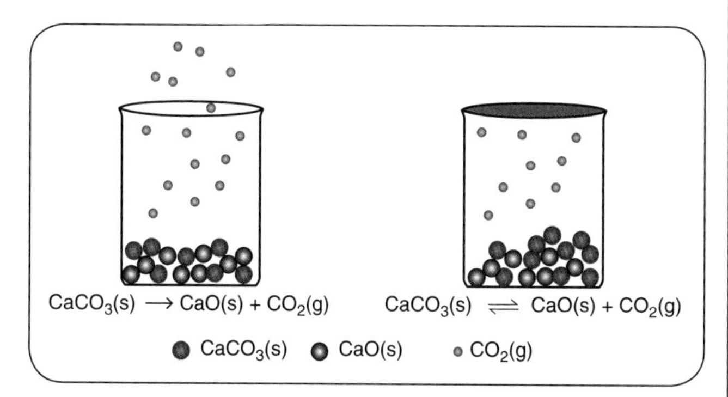

Figure 13.01

In **a** carbon dioxide can escape and the reaction goes to completion. In **b** the carbon dioxide cannot escape and there is an equilibrium between calcium carbonate and the decomposition products.

The equilibrium state can be approached from either the direction of reactants only or products only. For example, the equilibrium $H_2(g) + I_2(g) \rightleftharpoons 2HI(g)$ can be approached from either $H_2(g) + I_2(g)$ alone or $HI(g)$ alone (Figure 13.02).

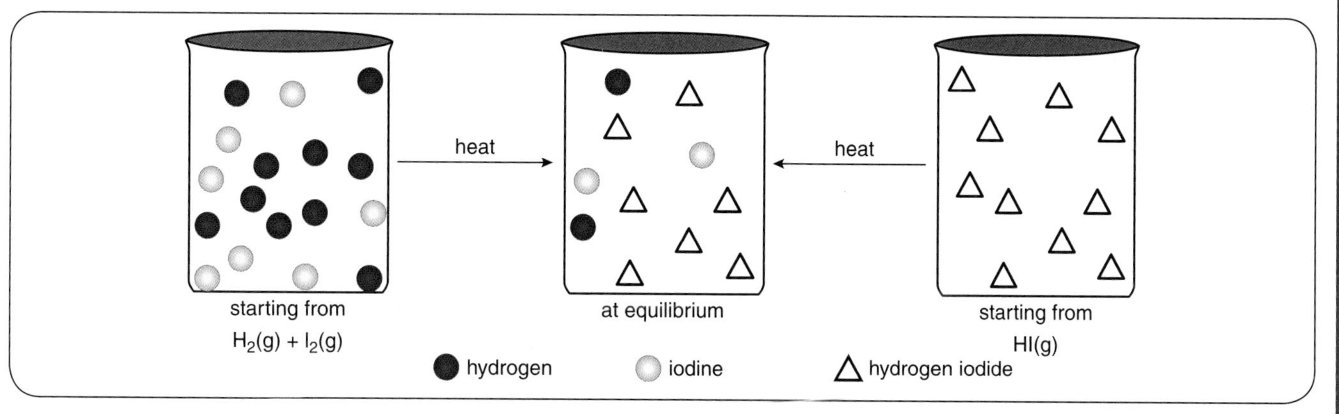

Figure 13.02

We use the term **position of equilibrium** to refer to the relative concentrations of reactants and products at equilibrium. The position of equilibrium is altered if we add more reactants or products or change the temperature or pressure.

TERMS

Position of equilibrium: The relationship between the equilibrium concentrations of reactants and products.

Changing the concentration of reactants or products

When the concentration of one or more reactants is increased:

- the position of equilibrium moves to the right to reduce the effect of increasing the concentration of reactant
- more products are formed until equilibrium is restored

When the concentration of one or more products is increased:

- the position of equilibrium moves to the left to reduce the effect of increasing the concentration of product
- more reactants are formed until equilibrium is restored

Example:

The reaction of iodine monochloride, ICl, with chlorine to form iodine trichloride, ICl_3:

$$ICl(l) + Cl_2(g) \rightleftharpoons ICl_3(s)$$

brown yellow

- Increasing the concentration of $ICl(l)$ or $Cl_2(g)$ moves the position of equilibrium to the right. The amount of yellow solid increases.
- Decreasing the concentration of chlorine moves the position of equilibrium to the left to reduce the effect of decreasing the concentration. The amount of brown solid increases.

Changing the pressure

Change in pressure only affects reactions where gases are reactants or products.

- Increasing the pressure moves the position of equilibrium to the side with the smaller volume of gases (smaller number of moles of gas molecules).
- Decreasing the pressure moves the position of equilibrium to the side with the larger volume of gases (larger number of moles of gas molecules in the stoichiometric equation).

In the reaction:

$2SO_2(g)$	+	$O_2(g)$	$\rightleftharpoons$	$2SO_3(g)$
2 molecules		1 molecule		2 molecules
2 moles		(1 volume)		(2 volumes)

- Increasing the pressure, shifts the position of equilibrium to the right which has a smaller volume (smaller number of molecules). More $SO_3(g)$ is formed.
- Decreasing the pressure, shifts the position of equilibrium to the left in the direction of the larger volume (larger number of molecules). More $SO_2(g)$ and $O_2(g)$ are formed.

Note that if there are equal numbers of moles of gaseous reactants and products, a change in pressure does not affect the position of equilibrium.

Changing the temperature

In an equilibrium reaction, if the forward reaction is exothermic, the backward reaction is endothermic by an equal amount, for example:

$\Delta H = -9.6$ kJ/mol forward reaction exothermic

$$H_2(g) + I_2(g) \rightleftharpoons 2HI(g)$$

$\Delta H = +9.6$ kJ/mol backward reaction endothermic

For an endothermic reaction, the position of equilibrium shifts to the right when the temperature increases. This is because:

- an increase in temperature increases the energy of the surroundings
- the reaction goes in the direction that opposes the increase in energy
- so the reaction goes in the direction in which energy is absorbed (the endothermic direction)

For an exothermic reaction, the position of equilibrium shifts to the left when the temperature increases. This is because:

- the reaction goes in the direction that opposes the increase in energy
- so the reaction goes in the direction in which energy absorbed (the endothermic / backward direction)

TIP

Remember that catalysts have no effect on the position of equilibrium. This is because they increase the rate of the forward and the back reaction equally.

Sample answer

Calcium carbonate decomposes when heated. When calcium carbonate is heated in a closed container the following equilibrium occurs:

$$CaCO_3(s) \rightleftharpoons CaO(s) + CO_2(g)$$

Describe and explain:

a The effect of decreasing the pressure on the position of equilibrium. **[2]**

The position of equilibrium moves to the left **[1]** in the direction of fewer molecules of gas in the equation / lower volume of gas **[1]**.

b The effect of increasing the temperature on the position of equilibrium. **[3]**

Reaction is endothermic **[1]** because heat is needed for the reaction to occur **[1]**. For an endothermic reaction, position of equilibrium moves to the right if temperature is increased **[1]**.

c What happens to the reaction if the container develops a leak. **[2]**

The reaction goes to completion **[1]** because carbon dioxide escapes **[1]**.

Worked example 13.02

Ammonia is synthesised by the Haber process. The reaction takes place in a closed reaction vessel. The reaction is exothermic.

$$N_2(g) + 3H_2(g) \rightleftharpoons 2NH_3(g)$$

Describe what happens to the position of equilibrium when

a the pressure is increased

b ammonia is removed

c the temperature is decreased

d a catalyst is added. **[4]**

Remember that in an equilibrium reaction, the reaction always tries to oppose any change in conditions so:

a The position of equilibrium moves to the right. **[1]** *This is in the direction of fewer molecules of gas in the equation / lower volume of gas.*

b The position of equilibrium moves to the right. **[1]** *This is in the direction of increasing the concentration of ammonia / in order to counteract the effect of removing the ammonia.*

c The position of equilibrium moves to the right. **[1]** *This is because decreasing the temperature favours the exothermic reaction.*

d No effect. **[1]** *Catalysts only affect rate of reaction (not position of equilibrium) and the rate is the same in the backward and forward direction in an equilibrium reaction.*

TIP

Remember that in an equilibrium reaction, the reaction always tries to oppose any change in conditions, e.g. if you increase the concentration, the reaction tries to reduce it, and if you add heat energy, the reaction tries to absorb heat energy.

Progress check

13.03 State three characteristics of chemical equilibrium. [3]

13.04 If a reaction is exothermic, how does increase in temperature affect the position of equilibrium? [1]

13.05 The equation describes the synthesis of methanol, CH_3OH:

$$CO(g) + 2H_2(g) \rightleftharpoons CH_3OH(g)$$

Describe the effect of the following on the position of equilibrium:

a adding more carbon monoxide [1]

b decreasing the pressure [1]

c removing the methanol. [1]

13.06 The equation describes the synthesis of sulfur trioxide, SO_3:

$$2SO_2(g) + O_2(g) \rightleftharpoons 2SO_3(g)$$

The reaction is exothermic.

Describe three ways by which you could shift the position of equilibrium to the right by changing the equilibrium conditions. [3]

Exam-style questions

Question 13.01

When pink cobalt chloride solution, $CoCl_2.6H_2O$, is heated, it turns blue.

a Write a symbol equation for this reaction. [2]

b How can you reverse this reaction? [1]

c Write a symbol equation for the reverse reaction. [1]

d Pink cobalt chloride is a hydrated salt. What is the name given to a salt without any water of crystallisation? [1]

Question 13.02

Bromine dissolves in water to form the equilibrium mixture shown below:

$$Br_2(l) + H_2O(l) \rightleftharpoons Br^-(aq) + BrO^-(aq) + 2H^+(aq)$$

Bromine is red-brown but the products are colourless.

Describe what you would observe and give an explanation in each case for the following:

a A few drops of concentrated hydrochloric acid are added to the reaction mixture. [3]

b A few drops of concentrated alkali are added to the reaction mixture. [4]

c A few drops of concentrated aqueous sodium bromide is added to the reaction mixture. [3]

Question 13.03

A mixture of hydrogen and iodine is put into a sealed tube of fixed volume. The tube is then heated.

$$H_2(g) + I_2(g) \rightleftharpoons 2HI(g)$$

a When the hydrogen and iodine were first in the tube, the position of equilibrium had not been reached. Compare the rate of the forward reaction and the backward reaction as equilibrium is approached and at equilibrium. Explain your answers. [5]

b What effect does increasing the pressure have on the position of equilibrium? Explain your answer. [2]

c What effect does adding more hydrogen have on the position of equilibrium? Explain your answer. [2]

d Increasing the temperature decreases the concentration of hydrogen iodide in the equilibrium mixture. Is the reaction exothermic or endothermic? Explain why. [2]

Question 13.04

When bismuth trichloride reacts with water to form bismuth oxychloride and hydrochloric acid the following equilibrium is achieved:

$$BiCl_3(aq) + H_2O(l) \rightleftharpoons BiOCl(s) + 2HCl(aq)$$

$BiCl_3$ is colourless but $BiOCl$ is a white solid.

Describe and explain what happens when:

a the concentration of acid is increased [3]

b water is added [3]

c more of the white solid is added [3]

Revision checklist

You should be able to:

- Describe reversible reactions involving hydrated and anhydrous salts
- Explain the characteristic features of an equilibrium reaction
- Explain how changing concentration or pressure affects the position of equilibrium
- Explain how changing the temperature affects the position of equilibrium for exothermic and endothermic reactions

Unit 14

Redox reactions

Learning outcomes

By the end of this unit you should:

- Be able to define oxidation and reduction in terms of oxygen gain or loss
- Be able to define redox in terms of electron transfer
- Be able to describe the use of oxidation state in naming ions
- Be able to identify redox reactions by changes in oxidation state
- Know and use the terms oxidising agent and reducing agent
- Be able to identify oxidising and reducing agents from simple equations
- Be able to relate colour changes using acidified potassium manganate(VII) and potassium iodide to redox reactions and oxidising and reducing agents

14.01 Oxidation and reduction

TERMS

Oxidation: The gain of oxygen, loss of electrons or increase in oxidation state.

Reduction: The loss of oxygen, gain of electrons or decrease in oxidation state.

When copper oxide reacts with hydrogen both **oxidation** and **reduction** are occurring at the same time:

$$CuO(s) + H_2(g) \rightarrow Cu(s) + H_2O(l)$$

- Hydrogen gains oxygen atoms from copper oxide. The hydrogen has been oxidised.
- Copper oxide is losing oxygen to the hydrogen. The copper oxide has been reduced.
- Oxidation and reduction have occurred at the same time. This type of reaction is called a **redox reaction**.

TERMS

Redox reaction: A reaction where oxidation and reduction occur together.

TIP

Another definition of oxidation and reduction which is useful in organic chemistry (see Unit 27) is that the gain of hydrogen is reduction and the loss of hydrogen is oxidation.

Worked example 14.01

Explain why reactions A, B and C are redox reactions. In each case identify the compound which has been oxidised and the compound which has been reduced. **[7]**

A $2NO + O_2 \rightarrow 2NO_2$
B $CuO + H_2 \rightarrow Cu + H_2O$
C $CO_2 + C \rightarrow 2CO$

This question is about gain and loss of oxygen. Do not try to answer it in terms of electrons because there are no ions or electrons shown.

They are redox because all three reactions involve both oxidation and reduction **[1]**. *This is suggested in the stem of the question!*

A *The NO is oxidised because it gains oxygen to form NO_2.* So NO is oxidised **[1]**.

The oxygen must be reduced because oxidation and reduction occur together. So O_2 is reduced **[1]**.

> B *The* H_2 *is oxidised because it gains oxygen to form* H_2O*.* So H_2 is oxidised **[1]**.
>
> *The CuO is reduced because it loses oxygen to form Cu.* So CuO reduced **[1]**.
>
> C *The C is oxidised because it gains oxygen to form CO.* So C is oxidised **[1]**.
>
> *The* CO_2 *is reduced because it loses one oxygen atom in forming CO.* So CO_2 is reduced **[1]**.

Progress check

14.01 What is meant by a reduction b redox reaction? [2]

14.02 Do the underlined elements get oxidised or reduced during these reactions. In each case explain your answer.

a <u>CuO</u> + H_2 → Cu + H_2O [1]

b N_2 + <u>O_2</u> → 2NO [1]

c CO + <u>$2H_2$</u> → CH_3OH [1]

d PbO + <u>C</u> → Pb + CO [1]

14.02 Electron transfer in redox reactions

- Oxidation is loss of electrons.
- Reduction is gain of electrons.

Sodium reacts with chlorine to form the ionic compound sodium chloride:

$$Na(s) + Cl_2(g) \rightarrow NaCl(s)$$

In this reaction, a sodium atom loses an electron:

$$Na \rightarrow Na^+ + e^-$$

The sodium has been oxidised because the sodium atom has lost an electron.

Each chlorine atom in the chlorine molecule gains one electron:

$$Cl_2 + 2e^- \rightarrow 2Cl^-$$

The chlorine has been reduced because each chlorine atom has gained an electron.

Equations like this showing the oxidation and reduction reaction separately are called half equations. In Unit 10 you saw that the reactions at the electrodes in electrochemical cells are expressed by half equations.

- Oxidation reactions occur at the anode, e.g. $2I^- \rightarrow I_2 + 2e^-$
- Reduction reactions occur at the cathode, e.g. $Cu^{2+} + 2e^- \rightarrow Cu$

Another example is the reaction of halogens with halides (see Unit 20):

$$Cl_2(aq) + 2Br^-(aq) \rightarrow 2Cl^-(aq) + Br_2(aq)$$

Chlorine has gained electrons from bromine. Chlorine has been reduced:

$$Cl_2 + 2e^- \rightarrow 2Cl^-$$

Bromide ions have lost electrons to chlorine. Bromide ions have been oxidised:

$$2Br^- \rightarrow Br_2 + 2e^-$$

A useful way of remembering oxidation and reduction in terms of electrons is 'OIL RIG': <u>O</u>xidation <u>I</u>s <u>L</u>oss and <u>R</u>eduction <u>I</u>s <u>G</u>ain.

14.03 Oxidation state

TERMS

Oxidation state: A number describing the degree of oxidation (or reduction) of an atom in a compound.

An **oxidation state** (oxidation number) is a number given to each atom or ion in a compound to show the degree of oxidation.

There are rules for applying oxidation numbers. The simplest of these are given below.

- Oxidation number refers to a single atom or ion in a compound.
- The oxidation number of each atom in an element is 0, e.g. each Cl in Cl_2 = 0, Zn = 0.

The oxidation number of an ion is the charge on the ion, e.g. $Na^+ = +1$, $Mg^{2+} = +2$, $Al^{3+} = +3$, $Cl^- = -1$, $O^{2-} = -2$.

- The sum of all the oxidation numbers of atoms or ions in a compound is zero, e.g. in $CaCl_2$:

  oxidation number of $Ca^{2+} = +2$

  oxidation number of $2 \times Cl^- = 2 \times -1 = -2$

- Some elements form more than one type of ion, e.g. Fe^{2+} and Fe^{3+}. The oxidation number of these ions can usually be worked out from the formula of the oxide or chloride: in $FeCl_3$, Cl^- has an oxidation number of -1, so to balance 3 chloride ions:

  $3 \times -1 = -3$, the iron ion must be Fe^{3+}.

Oxidation state and naming compounds

The Roman numerals in the formulae of some compounds show the oxidation state of particular atoms. We use Roman numerals when more than one oxidation state is possible.

Examples:

$FeSO_4$: the iron ion is Fe^{2+}, so the compound is named iron(II) sulfate.

$CrCl_3$: the chromium ion is Cr^{3+}, so the compound is named chromium(III) chloride.

$KMnO_4$: the manganese atom has an oxidation state of +7, so the compound is named potassium manganate(VII).

$NaClO_3$: the chlorine atom has an oxidation state of +5, so the compound is named sodium chlorate(V).

TIP

You do not have to know how to work out the oxidation state of atoms in compound ions such as MnO_4^- and ClO_3^-. Note that the Roman numeral comes after the name in this type of ion.

Redox and oxidation state

Oxidation and reduction can be defined in terms of changes of oxidation state.

- Oxidation is an increase in oxidation state.
- Reduction is a decrease in oxidation state.

Example 1: The reaction of zinc with aqueous copper(II) ions, Cu^{2+}:

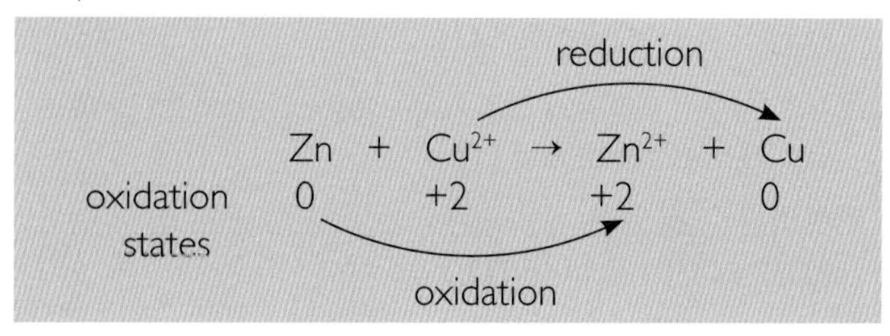

Zinc atoms have been oxidised because their oxidation state increases from 0 to +2.

Copper ions have been reduced because their oxidation state decreases from +2 to 0.

Example 2: The reaction of chlorine with bromide ions:

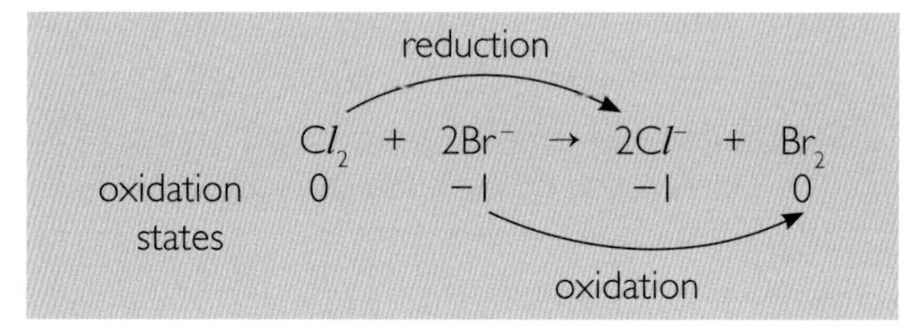

Bromide ions have been oxidised because the oxidation state of each bromide ion increases from −1 to 0.

Chlorine atoms have been reduced because the oxidation state of each chlorine atom decreases from 0 to −1.

14.04 Oxidising agents and reducing agents

An **oxidising agent** is a substance which oxidises another substance during a redox reaction.

- The oxidising agent gains electrons and gets reduced.
- The oxidation state of a particular atom or ion in the oxidising agent decreases.

TERMS

Oxidising agent: A substance that oxidises another substance by accepting electrons.

Reducing agent: A substance that reduces another substance by donating (adding) electrons.

A **reducing agent** is a substance which reduces another substance during a redox reaction.

- The reducing agent loses electrons and gets oxidised.
- The oxidation state of a particular atom or ion in the reducing agent increases

Example 1: The reaction of zinc with aqueous copper(II) ions, Cu^{2+}:

$$Zn + Cu^{2+} \rightarrow Zn^{2+} + Cu$$

Zinc is the reducing agent because it reduces copper ions. Zinc gives its electrons to the copper ions, whose oxidation state decreases.

The Cu^{2+} ion is the oxidising agent because it oxidises zinc. The Cu^{2+} ions accept electrons from the zinc atoms, whose oxidation state increases.

Example 2: The reaction of chlorine with bromide ions:

$$Cl_2(aq) + 2Br^-(aq) \rightarrow 2Cl^-(aq) + Br_2(aq)$$

The Br^- ion is the reducing agent because it reduces chlorine. Bromide ions give their electrons to the chlorine atoms, whose oxidation state decreases.

Chlorine is the oxidising agent because it oxidises bromide ions. The chlorine atoms accept electrons from the bromide ions, whose oxidation state increases.

14.05 Colour tests for oxidising agents and reducing agents

Tests for common oxidising and reducing agents involve the observation of a colour change in the solution under test. For the test to work well the test solution usually has to be acidic, so a few drops of sulfuric acid are added to the reaction mixture when appropriate.

Test for oxidising agents

Potassium iodide is a reducing agent which is used to test for many oxidising agents. When potassium iodide is added to an acidified solution of an oxidising agent such as hydrogen peroxide or aqueous chlorine, the solution turns brown because iodine is formed.

Example: The reaction of potassium iodide with chlorine.

Potassium iodide contains the iodide ion, I^-:

	$Cl_2(aq)$	+	$2I^-(aq)$	→	$2Cl^-(aq)$	+	$I_2(aq)$
	colourless solutions				colourless		brown
oxidation state:	0		−1		−1		0

The I^- ion is the reducing agent because it reduces chlorine. Iodide ions give their electrons to the chlorine atoms, whose oxidation state decreases.

Chlorine is the oxidising agent because it oxidises iodide ions. The chlorine atoms accept electrons from the iodide ions, whose oxidation state increases.

Test for reducing agents

Potassium manganate(VII) is an oxidising agent which is used as a test for reducing agents. When acidified potassium manganate(VII) is added to a reducing agent such as zinc or sodium sulfite (Na_2SO_3), its colour changes from pink-purple to colourless.

Example: The reaction of potassium manganate(VII) with aqueous Fe^{2+} ions.

Note: you will not be asked to write equations involving potassium manganate(VII).

Potassium manganate(VII) contains the manganate(VII) ion, MnO_4^-. H^+ is the acid:

	$MnO_4^-(aq)$	+	$5Fe^{2+}(aq)$	+	$8H^+(aq)$
	pink-purple		colourless solutions		
	+7		+2		
→	$Mn^{2+}(aq)$	+	$5Fe^{3+}(aq)$	+	$4H_2O(l)$
	colourless		light yellow		colourless
	+2		+3		

Potassium manganate(VII) is the oxidising agent because it oxidises Fe^{2+} ions. The MnO_4^- ions accept electrons from the Fe^{2+} ions whose oxidation state increases from +2 to +3.

The Fe^{2+} ion is the reducing agent because it reduces potassium manganate(VII). Fe^{2+} ions give electrons to the MnO_4^- ions and the oxidation state of Mn decreases from +7 to +2.

Worked example 14.02

Describe a test and results to show that aqueous Cl_2 is a good oxidising agent by using aqueous iodide ions, I^-. Write an ionic equation for this reaction and explain this reaction in terms of oxidation and reduction. The products are chloride ions and iodine. **[7]**

Remember that potassium iodide is a reducing agent that is used to test for oxidising agents. You need to remember the colour change when it reacts.

Add aqueous chlorine to a solution of potassium iodide **[1]**: colour turns from colourless **[1]** to brown. **[1]**

Remember that the brown colour is due to the iodine produced in the reaction and that iodine is I_2 not just I.

$Cl_2 + 2I^- \rightarrow 2Cl^- + I_2$ (**1 mark** for reactants and products, **1 mark** for balance.)

The equation shows electrons, so the last part of the question should be answered in terms of electron transfer.

Iodide ions have been oxidised because they have lost electrons **[1]**. Chlorine has been reduced because it has gained electrons **[1]**.

Progress check

14.03 Which is the reducing agent in the following reactions?

a $Cl_2 + 2Br^- \rightarrow 2Cl^- + Br_2$ [1]

b $Zn + FeSO_4 \rightarrow Fe + ZnSO_4$ [1]

c $MnO_4^- + 5Fe^{2+} + 8H^+ \rightarrow Mn^{2+} + 5Fe^{3+} + 4H_2O$ [1]

14.04 Which of the following are oxidations and which are reductions?

a tin(IV) chloride to tin

b potassium dichromate(VI) to chromium(III) chloride

c magnesium to magnesium ions

d iron(II) sulfate to iron(III) sulfate. [4]

Sample answer

a Potassium manganate(VII) is used to test for reducing agents. It reduces hydrogen sulfide to sulfur. Describe what you would observe during this reaction. **[3]**

The pink-purple colour of potassium manganate(VII) **[1]** decolourises **[1]** and a yellow precipitate of sulfur appears **[1]**.

b The equation describes this reaction:

$$2MnO_4^-(aq) + 5H_2S(g) + 6H^+(aq) \rightarrow 2Mn^{2+} + 5S(s) + 8H_2O(l)$$

i What is the change in oxidation state of the manganese in this reaction? **[2]**

+7 (in potassium manganate(VII)) **[1]** to +2 in Mn^{2+} **[1]**.

ii In this reaction potassium manganate(VII) undergoes reduction. Explain in three different ways how potassium manganate(VII) undergoes reduction. **[3]**

The oxidation state goes down (from +7 to +2) **[1]**.

MnO_4^- gains electrons **[1]** MnO_4^- loses oxygen **[1]**.

Exam-style questions

Question 14.01

Lead(II) oxide, PbO, reacts with carbon when heated to form lead and carbon monoxide.

a Explain the meaning of the symbol (II) in lead(II) oxide. [2]

b Construct a symbol equation for this reaction. [1]

c Which substance gets reduced in this reaction? Explain your answer. [2]

d Give the name for the process in which oxidation and reduction occur at the same time. [1]

Question 14.02

The equation shows the redox reaction between iodide ions and chlorine:

$$Cl_2(aq) + 2I^-(aq) \rightarrow 2Cl^-(aq) + I_2(aq)$$

a Write two half equations for this reaction and identify which substance gets oxidised and which gets reduced. [4]

b Identify the oxidising and the reducing agent in this reaction. Explain your answer. [4]

c Potassium iodide is used to test for oxidising agents. It reacts with the oxidising agent, hydrogen peroxide. Describe how you could tell that hydrogen peroxide is an oxidising reagent by using potassium iodide. Explain your answer. [3]

d The oxidation state of oxygen in hydrogen peroxide is −1.

Hydrogen peroxide reacts with aqueous Fe^{2+} ions:

$$2Fe^{2+} + H_2O_2 + 2H^+ \rightarrow 2Fe^{3+} + 2H_2O$$

Use ideas about oxidation states to explain which substance has been oxidised and which substance has been reduced in the reaction. [4]

Question 14.03

The equation describes the reaction of iron with sulfuric acid.

$$Fe + H_2SO_4 \rightarrow FeSO_4 + H_2$$

Acids contain hydrogen ions, H^+.

a Write two half equations for this reaction. [2]

b Identify which is the oxidising agent and which is the reducing agent in this reaction. Explain your answer. [4]

c When heated, aluminium reacts with iron(III) oxide:

$$Fe_2O_3 + 2Al \rightarrow Al_2O_3 + 2Fe$$

Which substance gets oxidised in this reaction? Give an explanation for your answer in terms of oxidation state or charge. [3]

Revision checklist

You should be able to:

- Define oxidation and reduction in terms of oxygen gain or loss
- Define redox reactions in terms of electron transfer
- Explain redox reactions in terms of change in oxidation state
- Define oxidising agent and reducing agent
- Describe colour changes when acidified potassium manganate(VII) is used as an oxidising agent
- Describe colour changes when aqueous potassium iodide is used as a reducing agent

Unit 15

Acids, bases and oxides

Learning outcomes

By the end of this unit you should:

- Be able to describe some chemical properties of acids and bases
- Be able to describe the relative acidity and alkalinity with reference to the pH scale
- Understand how universal indicator is used to measure pH
- Describe the use of litmus and methyl orange indicators
- Be able to describe and explain the importance of controlling acidity in soil
- Be able to define acids and bases in terms of proton transfer
- Understand the meaning of weak and strong acids and bases
- Be able to classify oxides as either acidic or basic
- Be able to classify oxides as neutral or amphoteric

15.01 Acids, bases and salts

Acids

- Acids have a sour taste and are corrosive.
- Acids turn damp blue litmus red.
- Acids have pH values below 7 (see below).
- A simple definition of an acid is a substance which neutralises a base to form a salt and water. For example:

$$\underset{\text{acid}}{2HCl(aq)} + \underset{\text{base}}{CaO(s)} \rightarrow \underset{\text{salt}}{CaCl_2(aq)} + \underset{\text{water}}{H_2O(l)}$$

- Acids have hydrogen atoms which can be replaced by a metal or ammonium ion, NH_4^+. In the reaction above, the hydrogen in the acid is replaced by the calcium ions.

Bases

- A simple definition of a base is a substance which neutralises an acid to form a salt and water. For example:

$$\underset{\text{base}}{2NaOH(aq)} + \underset{\text{acid}}{H_2SO_4(aq)} \rightarrow \underset{\text{salt}}{Na_2SO_4(aq)} + \underset{\text{water}}{2H_2O(l)}$$

- Bases are usually oxides or hydroxides of metals.

TERMS

Alkali: A soluble base.

- A base which is soluble in water is called an **alkali**. The hydroxides of the Group I metals and aqueous ammonia, $NH_3(aq)$, are alkalis.
- Alkalis turn damp red litmus blue.
- Alkalis have pH value above 7 (see below).

Salts

When an acid reacts with a metal, a metal oxide, a metal hydroxide or metal carbonate, a **salt** is formed.

TERMS

Salt: A compound formed when the hydrogen in an acid is replaced by a metal or ammonium ion.

Chlorides are salts formed from hydrochloric acid. For example, sodium chloride:

$$\underset{\text{hydrochloric acid}}{HCl(aq)} + \underset{\text{sodium hydroxide}}{NaOH(aq)} \rightarrow \underset{\text{sodium chloride}}{NaCl(aq)} + \underset{\text{water}}{H_2O(l)}$$

Sulfates are salts formed from sulfuric acid. For example, potassium sulfate:

$H_2SO_4(aq)$	+ $2KOH(aq)$	→ $K_2SO_4(aq)$	+ $2H_2O(l)$
sulfuric acid	potassium hydroxide	potassium sulfate	water

Nitrates are salts formed from nitric acid. For example, lithium nitrate:

$HNO_3(aq)$	+ $LiOH(aq)$	→ $LiNO_3(aq)$	+ $H_2O(l)$
nitric acid	lithium hydroxide	lithium nitrate	water

15.02 Chemical reactions of acids

Reaction with metals

Acids react with metals above hydrogen in the reactivity series (see Unit 21) to form a salt and hydrogen.

	metal +	acid	→	metal salt +	hydrogen
Example:	Zn	+ H_2SO_4	→	$ZnSO_4$ +	H_2
	zinc	sulfuric acid		zinc sulfate	

Remember that not all acids produce hydrogen when they react with metals. Nitric acid and concentrated sulfuric acid can be oxidising agents and hydrogen is not produced.

Reaction with metal oxides

Acids react with many metal oxides to form a salt and water.

	metal oxide +	acid	→	metal salt +	water
Example:	MgO	+ 2HC*l*	→	$MgCl_2$	+ H_2O
	magnesium oxide	hydrochloric acid		magnesium chloride	

Reaction with metal hydroxides and aqueous ammonia

Acids react with many metal hydroxides to form a salt and water.

	metal hydroxide +	acid	→	metal salt +	water
Example:	NaOH	+ HNO_3	→	$NaNO_3$	+ H_2O
	sodium hydroxide	nitric acid		sodium nitrate	

A solution of aqueous ammonia contains hydroxide ions. So ammonia also reacts to form a salt:

$2NH_3$	+ H_2SO_4	→ $(NH_4)_2SO_4$
ammonia	sulfuric acid	ammonium sulfate

Reaction with carbonates

Acids react with carbonates to form a salt, water and carbon dioxide:

metal carbonate	+ acid	→	salt	+ water	+ carbon dioxide
$CaCO_3$	+ 2HC*l*	→	$CaCl_2$	+ H_2O	+ CO_2
calcium carbonate	hydrochloric acid		calcium chloride		

Worked example 15.01

a What could you do to tell the difference between an acidic and alkaline solution? **[2]**

The best ways involve chemistry not taste! Either the litmus test or the pH test is suitable. Do not suggest indicators other than litmus or universal indicator because if the acid or base is very weak, a colour change may not be seen.

Acids turn blue litmus red **[1]**; alkalis turn red litmus blue **[1]**.

b Name three different compounds you could add to hydrochloric acid to make the salt calcium chloride. Write word equations for the formation of the salt. **[4]**

Since it is calcium chloride you are making, the substances must be calcium compounds. But not calcium metal because that is not a compound. When writing the equations, concentrate on the products. Do not forget that with a carbonate, salt + carbon dioxide + water are formed. And do not forget the arrow and + signs!

The compounds are calcium oxide, calcium hydroxide and calcium carbonate **[1]**.

Calcium oxide + hydrochloric acid → calcium chloride + water [1]

Calcium hydroxide + hydrochloric acid → calcium chloride + water [1]

Calcium carbonate + hydrochloric acid → calcium chloride + water + carbon dioxide [1]

15.03 Chemical reactions of bases

Reaction with acids

Metal oxides and hydroxides react with acids to form a salt and water.

Examples:

$$CaO + 2HCl \rightarrow CaCl_2 + H_2O$$

calcium oxide + hydrochloric acid → calcium chloride + water

$$KOH + HNO_3 \rightarrow KNO_3 + H_2O$$

potassium hydroxide + nitric acid → potassium nitrate + water

Reaction of alkalis with ammonium salts

Ammonium salts decompose when warmed with alkalis. A salt, water and ammonia are produced:

$$NH_4Cl + NaOH \rightarrow NaCl + H_2O + NH_3$$

ammonium chloride + sodium hydroxide → sodium chloride + water + ammonia

15.04 The pH scale

- The pH scale is a scale of numbers used to show how acidic or alkaline a solution is.
- pH values below 7 are acidic.
- pH values above 7 are alkaline.
- A pH of exactly 7 is **neutral**.
- The lower the pH, the more acidic is the solution.
- The higher the pH, the more alkaline is the solution.

TERMS

Neutral solution: A solution with a pH of 7.

Finding the pH

Using **universal indicator**:

- A drop of the solution to be tested is added to universal indicator paper (or solution).
- The colour of the indicator is matched against a colour chart showing the pH corresponding to different colours.

TERMS

Universal indicator: A mixture of indicators used to measure pH values.

Using a pH meter:

- A pH electrode connected to a pH meter is dipped into the solution to be tested.
- The reading on the meter gives a direct reading of the pH.
- A pH meter gives a more accurate value of pH than universal indicator.

Remember that the higher acidity the lower pH and the lower the acidity the higher the pH.

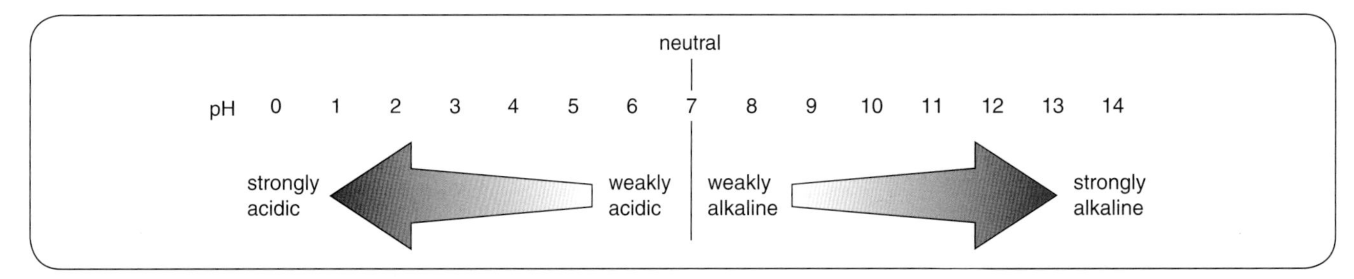

Figure 15.01 The pH scale

15.05 Neutralisation

Neutralisation is the combination of oxide or hydroxide ions (from a base) with the hydrogen ions (from an acid) to form a salt and water.

The reaction of an acid with an alkali is an example of a neutralisation reaction. We neutralise sodium hydroxide with hydrochloric acid by adding a few drops of an acid base indicator, such as methyl orange, to the alkali (sodium hydroxide). The methyl orange is yellow in alkaline solutions.

- Add the acid a little at a time to the alkali (see Figure 15.02).
- When the methyl orange indicator turns red, we know that the solution has been neutralised.
- The indicator can be removed by the addition of carbon powder then filtering.

Controlling soil acidity

Soil may become acidic because of:

- acid rain (see Unit 24)
- use of fertilisers containing ammonium salts.
- formation of acids when bacteria and fungi break down organic matter.

Most crop plants do not grow well if the soil is too acidic. So farmers add crushed or powdered limestone (calcium carbonate), lime (calcium oxide) or slaked lime (calcium hydroxide) to the soil to neutralise the excess acidity:

$$CaCO_3 + 2H^+ \rightarrow Ca^{2+} + H_2O + CO_2$$

calcium carbonate, from acids, calcium ions

$$CaO + 2H^+ \rightarrow Ca^{2+} + H_2O$$

calcium oxide, from acids, calcium ions

Farmers have to be careful not to add too much lime to the soil. When dissolved in water, lime and slaked lime form strongly alkaline solutions. Most plants do not grow well in alkaline conditions.

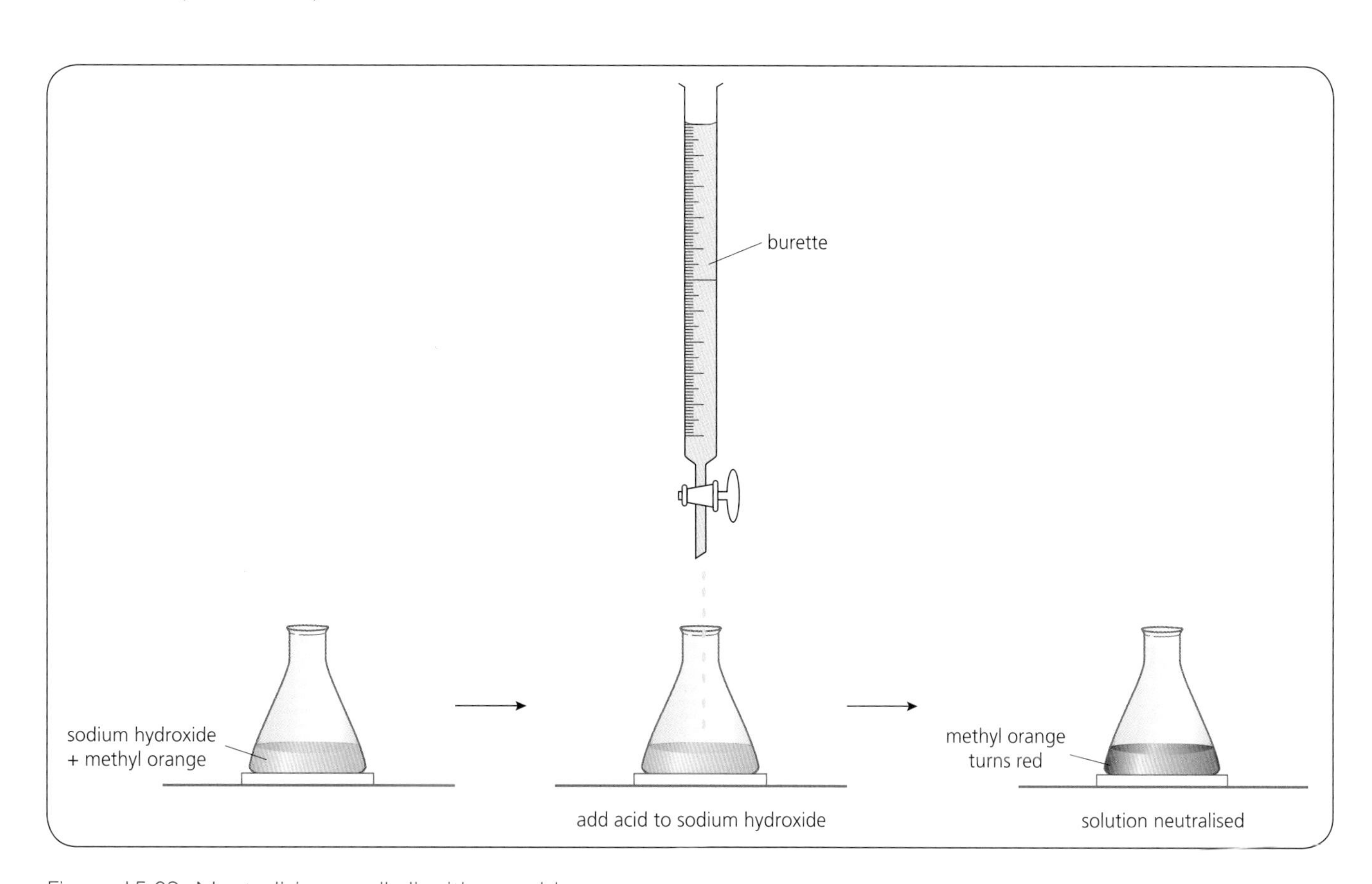

Figure 15.02 Neutralising an alkali with an acid

Progress check

15.01 Describe the following pH values as any of:

neutral, strongly acidic, strongly basic, weakly acidic, weakly basic,

a pH 5

b pH 13

c pH 7

d pH 2 [4]

15.02 Copy and complete these word equations:

a magnesium + sulfuric acid → [2]

b sodium carbonate + hydrochloric acid → [3]

c zinc oxide + nitric acid → [2]

d potassium hydroxide + hydrochloric acid → [2]

15.03 Copy and complete these chemical equations:

a $2NH_3 + H_2SO_4 \rightarrow$ .... [1]

b .... + 2......... → $K_2SO_4 + 2H_2O$ [2]

c $MgCO_3 + 2HCl \rightarrow$ .... + ... + ... [3]

d $NH_4Cl + NaOH \rightarrow NH_3$ + ... + ... [2]

e Ba + ... → $BaSO_4$ + ... [2]

15.04 Describe how you can find the pH of a solution if you do not have a pH meter. [2]

15.05 Describe how to neutralise an alkali with an acid.[3]

15.06 When neutralising acidic soil, explain why it is better to add calcium carbonate than calcium oxide. [4]

15.06 Proton transfer in acids and bases

When hydrogen chloride dissolves in water, it ionises (splits up into ions). Hydrogen ions are formed:

$HCl(g)$	+	(aq)	→	$H^+(aq)$	+	$Cl^-(aq)$
hydrogen chloride		water		proton		chloride ion

When an alkali dissolves in water, it ionises and forms hydroxide ions, OH^-:

$$NaOH(s) + (aq) \rightarrow Na^+(aq) + OH^-(aq)$$

We can define acids and bases in terms of the transfer of hydrogen ions.

When discussing acids and bases, we usually use the term proton instead of hydrogen ion.

TERMS

Acid: A proton donor.

Base: A proton acceptor.

- An **acid** gives (donates) one or more protons to a base.
- A **base** takes (accepts) one or more protons from an acid.

Hydrochloric acid is an acid because it transfers a proton to a base:

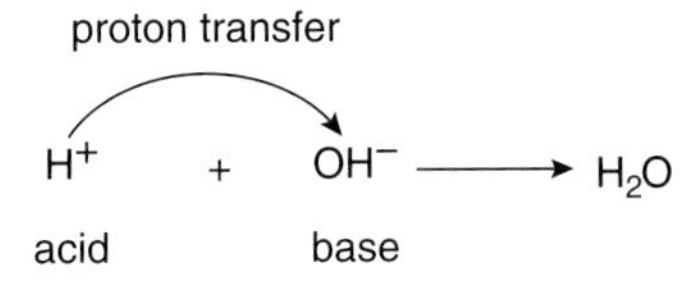

Ammonia is a base because it accepts a proton from an acid:

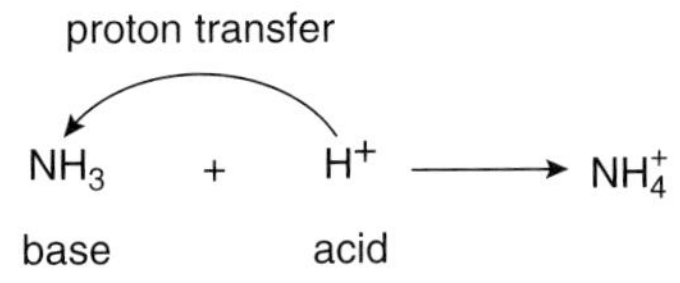

We can simplify the equations for proton transfer by ignoring the water when we show the acid dissolving. For example:

$$HCl(aq) \rightarrow H^+(aq) + Cl^-(aq)$$
$$CH_3COOH(aq) \rightleftharpoons H^+(aq) + CH_3COO^-(aq)$$

The ionic equation for the reaction of any acid with any alkali is:

H^+	+	OH^-	→	H_2O
from acid		from alkali		

15.07 Acid and base concentration

A concentrated solution of an acid or base contains more particles of acid or base per dm^3 than a dilute solution. It does not tell us anything about how well the acid ionises.

TIP

It is incorrect to use the words 'strong' or 'weak' when referring to the concentration of acids in mol/dm^3. You should use the terms 'concentrated' or 'dilute'.

- The lower the pH, the greater is the concentration of H^+ ions.
- The higher the pH, the greater is the concentration of OH^- ions.

15.08 Strong and weak acids and bases

- **Strong acids** ionise completely (dissociate completely) in solution. Hydrochloric, sulfuric and nitric acids are strong acids:

$$HCl \rightarrow H^+ + Cl^-$$

- **Strong bases** ionise completely in solution. Group I hydroxides are strong bases:

$$NaOH \rightarrow Na^+ + OH^-$$

TERMS

Strong (acid/base): An acid or base which ionises completely in solution.

Weak (acid/base): An acid or base which ionises partially in solution.

- **Weak acids** are only partially ionised (partly dissociated) in solution. There are many more molecules of unionised acid present than there are ions. There is an equilibrium between molecules of unionised acid and their ions.

Organic acids such as ethanoic acid, CH_3COOH, are weak acids:

$$CH_3COOH \rightleftharpoons H^+ + CH_3COO^-$$

- **Weak bases** are only partially ionised (partly dissociated) in solution. There are many more molecules of unionised base present than there are ions. There is an equilibrium between molecules of unionised base and their ions. Aqueous ammonia is a weak base:

$$NH_3(g) + H_2O(l) \rightleftharpoons NH_4^+(aq) + OH^-(aq)$$

We can distinguish strong acids from weak acids by comparing solutions of equal concentrations. At equal concentrations:

- strong acids have a lower pH than weak acids
- strong acids will react faster with metals, metal carbonates and metal oxides (at the same temperature)
- strong acids will conduct electricity better than weak acids

We can distinguish strong bases from weak bases by comparing solutions of equal concentrations. At equal concentrations:

- strong bases have a higher pH than weak bases
- strong bases will conduct electricity better than weak bases

Sample answer

Ammonia is a weak base. Ammonia reacts with water to form an equilibrium mixture expressed in the equation:

$$NH_3(g) + H_2O(l) \rightleftharpoons NH_4^+(aq) + OH^-(aq)$$

Describe each of the reactants and the OH^- ion as either acids or bases. Explain your answer using ideas of proton transfer. **[7]**

NH_3 is a base because it accepts a proton **[1]** from water **[1]**.

Water is an acid because it donates a proton **[1]** to ammonia **[1]**.

In the backward reaction hydroxide ion is a base **[1]** because it accepts a proton **[1]** from the acid NH_4^+ **[1]** to form water **[1]**.

15.09 Acidic and basic oxides

- **Basic oxides:** most metal oxides are basic oxides. They react with acids to form a salt and water:

CaO +	2HCl	→ $CaCl_2$	+ H_2O
calcium oxide	hydrochloric acid	calcium chloride	water

TERMS

Acidic oxide: An oxide that reacts with a base to form a salt and water.

Basic oxide: An oxide that reacts with an acid to form a salt and water.

- **Acidic oxides:** many non-metal oxides are acidic oxides. Acidic oxides react with bases to form salts. Examples include:

SO_2	+ 2NaOH	→ Na_2SO_3	+ H_2O
sulfur dioxide		sodium sulfite	

CaO	+ SiO_2	→ $CaSiO_3$
calcium oxide	silicon dioxide	calcium silicate

Some acidic oxides dissolve in water to form acids:

SO_3	+ H_2O	→ H_2SO_4
sulfur trioxide		sulfuric acid

Other acidic oxides include oxides of phosphorus; nitrogen dioxide, NO_2, and carbon dioxide.

15.10 Amphoteric and neutral oxides

- **Amphoteric oxides** react with acids as well as with bases. A salt and water are formed. Examples are aluminium oxide and zinc oxide:

$$Al_2O_3(s) + 6HCl(aq) \rightarrow 2AlCl_3(aq) + 3H_2O(l)$$
$$Al_2O_3(s) + 2NaOH(aq) \rightarrow 2NaAlO_2(aq) + H_2O(l)$$
sodium aluminate

$$ZnO(s) + 2HNO_3(aq) \rightarrow Zn(NO_3)_2(aq) + H_2O(l)$$
$$ZnO(s) + 2NaOH(aq) \rightarrow Na_2ZnO_2(aq) + H_2O(l)$$
sodium zincate

The hydroxides of aluminium and zinc behave in a similar way. Aluminates and zincates are soluble in aqueous alkaline solution. So when excess sodium hydroxide is added to a solution containing zinc or aluminium ions, the precipitate of hydroxide redissolves in excess sodium hydroxide (see Unit 17).

Worked example 15.02

a Oxides can be classified as acidic or basic. Explain the meaning of these two terms and describe what types of elements form basic and acidic oxides. **[4]**

If you remember that acids react with bases and that acids are formed when non-metals react with water, you can work the rest out because they are opposites!

An acidic oxide reacts with a base to form a salt and water **[1]**.

A basic oxide reacts with an acid to form a salt and water **[1]**.

Basic oxides are oxides of (reactive) metals **[1]**.

Acidic oxides are oxides of (some) non-metals **[1]**.

b Copy and complete the following symbol equations:

i $Mg(OH)_2 + \ldots HCl \rightarrow MgCl_2 + \ldots\ldots$ **[2]**

First balance the Cl (two on the left to balance the two on the right):

$Mg(OH)_2 + 2HCl \rightarrow MgCl_2 + \ldots H_2O$ **[1]**

Then balance the water:

$Mg(OH)_2 + 2HCl \rightarrow MgCl_2 + 2H_2O$ **[1]**

ii $ZnO + 2\ldots\ldots \rightarrow Zn(NO_3)_2 + \ldots\ldots$ **[2]**

The acid used to make nitrates is nitric acid, HNO_3. (You can get the formula from the NO_3 of the nitrate.)

$ZnO + 2HNO_3 \rightarrow Zn(NO_3)_2 + \ldots\ldots$ **[1]**

When an acid reacts with an oxide, water is formed as well as the salt.

$ZnO + 2HNO_3 \rightarrow Zn(NO_3)_2 + H_2O$ **[1]**

TERMS

Neutral oxide: An oxide that does not react with an acid or base.

Amphoteric oxide: An oxide that reacts with both acids and bases to form a salt and water.

- **Neutral oxides** do not react with acids or bases. Examples are nitrogen(I) oxide, N_2O, nitrogen(II) oxide, NO, and carbon monoxide, CO.

Progress check

15.07 Class the following oxides as either acidic or basic:

a magnesium oxide

b phosphorus(V) oxide

c carbon dioxide

d sodium oxide. [4]

15.08 Chloric(III) acid, $HClO_2$, is a weak acid.

Write symbol equations for:

a the ionisation of chloric(III) acid in water [1]

b the reaction of chloric(III) acid with sodium oxide, Na_2O [2]

c the reaction of chloric(III) acid with zinc. [2]

15.09 Aluminium oxide, Al_2O_3, is an amphoteric oxide. Write symbol equations for the reaction of aluminium oxide with

a potassium hydroxide to form the potassium aluminate, $KAlO_2$(aq) [2]

b hydrochloric acid. [2]

15.10

a What is the meaning of the terms strong acid and weak acid?

b How could you distinguish between a strong and weak acid by experiment, other than by measuring the difference in pH or electrical conductivity? [3]

Exam-style questions

Question 15.01

Calcium oxide, CaO, is a base which reacts with water to form calcium hydroxide, $Ca(OH)_2$.

a What is the meaning of the term base? [1]

b A solution of calcium hydroxide is alkaline. Suggest a value for the pH of calcium hydroxide and what effect does calcium hydroxide have on litmus? [2]

c Write word equations for:

i the reaction of calcium hydroxide with hydrochloric acid

ii the reaction of calcium oxide with nitric acid. [2]

d Farmers sometimes add calcium oxide to the soil. Explain why they do this. [3]

e Copy and complete the symbol equation for the reaction of calcium carbonate with hydrochloric acid.

$CaCO_3$ + ... $HCl \rightarrow CaCl_2$ + ... + ... [3]

Question 15.02

Sulfuric acid, H_2SO_4, is a typical acid.

a What would you observe when sulfuric acid is added to:

i blue litmus paper?

ii calcium carbonate? Explain your answer to part ii. [3]

b Write word equations for the reaction of sulfuric acid with:

i ammonia

ii sodium hydroxide. [2]

c Write a symbol equation for the reaction of sulfuric acid with iron. The salt formed is iron(II) sulfate. [2]

d Describe how you would use sulfuric acid to neutralise a solution of potassium hydroxide using a named acid–base indicator. [4]

e Describe how you could use universal indicator to find the pH of a solution. [2]

Question 15.03

Hydrochloric acid, HCl, is a strong acid but propanoic acid, CH_3CH_2COOH, is a weak acid.

a Construct equations for the ionisation of both hydrochloric and propanoic acid in water. [3]

b Explain why a solution of 1 mol/dm^3 hydrochloric acid reacts rapidly with magnesium but a solution of 1 mol/dm^3 propanoic acid reacts slowly. [4]

c Describe and explain two other ways by which you could distinguish 1 mol/dm^3 solutions of hydrochloric and propanoic acids. [5]

d Propanoic acid reacts with sodium to form the salt sodium propanoate.

Write a symbol equation for this reaction. [2]

Revision checklist

You should be able to:

- Describe the effect of acids and bases on litmus and methyl orange
- Describe the reactions of acids with metals, bases and carbonates
- Describe the reactions of bases with acids and with ammonium salts
- Describe solutions as acidic, neutral or alkaline using the pH scale
- Describe the use of universal indicator to measure pH
- Explain why calcium oxide and calcium carbonate are used to control soil pH
- Describe acids as proton donors and bases as proton acceptors
- Describe weak and strong acids and bases in terms of degree of ionisation
- Describe many metal oxides as basic and many non-metal oxides as acidic
- Describe neutral and amphoteric oxides

Making salts

Learning outcomes

By the end of this unit you should:

- Understand how to prepare salts by adding an acid to a reactive metal, a metal oxide or a metal carbonate
- Understand how to prepare salts by adding an acid to a metal hydroxide or aqueous ammonia
- Understand how to prepare insoluble salts by precipitation
- Understand how to make a salt from suitable reactants

16.01 Preparing soluble salts

Adding acid to an insoluble base

The base used is usually a metal oxide. The metal ions replace the hydrogen ions in the acid.

Example 1: Copper sulfate from copper oxide:

$$CuO(s) + H_2SO_4(aq) \rightarrow CuSO_4(aq) + H_2O(l)$$

Example 2: Zinc chloride from zinc oxide:

$$ZnO(s) + 2HCl(aq) \rightarrow ZnCl_2(aq) + H_2O(l)$$

The general method is as follows:

1. Add excess metal oxide to the acid in a beaker.
2. Warm the beaker gently to make sure reaction is complete.
3. Filter off the excess metal oxide. The filtrate is a solution of the salt.
4. Put the filtrate into an evaporating basin and leave to crystallise (see below).

Adding acid to a metal

This method is very similar to the one above. The metal reacts with the acid by a redox reaction and the metal ions replace the hydrogen ions in the acid.

The method is not used to prepare salts of very reactive metals such as sodium because the reaction is too violent.

Example 1: Iron(II) sulfate from iron:

$$Fe(s) + H_2SO_4(aq) \rightarrow FeSO_4(aq) + H_2(g)$$

Example 2: Zinc nitrate from zinc:

$$Zn(s) + 2HNO_3(aq) \rightarrow Zn(NO_3)_2(aq) + H_2(g)$$

The general method is as follows:

1. Add excess metal to the acid in a beaker.
2. Warm the beaker gently to make sure reaction is complete.
3. Filter off the excess metal. The filtrate is a solution of the salt.
4. Put the filtrate into an evaporating basin and leave to crystallise (see below).

Crystallisation

The last stage in the preparation of a soluble salt is crystallisation. The general method is as follows:

1. Gently heat the salt solution in an evaporating basin to concentrate it (or let the water evaporate slowly by leaving the solution in a warm place).
2. Evaporate the solvent until a saturated solution is reached (crystallisation point).
3. Leave the saturated solution to form crystals.
4. Filter off the crystals then dry them between filter papers.

TIP

When carrying out crystallisation:

- do not 'boil off' too much water: bigger crystals are produced if the evaporation is slow
- do not heat the crystals to dry them otherwise they will lose their water of crystallisation.

Worked example 16.01

Suggest how you could prepare dry crystals of the soluble salt zinc sulfate from zinc. **[7]**

To make a <u>*sulf*</u>*ate, the corresponding acid is* <u>*sulf*</u>*uric acid. Excess zinc should be used to make sure that all the sulfuric acid is reacted. You must also remember to filter off the excess zinc before partly evaporating the solution. Make sure that you do not just write 'evaporate'. It is better to refer to the crystallisation point.*

Add excess zinc **[1]** to (dilute) sulfuric acid **[1]**, filter off the excess zinc **[1]** then evaporate the filtrate to the crystallisation point **[1]**. Leave the solution to crystallise **[1]**.

Note that the question asks for dry crystals so you:

Filter off crystals **[1]** and dry them between filter papers **[1]**.

Adding acid to a solution of an alkali

A titration method is used to prepare a soluble salt from a solution of a soluble base or a soluble carbonate. The reaction is a neutralisation. The metal ions replace the hydrogen ions in the acid.

Example 1: Sodium sulfate from sodium hydroxide:

$$2NaOH(aq) + H_2SO_4(aq) \rightarrow Na_2SO_4(aq) + 2H_2O(l)$$

Example 2: Sodium chloride from sodium carbonate:

$$Na_2CO_3(aq) + 2HCl(aq) \rightarrow 2NaCl(aq) + H_2O(l) + CO_2(g)$$

This method is especially useful for making ammonium salts:

$$2NH_3(aq) + H_2SO_4(aq) \rightarrow (NH_4)_2SO_4(aq)$$

The general method is as follows:

1. Place the solution of alkali or carbonate in a flask and add a few drops of acid–base indicator.
2. Add acid from a burette until the indicator changes colour. Record the volume of acid added.
3. Repeat the experiment without the indicator in the flask by adding the volume of acid you recorded in step 2.
4. Put the filtrate into an evaporating basin and leave to crystallise (see above).

Worked example 16.02

Suggest how you could prepare a dry crystalline sample of the soluble salt potassium nitrate using a titration method. **[11]**

For this method you need to select an acid and a soluble base, e.g. Group I metal hydroxides are soluble. Metal oxides are insoluble so cannot be used. You should know how to carry out the titration method (see Unit 9).

Make a solution of potassium hydroxide **[1]** and add an indicator to the solution **[1]**. Add nitric acid from a burette **[1]** until the indicator changes colour **[1]**. Record the volume of acid added **[1]**. Repeat the experiment without an indicator **[1]**, adding the same amount of acid **[1]**.

Once you have the neutralised solution, the procedure is the same as for the other methods.

Evaporate the solution to crystallisation point **[1]**. Leave the solution to crystallise **[1]**. Filter off crystals **[1]**. Dry crystals between filter papers **[1]**.

Make sure that you know which of these methods to use for the preparation of a soluble salt. The first two methods are suitable if you are starting with an insoluble base. The titration method is for alkalis or carbonates which are soluble in water.

Progress check

16.01 Name three different compounds you could use to make the soluble salt sodium sulfate using sulfuric acid. [3]

16.02 When making the soluble salt zinc sulfate from zinc oxide, suggest why excess zinc oxide should be used. [1]

16.03 Why would you not prepare potassium chloride by the reaction of potassium with hydrochloric acid? [1]

16.04 Write word equations for the preparation of these salts from oxides:

a copper(II) sulfate [2]

b nickel chloride [2]

c zinc nitrate. [2]

16.05 Write word equations for the preparation of these salts from carbonates:

a iron(II) sulfate [2]

b magnesium chloride. [2]

16.06 What method would you use to prepare

a ammonium sulfate using sulfuric acid? [2]

b zinc chloride using hydrochloric acid? [1]

16.02 Preparing insoluble salts by precipitation

Solubility rules

TERMS

Precipitation reaction: A reaction in which a solid is formed when two solutions are mixed.

In some reactions a solid is formed when two solutions are mixed. These are **precipitation reactions.** The solid obtained is the precipitate. In order to make salts by precipitation we have to know the type of compounds which are insoluble in water.

Some rules for predicting solubility are given in the table.

Soluble compounds	Insoluble compounds
all salts of Group I elements	
all nitrates and ammonium salts	
most chlorides, bromides and iodides	chlorides, bromides and iodides of silver and lead
most sulfates	sulfates of calcium, barium and lead
group I carbonates	most other carbonates
group I hydroxides and barium hydroxide	most other hydroxides

Table 16.01

In questions involving salt preparation you are often given information about solubility of the reactants and products. This could be in the form of an equation with state symbols.

Preparing salts by precipitation

Example: Preparing lead(II) chloride:

$$Pb(NO_3)_2(aq) + 2NaCl(aq) \rightarrow PbCl_2(s) + 2NaNO_3(aq)$$

$Pb(NO_3)_2(aq)$	$2NaCl(aq)$	$PbCl_2(s)$	$2NaNO_3(aq)$
soluble salt	soluble salt	insoluble salt	soluble salt

To make an insoluble salt such as lead(II) chloride example we:

1. identify the ions present in the salt, e.g. lead and chloride
2. identify two soluble salts containing these ions, e.g. lead nitrate and sodium chloride
3. put one of the solutions into a beaker then add the other solution
4. filter off the precipitate
5. wash the precipitate with distilled water and dry it using filter paper or letting the water evaporate

TIP

When deciding the method to use to make a salt, you first have to find out if the salt is soluble or insoluble. If it is insoluble, it has to be made by the precipitation method.

Sample answer

Silver bromide is an insoluble salt. Suggest a method of preparing dry crystals of silver bromide. Give full experimental details and write an ionic equation for the reaction including state symbols. **[7]**

Add aqueous silver nitrate **[1]** to aqueous sodium bromide **[1]**. Filter off the silver bromide precipitate **[1]**. Wash the precipitate with water **[1]**. Dry precipitate between filter papers or in an oven at just above 100°C **[1]**.

$Ag^{+}(aq) + Br^{-}(aq) \rightarrow AgBr(s)$

(**1 mark** for correct reactants and products, **1 mark** for correct state symbols)

Progress check

16.07 Give the names of aqueous solutions you could mix to prepare the following insoluble salts:

a lead(II) chloride [2]

b barium sulfate [2]

c silver iodide. [2]

16.08 Write ionic equations, including state symbols, for the reactions in question 16.07, above. [6]

16.09 For each of the following compounds suggest whether they are soluble or insoluble in water:

a calcium sulfate

b lead chloride

c ammonium sulfate

d potassium nitrate

e iron(II) hydroxide. [5]

Exam-style questions

Question 16.01

Crystals of hydrated cobalt sulfate can be made from cobalt carbonate.

a State the name of the acid you should use to make this salt. [1]

b Write a word equation for the reaction. [2]

c The cobalt carbonate is in excess. Explain why. [1]

d Describe the method you would use to separate cobalt carbonate from aqueous cobalt sulfate. [1]

e Describe how you would prepare dry crystals of hydrated cobalt sulfate from an aqueous solution of cobalt sulfate. [4]

Question 16.02

This question is about the preparation of lead salts.

a Lead nitrate is soluble in water. Suggest a method of preparing dry crystals of lead nitrate from insoluble lead oxide. Give full experimental details. [8]

b Lead iodide is insoluble in water. Suggest a method of preparing dry crystals of lead iodide. Give full experimental details and write an ionic equation for the reaction including state symbols. [8]

Revision checklist

You should be able to:

- ☐ Describe the preparation of salts by adding an acid to a reactive metal, a metal oxide or a metal carbonate
- ☐ Describe the preparation of salts by adding an acid to a metal hydroxide or aqueous ammonia
- ☐ Demonstrating knowledge and understanding of the preparation of insoluble salts by precipitation
- ☐ Describe how to make a salt given suitable information

Unit 17

Identifying ions and gases

Learning outcomes

By the end of this unit you should:

- Be able to describe tests for the cations Al^{3+}, Ca^{2+}, Cu^{2+}, Cr^{3+}, Fe^{2+}, Fe^{3+} and Zn^{2+} using aqueous sodium hydroxide or aqueous ammonia
- Be able to describe a test for NH_4^+ ions
- Be able to describe flame tests for lithium, sodium, potassium and copper(II) ions
- Be able to describe tests for the anions carbonate, chloride, bromide, iodide, nitrate, sulfite and sulfate
- Be able to describe tests for the gases ammonia, carbon dioxide, chlorine, hydrogen, oxygen and sulfur dioxide

17.01 Tests for metal cations using sodium hydroxide or ammonia

- Many metal cations can be identified by observing the colour of the precipitate formed by the addition of dilute aqueous sodium hydroxide to an aqueous solution of the substance under test.
- If the sodium hydroxide is not in excess, a precipitate of a metal hydroxide is usually formed.
- In excess sodium hydroxide, some of the precipitates may dissolve, e.g. precipitates arising from solutions containing Zn^{2+}, Al^{3+} and Cr^{3+} ions.
- Aqueous ammonia contains hydroxide ions.
- On addition of aqueous ammonia most of the observations are similar to those observed with aqueous sodium hydroxide. There are some exceptions. For example, precipitates arising from solutions containing Zn^{2+} and Cu^{2+} ions dissolve in excess aqueous ammonia.

The table lists the tests and results for some cations.

You do not have to know the equations for the formation of the complex ion formed when copper hydroxide or zinc hydroxide dissolve in excess ammonia.

Cation tested (in aqueous solution)	Add a few drops of aqueous sodium hydroxide to the solution under test. Then add excess.	Add a few drops of aqueous ammonia to the solution under test. Then add excess.
aluminium, Al^{3+}	white precipitate. Excess: dissolves to form colourless solution	white precipitate. Excess: insoluble; white precipitate remains
calcium, Ca^{2+}	white precipitate. Excess: insoluble; white precipitate remains	no precipitate or very slight white precipitate
chromium(III), Cr^{3+}	green precipitate. Excess: forms greenish solution	grey-green precipitate. Excess: insoluble; grey-green precipitate remains
copper(II), Cu^{2+}	light blue precipitate. Excess: insoluble; light blue precipitate remains	light blue precipitate. Excess: dissolves to form a dark blue solution
iron(II), Fe^{2+}	green precipitate. Excess: insoluble; green precipitate remains	green precipitate. Excess: insoluble; green precipitate remains
iron(III), Fe^{3+}	red-brown precipitate. Excess: insoluble; red-brown precipitate remains	red-brown precipitate. Excess: insoluble; red-brown precipitate remains
zinc, Zn^{2+}	white precipitate. Excess: dissolves to form colourless solution	white precipitate. Excess: dissolves to form colourless solution

Table 17.01

More about the tests for cations

- If a precipitate is formed on addition of aqueous sodium hydroxide or ammonia, the hydroxide is insoluble in water. For example:

$$ZnCl_2(aq) + 2NaOH(aq) \rightarrow Zn(OH)_2(s) + H_2O(l)$$

- Ca^{2+} ions can be distinguished from Zn^{2+} and Al^{3+} ions because the precipitate of calcium hydroxide does not dissolve in excess sodium hydroxide but zinc hydroxide and aluminium hydroxide do dissolve.
- Zn^{2+} ions can be distinguished from Al^{3+} ions by the use of aqueous ammonia. Zinc hydroxide dissolves in excess aqueous ammonia to form a colourless solution but aluminium hydroxide does not dissolve.
- The hydroxides of most transition metal ions can be identified by their characteristic colours.

TIP

When conducting the test for Fe(II) ions, you must observe the colour of the precipitate straight away. If you leave the iron(II) hydroxide precipitate in the air for too long, it will oxidise to iron(III) hydroxide.

17.02 Test for ammonium ions, NH_4^+

- An aqueous solution containing ammonium ions is warmed.
- Ammonia gas is produced when warmed.
- Ammonia gas turns damp red litmus blue.

17.03 Flame test for metal cations

TERMS

Flame test: A test for particular metal ions by heating a sample of a compound containing the ion in a blue Bunsen flame. Characteristic colours are produced.

- A small sample of the compound under test is placed on the tip of a wire made from an unreactive metal such as nichrome or platinum.
- The compound under test is placed at the edge of a blue Bunsen flame.
- The colour given to the flame by the compound is observed.
- The test can be used to identify the ions of elements in Group I, and some Group II and transition metal ions.

Metal ion	Colour given to flame
lithium, Li^+	red
sodium, Na^+	yellow
potassium, K^+	lilac
copper, Cu^{2+}	blue-green

Table 17.02

Progress check

17.01 Describe tests and results for the cations in:

a sodium carbonate [2]

b calcium iodide [2]

c lithium sulfate [2]

d aluminium chloride. [2]

17.02 What colour precipitates are formed when a few drops of sodium hydroxide are added to solutions of:

a Cu^{2+} ions?

b Fe^{3+} ions?

c Zn^{2+} ions? [3]

17.03 An aqueous solution containing Cu^{2+} ions, is divided into two portions. Aqueous ammonia is added to the first portion.

a Describe what you would observe when aqueous ammonia is added until it is in excess. [4]

b The water from the second portion is evaporated. Some of the solid is placed on a wire loop and put in a Bunsen flame. What would you observe? [1]

17.04 Tests for anions

Anion under test	Test details	Result
carbonate, CO_3^{2-} (aqueous solution or solid)	add a dilute acid and test the gas given off	effervescence. Gas produced turns limewater milky
chloride, Cl^- (aqueous solution)	acidify with dilute nitric acid then add aqueous silver nitrate	white precipitate
bromide, Br^- (aqueous solution)	acidify with dilute nitric acid then add aqueous silver nitrate	cream precipitate
iodide, I^- (aqueous solution)	acidify with dilute nitric acid then add aqueous silver nitrate	yellow precipitate
nitrate, NO_3^- (aqueous solution)	add aqueous sodium hydroxide and aluminium foil then warm gently and test the gas given off	gas given off has pungent smell and turns red litmus blue
sulfate, SO_4^{2-} (aqueous solution)	acidify with dilute nitric acid then add aqueous barium nitrate	white precipitate
sulfite, SO_3^{2-} (aqueous solution)	add dilute acid then warm gently and test the gas given off	gas given off turns acidified aqueous potassium manganate(VII) colourless

Table 17.03

A solution that loses its colour is said to be colourless. It is incorrect to use the word 'clear' instead of 'colourless'.

More about the tests for anions

- Identifying carbonates:

Carbonates react with acids to produce carbon dioxide. For example:

$$CaCO_3(s) + 2HCl(aq) \rightarrow CaCl_2(aq) + CO_2(g) + H_2O(l)$$

The carbon dioxide released turns limewater milky.

- Identifying halides (chlorides, bromides and iodides):

The solution to be tested is first acidified with nitric acid to remove any carbonates.

Chlorides give a white precipitate of silver chloride:

$$AgNO_3(aq) + NaCl(aq) \rightarrow AgCl(s) + NaNO_3(aq)$$

The precipitate of silver chloride is white. With bromide ions, a cream precipitate of AgBr is formed and with iodide ions a yellow precipitate of AgI is formed.

- Identifying sulfates:

The solution to be tested is first acidified with nitric acid to remove any carbonates. The addition of aqueous barium nitrate produces a white precipitate of barium sulfate if a sulfate is present:

$$Ba(NO_3)_2(aq) + MgSO_4(aq) \rightarrow BaSO_4(s) + Mg(NO_3)_2(aq)$$

- Identifying sulfites:

Sulfites contain the ion SO_3^{2-}. When an aqueous sulfite is heated with a dilute acid, sulfur dioxide, SO_2, is released:

$$Na_2SO_3(aq) + 2HCl(aq) \rightarrow SO_2(g) + 2NaCl(aq) + H_2O(l)$$

The presence of sulfur dioxide can be confirmed by bubbling the gas through a solution of acidified potassium manganate(VII), which turns from purple to colourless if sulfur dioxide is present.

When acidifying the test reagent, remember what you are testing for. You should not acidify with hydrochloric acid when testing for halides or with sulfuric acid when testing for sulfates.

17.05 Identifying gases

Gas	Test for gas	Result
ammonia, NH_3	damp red litmus paper	turns litmus blue
carbon dioxide, CO_2	limewater	turns limewater milky
chlorine, Cl_2	damp litmus paper	bleaches litmus

Gas	Test for gas	Result
hydrogen, H_2	lighted splint	gives a squeaky 'pop' sound
oxygen, O_2	glowing splint	relights the splint
sulfur dioxide, SO_2	acidified aqueous potassium manganate(VII)	turns acidified aqueous potassium manganate(VII) from purple to colourless

Table 17.04

TIP

A common error is to confuse the tests for hydrogen and oxygen. Remember that a ligHted splint has the H for hydrogen and a glOwing splint has the O for oxygen.

Worked example 17.01

The label has fallen off two bottles of white solids. The solids could be either potassium carbonate or an ammonium salt. Describe tests for the cations and the ammonium ions present in order to deduce which solid is which. **[7]**

There are three types of tests for cations that you should know: addition of sodium hydroxide or ammonia to test for the colour and solubility of any precipitate formed, test for ammonium ions and the flame test. If the cation forms a soluble hydroxide, the flame test should be used. So for potassium ions, use the flame test.

Test for potassium ions using a flame test **[1]**. The Bunsen flame turns lilac **[1]**. Test for ammonium ions by warming gently **[1]**. If a gas is given off (ammonia) which turns red litmus blue **[1]** ammonium ions are present.

There are several types of test for anions. Some involve the addition of acid and the testing of a gas (carbonate / sulfite). Others involve precipitation (halides / sulfates). So for a carbonate you add acid and test for carbon dioxide and for a chloride you add silver nitrate.

Test for carbonate by adding hydrochloric acid **[1]** then testing the gas given off with limewater **[1]**. If the limewater goes cloudy **[1]** a carbonate is present.

More about the tests for gases

- When carbon dioxide is bubbled through limewater (a solution of calcium hydroxide), the limewater goes milky. This is due to the formation of a suspension of calcium carbonate:

$$Ca(OH)_2(aq) + CO_2(g) \rightarrow CaCO_3(s) + H_2O(l)$$

- Both chlorine and sulfur dioxide bleach damp litmus paper but of these two, only sulfur dioxide decolourises acidified potassium manganate(VII).

Sample answer

Two solutions M and N both form a green precipitate when a few drops of sodium hydroxide are added.

a When a large excess of concentrated sodium hydroxide is added to M, the precipitate dissolves. When excess sodium hydroxide is added to N, the precipitate does not dissolve. Identify the metal cations present in M and N. **[2]**

M contains Cr^{3+} ions **[1]**, N contains Fe^{2+} ions **[1]**.

b Describe the effect of adding a few drops of aqueous ammonia to M and N and then adding excess aqueous ammonia. **[4]**

With a few drops of ammonia M produces a grey-green precipitate **[1]**. With excess ammonia the precipitate does not dissolve **[1]**. With a few drops of ammonia N produces a grey-green precipitate **[1]**. With excess ammonia the precipitate does not dissolve **[1]**.

c The green precipitate from N is left in the air for 10 minutes. After 10 minutes the precipitate is beginning to turn a red-brown colour. Suggest why. **[2]**

It has been oxidised by oxygen in the air **[1]** to form Fe^{3+} ions **[1]**.

d Both solutions were acidified with nitric acid and aqueous silver nitrate was added. M produced a white precipitate and N produced a cream-coloured precipitate. Identify the anions present in M and N. **[2]**

M: chloride ions **[1]** N: bromide ions **[1]**.

Progress check

17.04 A metal reacts with hydrochloric acid. A gas is formed which pops with a lighted splint. The solution formed after leaving in the air gives a red-brown precipitate when a solution of sodium hydroxide is added.

- a Identify the cation present in the solution after exposure to air and the gas formed. [2]
- b Write a word equation for the reaction of the metal with the acid. [2]

17.05 A compound of metal T gives a white precipitate with aqueous sodium hydroxide but does not give a precipitate with aqueous ammonia. Hydrochloric acid is added to the compound. On heating, a gas is given off which turns acidified aqueous potassium manganate(VII) from purple to colourless.

- a Identify metal T. [1]
- b Identify the gas given off. [1]
- c Name the compound containing the ion of metal T. [1]
- d Describe what you would observe when aqueous silver nitrate is added to hydrochloric acid. [2]

Exam-style questions

Question 17.01

Substance Q is a liquid which dissolves in water to form a solution of pH 1.5.

- a What effect will Q have on blue litmus? Explain your answer. [2]
- b A solution of Q is added to solid R. A colourless gas is given off. Describe an experiment you could do to show that this gas is carbon dioxide. [3]
- c A small piece of solid R is heated in a blue Bunsen flame. The flame turns a green-blue colour. Which metal ion turns a Bunsen flame this colour? [1]
- d Identify solid R. [1]
- e Substance Q reacts with sodium carbonate to form a salt, S. S is a nitrate. Describe a test for nitrate ions. [4]

Question 17.02

G is a green gas. G is bubbled through a solution of acidified sodium sulfite.

- a G bleaches litmus paper. Identify G. [1]
- b Describe a test for sulfite ions and give the result. [4]
- c A product of the reaction between G and acidified sodium sulfite reacts with acidified barium nitrate solution to give a white precipitate. Give the chemical name of the precipitate. [1]
- d When G is bubbled through water, a mixture of ions is formed. One of these ions is the chloride ion. Describe a test for chloride ions and give the result. [3]
- e An aqueous solution of copper(II) chloride is divided into two parts. Aqueous sodium hydroxide is added to one portion until the hydroxide is in excess. Aqueous ammonia is added to the second portion until it is in excess. Describe the similarities and differences you would observe in these two reactions. [6]

Revision checklist

You should be able to:

- Describe tests for Al^{3+}, Ca^{2+}, Cu^{2+}, Cr^{3+}, Fe^{2+}, Fe^{3+} and Zn^{2+} ions using aqueous sodium hydroxide or aqueous ammonia
- Describe a test for NH_4^+ ions
- Describe flame tests for lithium, sodium, potassium and copper(II) ions
- Describe tests for carbonate, chloride, bromide, iodide, nitrate, sulfite and sulfate ions
- Describe tests for the gases ammonia, carbon dioxide, chlorine, hydrogen, oxygen and sulfur dioxide

The Periodic Table

Learning outcomes

By the end of this unit you should:

- Be able to describe the Periodic Table as a method of classifying elements
- Understand the use of the Periodic Table to predict the properties of elements
- Know that across a Period, there is a change from metallic to non-metallic character of the elements
- Be able to describe and explain the relationship between the Group number and the number of outer shell electrons
- Be able to identify trends in Groups in the Periodic Table from information given

18.01 The Periodic Table

TERMS

Periodic Table: An arrangement of elements in order of increasing proton number so that elements with the same number of electrons in their outer shell fall in the same vertical column.

Group: A vertical column of elements in the Periodic Table.

Period: A horizontal row of elements in the Periodic Table.

In the modern extended form of the **Periodic Table**:

- The elements are arranged in order of increasing proton number.
- A vertical column in the Periodic Table is called a **Group**.
- A horizontal row in the Periodic Table is called a **Period**.
- Elements are arranged in Groups numbered I–VIII.
- The first three Periods are called short Periods and the others are long Periods.

Predicting properties

- The elements within some Groups have similar chemical properties. For example, Group I elements react rapidly with water.
- There is a trend in physical properties down many Groups (see Units 20 and 21) and across a Period.
- We can use these trends to predict the properties of elements.

Refer to the Periodic Table (found at the back of this book) when looking up proton numbers, relative atomic masses and symbols.

18.02 Electronic structure and the Periodic Table

Groups

- The electron arrangement of an atom, especially the number of outer shell electrons determines the chemical properties of an element.
- Elements in the same Group have the same number of outer shell electrons.
- So for many Groups, the elements in the Group have similar chemical properties. This applies especially to Groups I, II, VII and VIII.
- The number of outer shell electrons is equal to the group number, e.g. magnesium in Group II has two electrons in its outer shell and chlorine in Group VII has seven electrons in its outer shell.
- Elements in Group VIII have eight electrons in their outermost shell except helium, which has two.

Periods

- In a short Period atoms of elements in the same Period have their outermost electrons in the same shell.
- The Period number of an element is the number of shells which contain electrons.
- As we go across a short Period, the number of outer shell electrons increases by one as the proton number increases by one but the number of occupied shells remains the same.

TIP

Remember that the first Period contains only two elements, hydrogen and helium. It is a common error to forget this.

Progress check

18.01 On what does the arrangement of the elements in the Periodic Table depend? [1]

18.02 Identify the element in Period 4 and Group 5 of the Periodic Table. [1]

18.03 Which Group of elements in the Periodic Table has a completely full outer electron shell? [1]

18.04 What is the relationship between the Group number and the number of electrons in the outer shell of an element? [1]

18.03 Metallic and non-metallic properties in the Periodic Table

- Metals occur on the left-hand side and in the middle of the Periodic Table.
- Non-metals occur on the right-hand side of the Periodic Table (see Figure 18.01).
- Across each Period there is a gradual change from metallic to non-metallic character.
- Down each Group the elements become more metallic in character.
- Between the metals and the non-metals are the elements which have some properties of both. These are called metalloids. Many (but not all) of the elements near the 'staircase line' shown on the Periodic Table in Figure 18.01, are metalloids.

The physical properties of metals and non-metals have been discussed in Unit 5. Some of the more important points related to electron distribution, bonding and reactivity are shown in the table below.

Property	Metals	Non-metals
electron arrangement	1 to 3 (or more in periods 5,6) electrons in the outer shell	4 to 7 electrons in the outer shell
bonding	metallic due to loss of outer shell electrons	covalent because of sharing outer shell electrons
type of oxide	basic (but a few are amphoteric)	usually acidic (but some are neutral)
reaction with acids	many react with acids	do not react with acids

Table 18.01

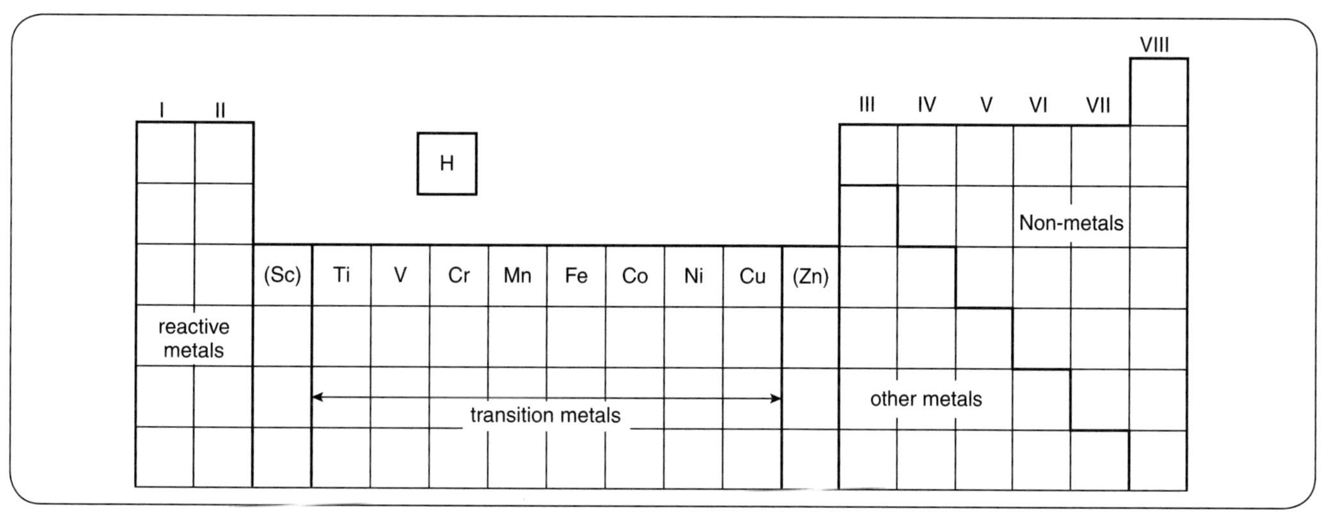

Figure 18.01 Metals and non-metals in the Periodic Table

Sample answer

Describe the arrangement of the elements in the Periodic Table in terms of proton number and how metallic and non-metallic character is related to the position of elements in the Periodic Table. **[7]**

Elements are arranged in horizontal Periods **[1]** and vertical Groups **[1]** in order of increasing proton (atomic) number **[1]**. Elements on the left of the table are metals **[1]**. Elements on the right of the table are non-metals **[1]**. There is a trend of decreasing metallic character across a Period **[1]** and a trend of increasing metallic character down a Group **[1]**.

18.04 Other trends in the Periodic Table

As we move from left to the right across a Period, the physical and chemical properties of the elements change. Periodicity is the regular occurrence of similar properties of the elements in the Periodic Table so that some Groups have similar properties or a trend in properties. Some examples are as follows:

- Across a Period, the boiling point increases then decreases. This reflects the different structures of the elements (Figure 18.02).

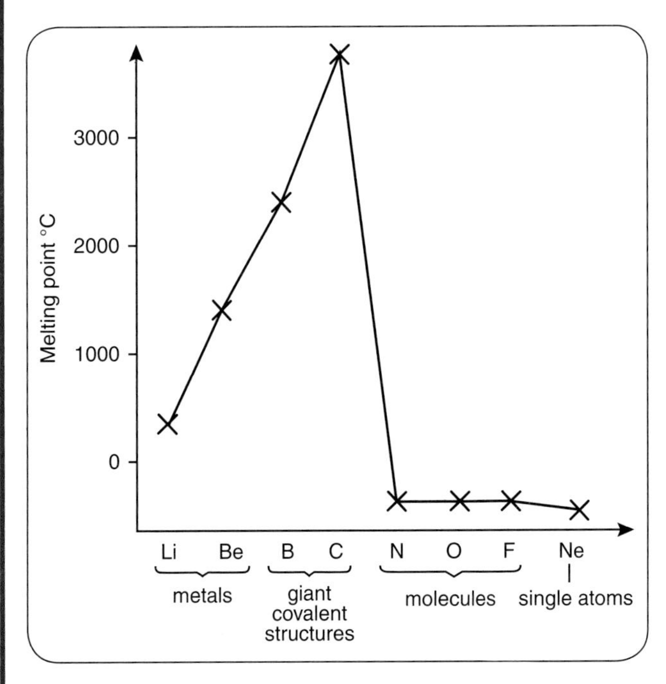

Figure 18.02

- Across a period the ability of the atoms to lose electrons decreases and their ability to gain electrons increases.
- The formulae of the chlorides show a pattern related to the number of electrons in the outer shell of their atoms:

$NaCl$, $MgCl_2$, $AlCl_3$, $SiCl_4$, PCl_3, SCl_2

The chlorides on the left of the Periodic Table tend to be ionic, those on the right tend to be covalent.

- Variation in chemical properties (which depend on the number of outer shell electrons in the atoms). For example, reaction with water:

  sodium (Group I) reacts rapidly and forms an alkaline solution

  magnesium (Group (II) reacts very slowly and forms a weakly alkaline solution

  aluminium (Group III) only reacts slowly when heated in steam

  silicon (Group IV), phosphorus (Group V) and sulfur (Group VI) do not react

  chlorine (Group VII) reacts slightly to form an acidic solution

- There are trends within each Group in both physical and chemical properties (see Units 19 and 20).

TIP

Some questions may ask you to identify trends across Periods as well as down a Group in order to deduce information about elements.

Worked example 18.01

Predict the properties of selenium (Se) from its position in the Periodic Table and the following information. Explain your predictions.

Melting points: O = −218 °C; S = +113 °C; Te = 450 °C **[6]**

This is a data analysis question where you have to predict trends from the data given. The major point to notice is that the elements in Group VI in descending order are O, S, Se, Te. So the properties of Se must be between those of S and Te. In this sort of question you are not expected to score all the marking points. In this case six out of the nine points shown here gets full marks.

Selenium is a solid **[1]** because both S and Te are solids **[1]**. Its melting point is between 113 and 450 °C **[1]** because the melting points decrease down the Group **[1]**. Selenium is a non-metal **[1]** because it is on the right-hand side of the Periodic Table **[1]** so it does not conduct electricity / other non-metallic properties **[1]**. It forms Se^{2-} ions **[1]** because sulfur and oxygen form ions with a −2 charge **[1]**.

Progress check

18.05 Describe the change in electrical conductivity across a Period. [2]

18.06 Where in the Periodic Table are elements having basic and acidic oxides found? [2]

18.07 Where in the Periodic Table are elements having amphoteric oxides most likely to be found? [1]

18.08 What is the relationship between the position of the elements in the Periodic Table and the number of electrons lost or gained when they form simple ions? [5]

18.09 How do the formulae of the chlorides of the elements in Period 3 vary across the Period? [3]

Sample answer

Some properties of the elements in Groups I to VII of Period 3 of the Periodic Table are given in Table 18.02. Deduce the missing figures and words in boxes A to I. **[10]**

A = 2,8,5 B = 2,8,6 C = 2,8,7 (**2 marks** for all three correct, **1 mark** for one or two correct)

D electrons shared **[1]** 4 (electrons shared) **[1]**

E $2e^-$ **[1]** gained **[1]**

F allow values between 900 and 2200 (actual value 1107) **[1]**

G insulator **[1]** H metal **[1]** I non-metal **[1]**

	Na	**Mg**	**Al**	**Si**	**P**	**S**	**Cl**
Electron arrangement	2,8,1	2,8,2	2,8,3	2,8,4	A	B	C
Electron gain or loss on forming ions	$1e^-$ lost	$2e^-$ lost	$3e^-$ lost	D	$3e^-$ gained	E	$1e^-$ gained
Boiling point/°C	883	F	2467	2355	280	445	−35
Electrical conductor?	yes	yes	yes	semi-conductor	no	G	no
Metal or non-metal	metal	metal	H	metalloid	I	non-metal	non-metal

Table 18.02

Exam-style questions

Question 18.01

Magnesium and phosphorus are in the same Period of the Periodic Table.

a To which Period and Group in the Periodic Table does phosphorus belong? [1]

b The oxides of calcium and phosphorus both react with water. Describe how you could distinguish between the two solutions formed. Explain your answer. [5]

c Which element, magnesium or phosphorus, will be a better electrical conductor? Explain your choice. [2]

d Which element, magnesium or phosphorus, will have the higher melting point?

Explain your answer. [4]

e The melting point of silicon is 1410 °C. The melting point of sulfur is 113 °C. Use this information to suggest the state of phosphorus at room temperature. Explain your answer. [2]

f Phosphorus is in the same Group as nitrogen, antimony and bismuth. Antimony and bismuth are metals. What is the trend in metallic nature in this Group? Explain your answer. [2]

Question 18.02

Germanium is in Group IV of the Periodic Table. One of its structures is similar to diamond. The elements in Group IV show trends in physical and chemical properties.

a Describe the structure and bonding of germanium. [3]

b Suggest why germanium has a high melting point. [2]

c By reference to the Periodic Table, suggest how the electrical conductivity varies down Group IV? Assume that carbon is in the form of diamond. Explain your answer. [2]

d The bond energy of the C–C bond in diamond is 350 kJ / mol. The bond energy of the Ge–Ge bond in germanium is 188 kJ / mol. Suggest a value for the bond energy of silicon. [1]

e Carbon and germanium form chlorides with the same formula.

i Predict the type of bonding in these chlorides. Explain your answer. [2]

ii Predict the formula of the chloride of germanium. [1]

Revision checklist

You should be able to:

- Describe the structure of the Periodic Table (Groups and Periods)
- Describe the change from metallic to non-metallic properties across a Period
- Describe how trends in the Periodic Table can be used to predict the properties of elements
- Describe and explain the relationship between number of outer shell electrons and Group number
- Identify trends and predict properties in the Periodic Table from information given

Metals

Learning outcomes

By the end of this unit you should:

- Be able to list the general physical properties of metals
- Be able to describe lithium, sodium and potassium as a collection of relatively soft metals showing a trend in melting point, density and reaction with water
- Be able to predict the properties of other elements in Group I
- Be able to describe the distinguishing properties of transition elements
- Know that transition elements have variable oxidation states
- Be able to identify representations of alloys from diagrams of structure
- Be able to explain in terms of their properties, why alloys are used instead of pure metal

19.01 The physical properties of metals

- All metals conduct electricity and heat, are malleable and ductile, and are shiny (see Unit 5).
- Most metals have higher melting and boiling points and higher density than non-metals.

19.02 The Group I metals

TERMS

Alkali metals: The elements in Group I of the Periodic Table.

- The Group I elements are also known as the **alkali metals**.
- The first three metals in Group I are lithium, Li; sodium, Na; and potassium, K.
- Their atoms have a single electron in their outer electron shell:

Li 2,1 Na 2,8,1 K 2,8,8,1

- When they form ions, they lose this electron:

Li^+ 2 Na^+ 2,8 K^+ 2,8,8

Physical properties of Group I metals

- They have shiny, silvery surfaces when freshly cut.
- They are good conductors of electricity and heat.
- They are all soft metals. The metals get softer down the Group.
- They all have low melting points for metals. For example, the melting point of sodium is 98 °C. The melting points decrease down the Group.
- They all have low densities. Lithium, sodium and potassium are less dense than water. There is a general trend of increasing density down the Group:

Density in g/cm^3: Li 0.53; Na 0.97; K 0.86; Rb 1.53; Cs 1.88

Sodium and potassium upset the pattern.

Chemical properties of Group I metals

- The Group I metals react readily with oxygen and water vapour in the air. So they are kept under oil to prevent the metals from reacting. Their reactivity increases down the Group.
- They all react rapidly with cold water. The reactivity increases down the Group.

Table 19.01 compares what we observe when lithium, sodium and potassium react with water.

Lithium	steady reaction slow stream of bubbles	disappears slowly	moves slowly over the surface	remains solid no flames
Sodium	rapid reaction bubbles produced rapidly	disappears quickly	moves quickly over the surface	melts into a ball. no flames
Potassium	violent reaction bubbles produced very rapidly	disappears very quickly	moves very quickly over the surface	melts into a ball; violet flame; may explode

Table 19.01

- The equations for the reactions are similar:

$$2Li + 2H_2O \rightarrow 2LiOH + H_2$$
$$2Na + 2H_2O \rightarrow 2NaOH + H_2$$

- The aqueous hydroxides, LiOH, NaOH and KOH are highly alkaline and turn red litmus blue.
- Hydrogen gas is given off in the reaction. With potassium, the reaction is violent enough to make the hydrogen burn. The violet flame observed is characteristic of potassium.

Remember that observations are what you see, hear and feel. The statement that 'a gas is given off' is not an observation but, 'bubbles are seen' is an observation.

Worked example 19.01

The table shows some properties of the elements in Group I of the Periodic Table.

Element	Melting point/°C	Density in g/cm³	Reactivity with oxygen
lithium	180	0.53	burns steadily
sodium	98	0.97	burns rapidly
potassium	64	0.86	
rubidium			explodes when burnt

Table 19.02

a Suggest a value for the melting point of rubidium. **[1]**

The trend is towards a lower melting point. A value a little way below 64 °C is suitable but not too close or too far below.

Anything between 30 and 55 °C is a good answer **[1]**.

b Why is it easier to predict the melting point of rubidium than its density? **[2]**

You have to look at the trend in both.

Melting point goes down all the time **[1]**. The density goes up then down so there is not enough information to show a trend **[1]**.

c Describe the reactivity of potassium burning in oxygen. **[1]**

The trend in reactivity is getting more reactive. You have to write something between 'rapidly' and 'explosive'.

Burns very rapidly **[1]**.

d Copy and complete the symbol equation for the reaction of potassium with water. .... $K + 2H_2O \rightarrow$ .... KOH + ... **[2]**

First select the second product (hydrogen) remembering that hydrogen is H_2 not just H.

.... $K + H_2O \rightarrow$ .... $KOH + H_2$

Then balance the potassium. There are two O atoms on the left so there must be two on the right. You get this by putting a '2' in front of the KOH. Then balance the other K.

$2K + 2H_2O \rightarrow 2KOH + H_2$ (**2 marks** if all correct. If not, then **1 mark** for H_2)

Progress check

19.01 Write a word equation for the reaction of caesium with water. [2]

19.02 Give the electronic structure of a a potassium atom b a potassium ion. [2]

19.03 State three trends in the properties of the alkali metals. [3]

19.04 State three observations about the reaction of sodium with water. [3]

19.05 Write a symbol equation for the reaction of rubidium with water. [2]

19.03 Transition elements

TERMS

Transition element: A block of metals in the middle of the Periodic Table which have characteristic properties such as high melting points and the formation of coloured compounds.

The **transition elements** are in the block of metallic elements between Groups II and III of the Periodic Table. The first row of transition elements starts with titanium, Ti, and ends with copper, Cu (see Figure 18.01).

Note that Sc and Zn are not transition elements. They do not form coloured compounds and only form one type of ion.

Transition metals have much higher melting points and densities than the Group I and Group II metals. They are also harder and stronger.

Properties of transition elements and their ions

Physical properties

- They have very high melting and boiling points. The melting point of chromium is 1857 °C compared with 98 °C for sodium.
- They have very high densities. The density of chromium is 7.2 g/cm^3 compared with 0.97 g/cm^3 for sodium.
- They form coloured compounds. Copper(II) sulfate is blue and iron(III) sulfate is yellow.
- They are strong and hard in comparison with the Group I metals.
- They have all the other typical metallic properties (see Unit 5).

Chemical properties

- Transition elements and their compounds are good catalysts. Iron catalyses the synthesis of ammonia and vanadium(V) oxide catalyses the synthesis of sulfur trioxide (see Unit 25).
- Transition elements are less reactive than Group I elements. They do not react with cold water. Most, however, react with steam. For example:

$$3Fe + 4H_2O \rightarrow Fe_3O_4 + 4H_2$$

- Transition elements have more than one oxidation state in their compounds. For example, iron compounds can have oxidation states of +2 and +3. This is shown by Roman numerals: iron(II) sulfate, iron(III) sulfate.

Remember that it is the compounds of transition elements that are coloured and not the elements. The elements have the typical lustre of metals.

Sample answer

Iron is a transition element. Lithium is an element in Group I of the Periodic Table. What differences are there in the physical and chemical properties of these two elements? **[5]**

Lithium has a low density but iron has a high density **[1]**.

Lithium has a low melting point but iron has a high melting point **[1]**.

Lithium is soft but iron is hard **[1]**.

Lithium reacts rapidly with cold water but iron does not **[1]**.

Lithium is not a good catalyst but iron is a good catalyst **[1]**.

NOTE: differences in colour and charge refer to the compounds and not the elements.

Progress check

19.06 Give two chemical properties of transition elements. [2]

19.07 Give three physical properties of transition elements. [3]

19.08 Compound X is yellow in colour and is a good catalyst. Compound Y is white in colour and is not a catalyst. Which one of these compounds contains a transition element ion ? [1]

19.09 Give two properties of transition element compounds which are not shown by compounds of Group I elements. [2]

19.04 Alloys

An alloy is a mixture made up of either two or more different metals or a metal and non-metals. Many alloys are mixtures of transition elements. The metal added to create the alloy becomes part of the lattice structure of the other metal (Figure 19.01). For example, brass is a mixture of copper and zinc. The zinc ions are larger than the copper ions and make the layers less regular.

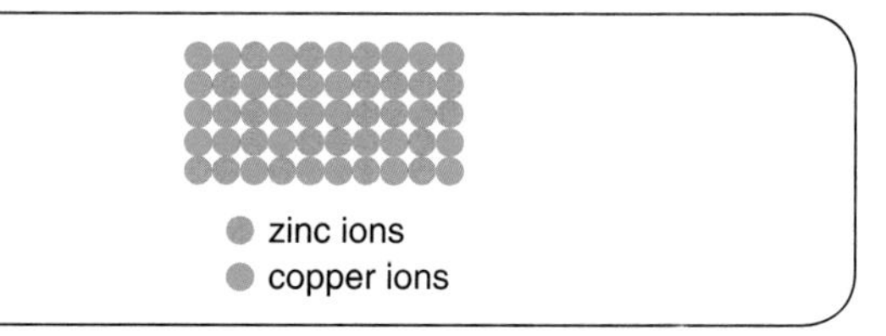

Figure 19.01

Alloys are used instead of pure metals for a number of reasons:

- They are stronger. Brass is stronger than copper and does not corrode.
- Alloys of iron with chromium or nickel are more resistant to corrosion.
- Alloys of iron with tungsten are very hard and do not change their shape at high temperatures.
- Aluminium alloyed with copper, manganese and silicon are used for making aircraft bodies because the alloy is stronger than aluminium alone and still has a low density comparable to pure aluminium.

Progress check

19.10 What does the term *alloy* mean? [3]

19.11 Why is an alloy of iron and nickel used for making knives rather than using pure iron? Give two reasons. [2]

Exam-style questions

Question 19.01

a What differences would you observe in the reaction of lithium with water compared with the reaction of potassium with water? [4]

b The solution formed when lithium hydroxide reacts with water is alkaline. How can you show that the solution is alkaline using a named indicator? [2]

c Write a word equation for the reaction of potassium with water. [2]

Question 19.02

Iron is a transition element.

a Give two physical properties of iron that are similar to sodium. [2]

b Give three physical properties of iron and its ions that are different from sodium and its ions. [3]

c Iron is usually used in the form of alloys rather than as pure iron. Give two reasons why alloys are used instead of the pure metal. [2]

d Copy and complete the symbol equation for the reaction of sodium with water. [2]

$2Na + 2H_2O \rightarrow$ ......NaOH + ... [2]

Question 19.03

Brass is an alloy of copper and zinc. Explain what is meant by the term alloy and describe the general metallic properties which copper and zinc have in common with **all** other metals. [7]

Revision checklist

You should be able to:

- Recall the physical properties of metals
- Describe the trends in density, melting point and hardness of lithium, sodium and potassium
- Describe the reaction of the Group I elements with water
- Predict the properties of other Group I elements from trends in data
- Describe the typical properties of transition elements
- Know that transition elements have variable oxidation states
- Identify alloys from diagrams of their structure
- Explain why alloys are used instead of pure metals

Unit 20

Halogens and noble gases

Learning outcomes

By the end of this unit you should:

- Know that chlorine, bromine and iodine are diatomic molecules and show a trend in colour and reaction with halide ions
- Predict the properties of other Group VII elements
- Know that the noble gases are unreactive
- Know some uses of the noble gases

20.01 The halogens

TERMS

Halogens: The elements in Group VII of the Periodic Table.

Halide: Compound containing an ion formed when a halogen atom has gained an electron to complete its outer shell of electrons.

Diatomic: Molecule containing two atoms.

- The **halogens** are all non-metals.
- All halogen atoms have seven electrons in their outer shell.
- The halogens form **halide** ions by gaining one electron to make their outer electron shell complete, e.g. chlorine (2,8,7) and chloride ion (2,8,8).
- The elements have a simple molecular structure of **diatomic** molecules (molecules made up of two atoms).
- They all have low melting and boiling points.

Table 20.01 shows some physical properties of the elements chlorine, bromine and iodine.

Halogen	State at 25 °C	Colour	Melting point/°C	Boiling point/°C
chlorine, Cl_2	gas	yellow-green	−101	−35
bromine, Br_2	liquid	red-brown	−7	+59
iodine, I_2	solid	grey-black	+114	+184

Table 20.01

There are trends in physical properties down the group:

- The colours get darker and more intense down the group.
- The melting points and boiling points increase down the group.
- The density (of the liquid at the boiling point) increases down the group.
- We can predict the properties of other Group VII elements because of these consistent trends down the Group.

TIP

Make sure that you know the difference in the colour of iodine solid, solution and vapour. Solid iodine is grey-black but iodine vapour is purple. An aqueous solution of iodine is brown. Iodine is relatively insoluble in water but soluble in aqueous potassium iodide.

Progress check

20.01 What colour is a chlorine gas b aqueous bromine c iodine vapour ? [3]

20.02 Describe the trend in:

a reactivity of the halogens up the group [1]

b density of the halogens down the group [1]

20.03 Write the electronic structure of

a a chlorine atom

b a fluoride ion. [2]

20.02 Displacement reactions of halogens

A displacement reaction is one in which one type of atom or ion replaces another in a compound.

When an aqueous solution of chlorine is added to an aqueous solution of potassium iodide, the chlorine displaces the iodide ion in the potassium iodide:

$Cl_2(aq)$ +	$2KI(aq)$	→ $2KCl(aq)$	+ $I_2(aq)$
chlorine	potassium iodide	potassium chloride	iodine
(colourless)	(colourless)	(colourless)	(brown)

Although concentrated aqueous solutions of chlorine may appear very light yellow-green, most solutions used in the school laboratory are colourless.

- Halogens are diatomic molecules which are elements.
- Halides are ionic compounds containing halide ions, e.g. chloride, Cl^-; bromide, Br^-; iodide, I^-.

The table below shows what happens when different halogens are added to different halides.

- If a reaction occurs, the names of the products are given.
- NR indicates no reaction.

Halogen (aqueous solution)	Halide (aqueous solution)		
	potassium chloride	potassium bromide	potassium iodide
chlorine (colourless)	NR	potassium chloride + bromine (red-brown)	potassium chloride + iodine (brown)
bromine (red brown)	NR	NR	potassium bromide + iodine (brown)
iodine (brown)	NR	NR	NR

Table 20.02

The results show the following:

- When a reaction takes place, the solution changes colour.
- A halogen higher in the group displaces a halogen lower in the group from its halide solution.
- This is because halogens higher in the Group are more reactive than those lower in the Group.
- The reactions are redox reactions.
- The reason why halogens higher in the Group are more reactive is that they are better oxidising agents.
- There is no reaction if a halogen lower in the group is added to a halide ion higher in the Group.

More about the reactions between halogens and halides

The reactions of halogens with halide ions are redox reactions. For example, the reaction:

$$Cl_2(aq) + 2KBr(aq) \rightarrow 2KCl(aq) + Br_2(aq)$$

Can be divided into two half equations:

$Cl_2(aq) + 2e^- \rightarrow 2Cl^-(aq)$ reduction of chlorine

and

$2Br^-(aq) \rightarrow Br_2(aq) + 2e^-$ oxidation of bromide ions

Worked example 20.01

a Use the data below to deduce the state, colour and boiling point of fluorine and astatine. **[6]**

Halogen	Colour	State at 25 °C	Boiling point / °C
fluorine			
chlorine	green	gas	−35
bromine	red	liquid	+59
iodine	grey-black	solid	+184
astatine			

Table 20.03

The answer depends on identifying the trends. The colour is getting darker down the Group, the state is going from gas to liquid to solid and the boiling point is increasing.

Fluorine: yellow / lighter in colour than chlorine **[1]**, state is gas **[1]**, boiling point: allow values between −40 and −250 °C (actual = −188 °C) **[1]**.

Astatine: black / deeper colour than iodine **[1]**, solid **[1]**, boiling point: allow 190–390 °C (actual = 337 °C) **[1]**.

b Explain why aqueous iodine reacts with potassium astatide? **[2]**

Halogens higher in the Group are more reactive than those lower in the Group and displace the halogen from the halide.

Iodine is more reactive than astatine **[1]** so displaces astatine **[1]** from potassium astatide.

c Copy and complete the chemical equation for this reaction. **[2]**

I_2 + ... KAt → ...... + 2KI

Remember that all the halogens are diatomic, so the formula for astatine is At_2. The number of K and At atoms must be the same on each side of the equation. One mark is for the correct formula of astatine, the second mark is for correct balance.

I_2 + 2KAt → At_2 + 2KI **[1]** for At_2, **[1]** for correct balance.

Sample answer

Will the following mixtures react or not? In each case give a reason for your answer. Write word equations for any reactions occurring and describe any observations you would make. **[8]**

a aqueous bromine and aqueous potassium chloride

b aqueous bromine and aqueous potassium iodide

For **a**, it does not react because bromine is less reactive than chlorine **[1]** so cannot displace the chlorine from potassium chloride **[1]**. The solution is the colour of bromine **[1]**.

For **b**, it reacts because bromine is more reactive than iodine **[1]** so bromine displaces iodine from potassium iodide **[1]**. The red-brown solution **[1]** turns (darker) brown **[1]**. The word equation is:

bromine + potassium iodide → iodine + potassium bromide **[1]**.

Progress check

20.04 Write a word equation for the reaction of aqueous potassium bromide with aqueous chlorine. [2]

20.05 a Write a chemical equation for the reaction of aqueous chlorine with aqueous potassium bromide. [2]

b Convert this equation into an ionic equation. [2]

20.03 The elements of Group VIII

TERMS

Noble gases: Elements in Group VIII of the Periodic Table.

Monatomic: Existing in the natural state as single atoms.

- The Group VIII (Group 0) elements are all gases, so they have very low melting and boiling points. They are called the **noble gases**.
- They are all colourless.
- They are **monatomic** – they exist in nature as single atoms.
- They have a complete outer electron shell of eight electrons (or two in the case of helium).
- They are inert (unreactive) because they cannot gain, lose or share electrons so cannot form bonds with other elements.

Some uses of the noble gases

- Helium is used for filling balloons because it is less dense than air and does not burn, unlike hydrogen.
- A mixture of helium and oxygen is used in breathing apparatus for deep sea divers.
- Neon, argon and xenon are used in advertising signs.
- Argon is used in providing an inert atmosphere in welding.
- Argon is used to fill electric light bulbs because it is inert.
- Neon, argon and xenon are used in advertising signs.
- Krypton is used in lasers for eye surgery and in lamps for car headlights.
- Xenon is used in lamps for lighthouses and hospital operating theatres.

Progress check

20.06 Compare the structures of chlorine and argon. [3]

20.07 Why is argon used in welding? [1]

20.08 What is meant by the noble gas electronic structure? [1]

Exam-style questions

Question 20.01

This question is about the noble gases. These gases are monatomic.

a Why are the noble gases unreactive? [2]

b What does the term monatomic mean? [1]

c Give two reasons why helium is used in balloons. [2]

d Explain why argon is used in electric light bulbs. [2]

e Deduce the electronic structure of argon. [1]

f The densities of the liquid noble gases in g/cm^3 are:

He 0.15, Ne 1.20, Ar 1.40, Kr ----, Xe 3.52

Deduce the density of liquid Kr. [1]

Question 20.2

Chlorine is a diatomic molecule.

a What do you understand by the term diatomic? [1]

b Explain why an aqueous solution of iodine does not react with an aqueous solution of potassium chloride. [1]

c What would you observe when an aqueous solution of chlorine reacts with an aqueous solution of potassium iodide? [2]

d Write a word equation for the reaction in part c. [1]

e Write a chemical equation for the reaction in part c. [2]

f Write an ionic equation, including state symbols, for the reaction in part c and explain which substance is the oxidising agent and which is the reducing agent. [6]

Revision checklist

You should be able to:

- [ ] Describe chlorine, bromine and iodine as diatomic molecules
- [ ] Describe the trend in colour and density of the halogens
- [ ] Predict the properties of other Group VII elements
- [ ] Describe the trend in the reactivity of halogens with halide ions
- [ ] Describe the noble gases as unreactive monatomic gases
- [ ] State the uses of the noble gases (inert atmosphere, balloons)

The reactivity series

Learning outcomes

By the end of this unit you should:

- Be able to place given metals in order of their reactivity by reference to their reactions with water or steam, with oxygen or with dilute hydrochloric acid
- Be able to place given metals in order of their reactivity by reference to the reduction of their oxides with carbon
- Be able to deduce an order of reactivity from a given set of experimental results
- Be able to describe the reactivity series as related to the tendency of a metal to form its positive ion with reference to aqueous ions, oxides and metals
- Be able to describe and explain the action of heat on the hydroxides, carbonates and nitrates of the listed metals
- Be able to account for the apparent unreactivity of aluminium in terms of the oxide layer which adheres to the metal

21.01 The reactivity of metals with oxygen

Table 21.01 shows the relative reactivity of different metals with oxygen.

Sodium	reacts violently, burning with a yellow flame
Gold	does not react
Iron	burns only when in the form of a powder or iron wool
Copper	Blackens in its surface but does not burn

Table 21.01

We can put these metals in order of their reactivity as:

gold → copper → iron → sodium
(least reactive) (most reactive)

When metals react with oxygen they form oxides:

$$\underset{\text{copper}}{2Cu} + \underset{\text{oxygen}}{O_2} \rightarrow \underset{\text{copper(II) oxide}}{2CuO}$$

21.02 The reactivity of metals with water

Table 21.02 shows the relative reactivity of different metals with water or steam, based on observations.

Potassium	reacts violently with cold water; hydrogen produced, burns and may explode.
Sodium	reacts violently with cold water but hydrogen produced does not catch fire
Calcium	reacts rapidly with cold water.
Magnesium	reacts very slowly with cold water but rapidly with steam.
Zinc	only reacts when heated strongly with steam
Iron	only reacts at red heat with steam
Copper	does not react with cold water or steam

Table 21.02

- The metals which react with cold water form a metal hydroxide and hydrogen as products. For example:

$$Ca + 2H_2O \rightarrow Ca(OH)_2 + H_2$$

- The metals which react with steam form a metal oxide and hydrogen as products. For example:

$$Zn + H_2O \rightarrow ZnO + H_2$$

21.03 The reactivity of metals with hydrochloric acid

Table 21.03 shows the relative reactivity of different metals with dilute hydrochloric acid based on observations.

Metal	Observation
Potassium	explosive reaction
Sodium	violent reaction and likely to explode
Calcium	very rapid reaction with a lot of bubbles produced rapidly
Magnesium	rapid reaction with bubbles produced at a steady rate
Zinc	slow reaction with bubbles produced slowly
Iron	very slow reaction with bubbles produced very slowly
Copper	no reaction even with concentrated acid

Table 21.03

We can use our observations of the rate of reaction of metals with water and hydrochloric acid to place metals in a **reactivity series** with the most reactive at the top (Figure 21.01).

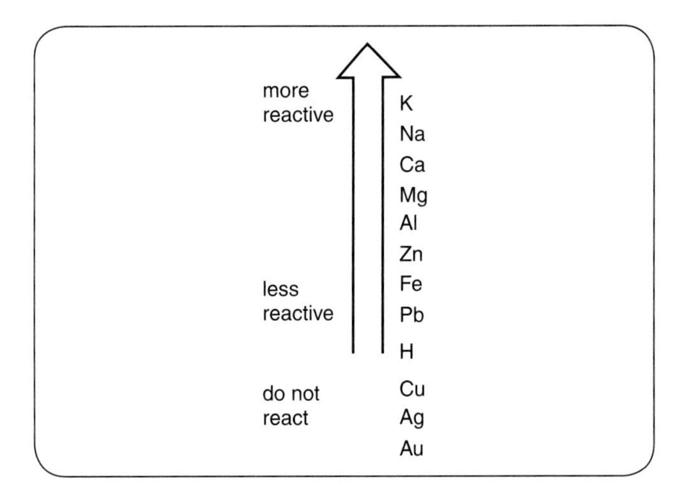

Figure 21.01

TERMS

Reactivity series: A list of elements (usually metals) in order of their reactivity, with the most reactive first.

In Figure 21.01 hydrogen is included. When acids react with metals, the hydrogen in the acid is replaced by a metal ion to form a salt. When we include hydrogen in this reactivity series, we see that:

- metals above hydrogen react with dilute hydrochloric acid to form a salt and hydrogen, e.g.

$$Fe + 2HCl \rightarrow FeCl_2 + H_2$$

- metals below hydrogen do not react with acids

TIP

In some questions you may be expected to deduce an order of reactivity from experimental results involving unfamiliar metals or metal oxides.

Worked example 21.01

Table 21.04 shows some observations made when different metals react with oxygen.

Metal	Observation
Copper	does not burn but its surface turns black
Iron	burns only when it is in powder form or as iron wool
Gold	does not react
Magnesium	a thin ribbon burns very rapidly

Table 21.04

a Put these elements in order of their reactivity and describe the observations you could make when they react with hydrochloric acid. **[4]**

The order of reactivity is the same as how fast the metal reacts. You need to look for words which describe the rate of reaction such as 'very rapidly' or 'burns slowly'. It should also be made clear which has the highest and which has the lowest reactivity:

magnesium (most reactive) > iron > copper > gold (least reactive) [1]

The order of reactivity with hydrochloric acid is likely to be the same as that with oxygen. Use the information above and what you know about the order of reactivity to suggest ***observations****.*

With HC*l*: with magnesium, bubbles are produced at a rapid rate [1]; iron reacts slowly and bubbles come off slowly [1]. With copper and gold there are no bubbles [1] because they do not react with acid.

b Write word equations for the reaction of magnesium with steam, and iron with hydrochloric acid. **[4]**

Remember that when metals react with steam, the oxide is formed and when they react with hydrochloric acid, a metal chloride is formed.

magnesium + steam →
magnesium oxide + hydrogen
(1 **mark** for each product)

iron + hydrochloric acid →
iron(II) chloride + hydrogen
(1 **mark** for each product)

21.04 Reducing metal oxides with carbon

Some metal oxides can be reduced by heating with carbon. For example, copper(II) oxide is reduced to copper.

$$2CuO + C \rightarrow 2Cu + CO_2$$

Figure 21.02 shows the metal reactivity series.

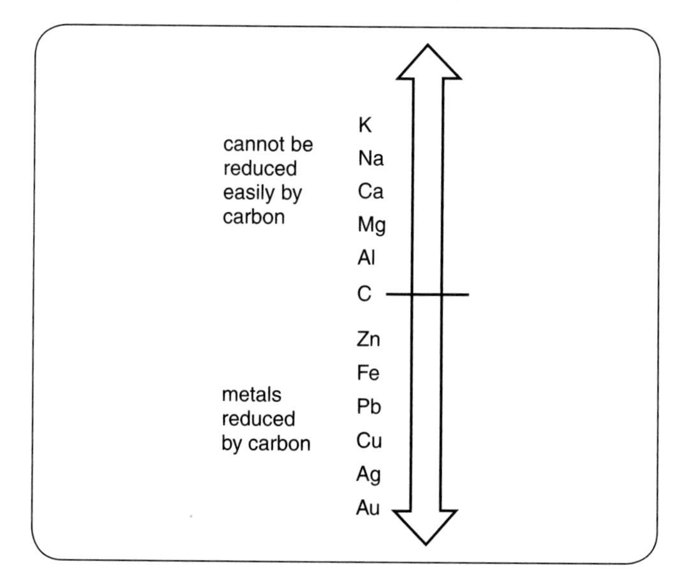

Figure 21.02

When metal oxides are heated with carbon, we see the following:

- The metals above carbon are too reactive. The carbon is not a strong enough reducing agent to remove the oxygen from the metal oxide. These metals are generally extracted from their oxides by electrolysis.
- The metals below carbon in the reactivity series can be reduced by heating with carbon because carbon is a better reducing agent than the metal. So carbon can remove the oxygen from the metal oxide.
- In general, the lower a metal is below carbon in the reactivity series, the less energy is required to remove oxygen from its oxide by heating with carbon.

Worked example 21.02

Here is a list of elements in order of reactivity. Carbon is also included.

most reactive $\xrightarrow{\text{Ca, Mg, C, Zn, Fe, Pb}}$ least reactive

a Which of these metals could be extracted from their oxides by electrolysis? Give a reason for your answer. **[3]**

Remember that metals above carbon in the reactivity series hold onto their oxygen too tightly for carbon to remove it.

Calcium and magnesium **[1]** because they are more reactive than carbon **[1]**. The oxygen is too strongly bonded to the metal for carbon to remove it **[1]**.

b Copper oxide can be reduced by heating with carbon. One of the products is carbon dioxide. Copy and complete the chemical equation for this reaction. **[3]**

2CuO + C → .... + ...

Start with the formula for carbon dioxide. You should know this! If copper oxide is reduced, then copper must be the other product since oxygen has been removed. The last mark is for the balancing.

$$2CuO + C \rightarrow 2Cu + CO_2$$

[1] [1] [1]

Progress check

21.01 Iron reacts with steam but calcium reacts with cold water. Give the names of the products formed in each reaction. [4]

21.02 Here is a list of elements in order of reactivity. Carbon is also included.

most reactive $\xrightarrow{\text{Ca, Mg, C, Zn, Fe, Cu, Ag}}$ least reactive

Which of these metals cannot be extracted from their oxides by heating with carbon? [1]

21.03 Which of the metals in question 21.02 will react with cold water? [1]

21.04 Which metals in question 21.02 do not react with hydrochloric acid? [1]

21.05 Write a chemical (symbol) equation for the reaction of zinc with hydrochloric acid, HCl, to form zinc chloride, $ZnCl_2$, and hydrogen. [2]

21.05 The reaction of metals with aqueous metal ions

Table 21.05 shows the results of adding some metals to some metal sulfates in aqueous solution. The solutions contain the metal ions shown.

Aqueous metal ion	Metal		
	magnesium	zinc	copper
magnesium ions, Mg^{2+}	no reaction	no reaction	no reaction
zinc ions, Zn^{2+}	reaction	no reaction	no reaction
copper ions, Cu^{2+}	reaction	reaction	no reaction

Table 21.05

- The results show that, metals higher in the reactivity series react with metal ions from metals lower in the reactivity series.
- These are displacement reactions. The more reactive metal replaces the less reactive metal in the compound (sulfate). For example:

$$Zn(s) + CuSO_4(aq) \rightarrow ZnSO_4(aq) + Cu(s)$$

- If a metal lower in the reactivity series is added to a solution of a salt of a metal higher in the reactivity series no reaction occurs. For example, adding zinc to magnesium ions.
- Metals higher in the reactivity series are better reducing agents and are better at losing electrons to form positive ions.

Explaining metal – metal ion displacement reactions

A metal higher in the reactivity series can replace a metal below it from a solution of its ions: by a **displacement reaction.** We can use the reactivity series to predict whether or not a reaction is likely to happen.

TERMS

Displacement reaction: A reaction in which one atom or ion (usually a metal) replaces another atom or ion in a compound.

Example: Will zinc react with aqueous copper(II) sulfate?

- Zinc is higher than copper in the reactivity series.
- So zinc loses electrons and forms ions more readily than copper. Zinc is a better reducing agent.
- Copper ions are better at accepting electrons than zinc ions.
- The reaction taking place will be one where zinc loses electrons and copper ions gain electrons.
- When zinc is added to a solution of copper(II) sulfate, a pink deposit of copper is seen and the blue colour of the copper sulfate fades:

$$Zn(s) + Cu^{2+}(aq) \rightarrow Zn^{2+}(aq) + Cu(s)$$

Sample answer

Explain why magnesium will react with an aqueous solution of zinc chloride but silver will not. In your answer refer to redox reactions and the ease with which metals form ions. **[6]**

Magnesium is higher than zinc in the reactivity series **[1]** because magnesium forms ions more readily than zinc **[1]**.

Magnesium is the better reducing agent **[1]** and zinc is the better oxidising agent. Silver is lower than zinc in the reactivity series **[1]** so zinc forms ions more readily than silver **[1]**. Zinc is the better reducing agent **[1]** so silver cannot release electrons to form ions.

21.06 Explaining the reaction of metal oxides with carbon

Metals can be used to reduce metal oxides. For example, when we heat aluminium powder with iron(III) oxide, the aluminium reduces the iron(III) oxide to iron:

$$Fe_2O_3 + 2Al \rightarrow 2Fe + Al_2O_3$$

We can explain the use of a more reactive metal to reduce the oxide of a less reactive metal using the same ideas as for the metal – metal ion displacement reactions.

- Aluminium is higher than iron in the reactivity series than iron.
- So aluminium loses electrons and forms ions more readily than iron. Aluminium is a better reducing agent.
- Iron(III) ions are better at accepting electrons than aluminium ions.
- The reaction taking place will be one where aluminium loses electrons and iron(III) ions gain electrons:

$$2Al(s) + 2Fe^{3+}(aq) \rightarrow 2Al^{3+}(aq) + Fe(s)$$

Progress check

21.06 Explain why calcium is extracted by electrolysis but iron is extracted by heating with carbon. [2]

21.07 Magnesium is more reactive than iron. Describe how you could use magnesium to extract iron from iron(III) oxide and write a chemical equation for this reaction. [4]

21.08 Use ideas about ease of ionisation of metals to explain why magnesium reacts with an aqueous solution of zinc sulfate. [3]

21.07 Thermal decomposition reactions

Some compounds decompose when heated. Two or more products are formed. This type of reaction is called **thermal decomposition**.

TERMS

Thermal decomposition: The breakdown of a compound into two or more products by heating.

Thermal decomposition of metal hydroxides

Most metal hydroxides decompose on heating. A metal oxide and water are formed. For example:

$$\underset{\text{zinc hydroxide}}{Zn(OH)_2(s)} \xrightarrow{\text{heat}} \underset{\text{zinc oxide}}{ZnO(s)} + H_2O(g)$$

Group II hydroxides decompose in a similar way. Group I metal hydroxides, apart from lithium hydroxide, do not decompose.

Thermal decomposition of carbonates

Many carbonates and hydrogen carbonates decompose when heated to form the metal oxide and carbon dioxide. For example:

$$MgCO_3(s) \xrightarrow{\text{heat}} MgO(s) + CO_2(g)$$

magnesium carbonate → magnesium oxide + carbon dioxide

Group I carbonates apart from lithium carbonate do not decompose on heating.

- The more reactive the metal, the more difficult it is to decompose the carbonate. So copper carbonate decomposes readily on heating but potassium carbonate does not decompose.

Thermal decomposition of nitrates.

- All nitrates decompose when heated.
- Nitrates of Group I metals decompose to form the metal nitrite and oxygen. For example:

$$2NaNO_3(s) \xrightarrow{heat} 2NaNO_2(s) + O_2(g)$$

sodium nitrate ⟶ sodium nitrite + oxygen

- The nitrates of most other metals decompose to form the metal oxide, nitrogen dioxide and oxygen:

$$2Cu(NO_3)_2 \xrightarrow{heat} 2CuO(s) + 4NO_2(g) + O_2(g)$$

copper(II) nitrate ⟶ copper(II) oxide + nitrogen dioxide + oxygen

TIP

Make sure that you know how the thermal decomposition of Group I nitrates differ from those of other nitrates. You also need to be able to write or complete balanced equations for these reactions.

21.08 The apparent lack of reactivity of aluminium

- Aluminium is high in the reactivity series but appears unreactive with water or acids.
- This is because the surface of freshly-made aluminium gets coated with a layer of aluminium oxide by reaction with oxygen in the air:

$$4Al + 3O_2 \rightarrow 2Al_2O_3$$

- This layer is fairly unreactive and is resistant to corrosion.
- The tough oxide layer adheres (sticks) to the surface of the aluminium very strongly and does not flake off.
- So the aluminium below is protected from reaction with acids and water.

Progress check

21.09 Write a chemical equation for the thermal decomposition of sodium nitrate. [2]

21.10 Describe and explain the ease of decomposition of the carbonates of the Group II metals. [3]

21.11 Write a chemical equation for the thermal decomposition of calcium nitrate.

Exam-style questions

Question 21.01

Magnesium, sodium and zinc react with water or steam to produce hydrogen.

a Describe any observations you would make when these metals are added to cold water. [6]

b Describe and explain how the conditions required for hydrogen to be produced can be used to put these metals in order of reactivity. [3]

c Write a word equation for the reaction of zinc with steam. [1]

d Copy and complete the symbol equation for the reaction of sodium with water.

$\ldots Na + 2H_2O \rightarrow \ldots\ldots NaOH + \ldots\ldots$ [2]

e Explain why zinc can be extracted from its oxide by reduction with carbon but magnesium is extracted from magnesium compounds by electrolysis. [3]

f A student wanted to compare the reactivity of magnesium, sodium and zinc by adding the metals to hydrochloric acid.

i How could you use the observations of the reaction of these metals with hydrochloric acid to put the metals in order of reactivity? [3]

ii Explain why you should not add sodium to hydrochloric acid. [1]

Question 21.02

Different metals were added to different metal chlorides in aqueous solution. The results are shown in the Table 21.06 below. A reaction is represented by a tick.

	Cobalt	Nickel	Tin	Zinc
Cobalt(II) chloride		✗	✗	✓
Nickel(II) chloride	✓		✗	✓
Tin(II) chloride	✓	✓		✓
Zinc chloride	✗	✗	✗	

Table 21.06

a Deduce the order of reactivity of the metals. Put the most reactive first. [1]

b Explain using ideas about the ease of formation of ions and redox reactions why tin does not react with zinc chloride. [3]

c Which ions can be reduced by cobalt metal? [1]

d Which ion is the best oxidising agent? [1]

e Tin(II) hydroxide undergoes thermal decomposition.

i What is meant by the term thermal decomposition? [1]

ii Construct a symbol equation for the thermal decomposition of tin(II) hydroxide. [1]

Question 21.03

Magnesium reacts with black copper(II) oxide when heated. Describe what you would observe during this reaction. Identify the oxidising agent and reducing agent and explain why the reaction occurs in terms of the tendency of metal ions to lose electrons. [5]

Revision checklist

You should be able to:

- Place metals in order of their reactivity with water, steam, oxygen or dilute hydrochloric acid
- Place metals in order of their reactivity by the reaction of the metal oxides with carbon
- Deduce an order of reactivity from a given set of experimental results
- Understand that a more reactive metal will displace a less reactive metal from an aqueous solution of its ions
- Understand how the ease with which a metal or the metal in an oxide forms positive ions is related to the reactivity series
- Describe the thermal decomposition of metal hydroxides, carbonates and nitrates
- Explain the apparent unreactivity of aluminium by reference to its oxide layer

Unit 22

Metals: extraction and uses

Learning outcomes

By the end of this unit you should:

- Be able to describe the ease of obtaining metals from their ores by relating the elements to the reactivity series
- Be able to describe and state the essential reactions in the extraction of iron from haematite
- Be able to describe the conversion of iron into steel using basic oxides and oxygen
- Be able to describe how the properties of iron are changed by the controlled use of additives to form steel alloys
- Be able to describe the extraction of zinc from zinc blende
- Know that aluminium is extracted from bauxite by electrolysis
- Be able to describe how aluminium is extracted form bauxite
- Be able to name the uses of mild steel and stainless steel
- Be able to relate the uses of copper to its use in electrical wiring and in cooking utensils
- Be able to relate the uses of aluminium to its use in aircraft manufacture and its use in food containers
- Be able to explain the uses of zinc for galvanising and for making brass
- Be able to discuss the advantages and disadvantages of recycling aluminium and iron and steel

22.01 The extraction of metals

TERMS

Ore: A rock containing a metal or metal compound in sufficient quantity that a metal can be extracted from it.

Few metals are found as elements in rocks. Most elements are extracted from compounds in particular types of rock. These rocks are called **ores**. In the last unit we learnt that:

- the position of the metal in the reactivity series influences the method of obtaining the metal from a metal compound
- metals which are lower in the reactivity series (zinc and lower) can be extracted by heating with carbon
- metals high in the reactivity series (aluminium and higher) are extracted by electrolysis

22.02 The extraction of iron

- The raw materials for extracting iron are iron ore, coke (carbon), limestone (calcium carbonate) and air.
- The commonest ore of iron is haematite, which is largely iron(III) oxide, Fe_2O_3.
- The iron is extracted in a blast furnace (see Figure 22.01).
- The main reducing agent is carbon monoxide, which is formed by reactions within the furnace.

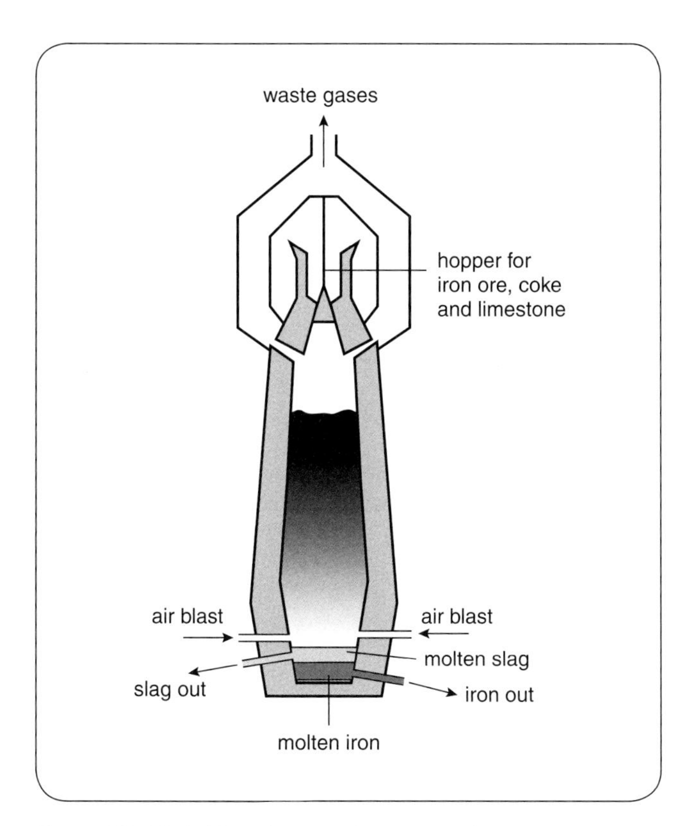

Figure 22.01 A blast furnace for extracting iron

The reactions producing iron are as follows:

- Coke (carbon) burns in a hot air blast to form carbon dioxide:

$$C(s) + O_2 \rightarrow CO_2(g)$$

- The carbon dioxide reacts with more coke to form carbon monoxide:

$$CO_2(g) + C(s) \rightarrow 2CO(g)$$

- The carbon monoxide reduces iron(III) oxide to iron:

$$Fe_2O_3(s) + 3CO(g) \rightarrow 2Fe(l) + 3CO_2(g)$$

The iron flows to the bottom of the furnace where it is removed.

Iron ore contains silicon(IV) oxide as an impurity. Limestone is added to the furnace to remove this impurity. The reactions involved are as follows:

- Limestone (calcium carbonate) decomposes on heating:

$$CaCO_3(s) \rightarrow CaO(s) + CO_2(g)$$

- The calcium oxide formed from this decomposition reacts with the silicon dioxide to form calcium silicate (slag), $CaSiO_3$.

$$CaO(s) + SiO_2(l) \rightarrow CaSiO_3(l)$$

The slag flows to the bottom of the furnace and being less dense than iron, floats on top of the iron.

Worked example 22.01

Iron is extracted from iron ore in a blast furnace.

a Explain the use of coke (carbon) and air in the blast furnace. **[7]**

The word 'explain' tells you that you have to write about the chemical reactions involved. Reactants and products should be mentioned.

Coke burns **[1]** in air **[1]** to form carbon dioxide **[1]**. This provides heat for the furnace **[1]**. The carbon dioxide then reacts with more coke **[1]** to form carbon monoxide **[1]**, which reduces the iron ore **[1]**.

b Explain why limestone (calcium carbonate) is added to the blast furnace. Include a relevant word equation in your answer. **[4]**

The word 'explain' tells you that you have to write about the chemical reactions involved. Remember that calcium carbonate does not react with silicon dioxide directly. It is first broken down to calcium oxide.

The limestone decomposes **[1]** to form calcium oxide **[1]**, which reacts with the sand (silicon dioxide) **[1]** to form slag **[1]**.

22.03 The conversion of iron into steel

Iron from the blast furnace contains impurities of carbon, phosphorus, silicon and sulfur. These have to be removed, otherwise the iron is too brittle. Steel is often made in a basic oxygen converter. This is a large vessel into which oxygen is blown.

The procedure for removing impurities and making steel is as follows:

- The molten iron is placed in the converter.
- Oxygen and powdered calcium oxide are blown onto the iron.

- The oxygen oxidises the carbon, phosphorus, silicon and sulfur in the iron to their oxides. They are all acidic oxides.
- Carbon dioxide and sulfur dioxide are gases, so they escape from the converter.
- Silicon and phosphorus oxides react with the calcium oxide to form a slag which is largely calcium silicate.

$SiO_2(l)$	+	$CaO(s)$	→	$CaSiO_3(l)$
silicon dioxide		calcium oxide		calcium silicate
(acidic oxide)		(basic oxide)		(slag)

- The slag floats on the molten iron and is removed.
- The carbon content of the steel depends on the amount of oxygen blown onto the molten iron. The longer the oxygen 'blow', the more carbon is removed.
- After the removal of the required amount of carbon, metals such as chromium are added to make particular types of steel alloys. When iron is converted to steel a specific amount of carbon is removed and/or a controlled amount of other metals, e.g. nickel or chromium are added. By controlling the amount of other metals added to the iron, alloys with particular strength, hardness or resistance to corrosion can be made.

Worked example 22.02

Describe and explain how impure iron from the blast furnace is converted to steel by removing the impurities, carbon and silicon. [7]

You should state what substances are added to the iron and then explain the purpose of these substances in detail and not just write that 'they get rid of impurities'. For the explanation write about a chemical reaction as well as how the substances are removed. If you can get seven of the ten marking points below you will get full marks for this question.

Oxygen is blown into the molten iron [1] and calcium oxide is added [1]. The oxygen reacts with carbon [1] to form carbon dioxide [1]. Carbon dioxide is a gas [1] so it can escape into the air [1]. Silicon reacts with oxygen to form silicon dioxide [1]. The silicon dioxide then reacts with calcium oxide [1] to form slag [1], which is removed from the surface of the molten iron [1].

Progress check

22.01 Give the names of the raw materials used in the extraction of iron in the blast furnace. [4]

22.02 Why is air blown into the blast furnace used to extract iron from haematite? [1]

22.03 Write a word equation for the reduction of iron(III) oxide in the blast furnace using carbon monoxide. [1]

22.04 Why are oxygen and powdered calcium oxide blown into the converter used to make steel from iron? [2]

22.05 Which of the following metals could be extracted by heating with carbon?

Magnesium, zinc, lead, aluminium, iron, calcium [1]

22.04 The extraction of zinc

- The raw materials for extracting zinc are the zinc ore called zinc blende, coke (carbon) and air.
- Zinc blende contains zinc sulfide, ZnS.
- The zinc blende is first converted to zinc oxide by heating the ore strongly in air.
- The zinc is extracted in a blast furnace.
- The main reducing agent is carbon monoxide, which is formed by reactions within the furnace.

The reactions producing zinc are as follows:

- Coke (carbon) burns in a hot air blast to form carbon dioxide:

$$C(s) + O_2 \rightarrow CO_2(g)$$

- The carbon dioxide reacts with more coke to form carbon monoxide:

$$CO_2(g) + C(s) \rightarrow 2CO(g)$$

- The carbon monoxide reduces zinc oxide to zinc:

$$ZnO(s) + CO(g) \rightarrow Zn(g) + CO_2(g)$$

- The zinc vapour rises to the top of the furnace where it condenses.

> **TIP** You do not need to know the details of the blast furnaces for making iron or zinc but you should be able to label a diagram provided and have knowledge of the main chemical reactions taking place.

22.05 The extraction of aluminium

Aluminium is extracted by electrolysis. Aluminium ore (bauxite) contains a high percentage of aluminium oxide. The aluminium oxide is first purified. The electrolysis is then carried out in cells having many carbon (graphite) electrodes.

The electrolysis of aluminium oxide

- Aluminium oxide needs to be molten for electrolysis to occur. But aluminium oxide has a very high melting point (2040 °C) and it is difficult and expensive to keep the aluminium oxide molten at this temperature.
- So the mineral cryolite, Na_3AlF_6, is added. When molten, the cryolite dissolves the aluminium oxide and lowers the melting point of the electrolyte to about 900 °C.

Figure 21.02 shows a diagram of the cell used to electrolyse molten aluminium oxide.

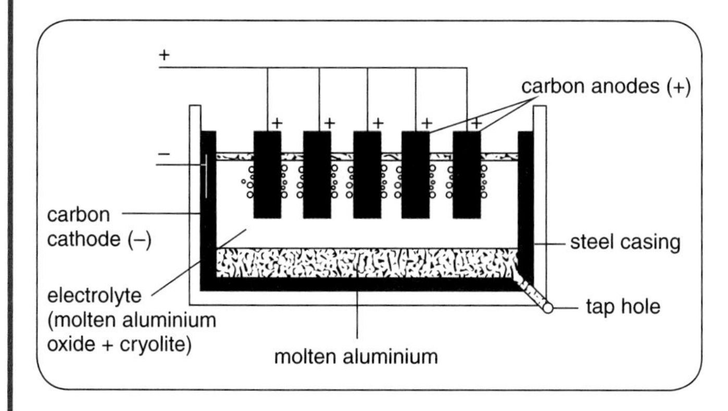

Figure 22.02

The reactions at the electrodes are as follows:

- Cathode: $Al^{3+} + 3e^- \rightarrow Al$

Aluminium ions are reduced to aluminium. The molten aluminium sinks to the bottom of the cell and is removed.

- Anode: $2O^{2-} \rightarrow O_2 + 4e^-$

Oxide ions are oxidised to oxygen. As the hot oxygen gas bubbles off, it reacts with the graphite anodes to form carbon dioxide. The graphite anodes 'burn away' and have to be replaced periodically.

- The overall reaction is:

$$2Al_2O_3 \rightarrow 4Al + 3O_2$$

> **TIP** You do not have to remember details of the purification of bauxite or technical details of the electrolysis cell for producing aluminium. You should concentrate on the processes, the use of cryolite and the equations.

Sample answer

Describe how aluminium is extracted from aluminium oxide by electrolysis. In your answer refer to:

- the type of electrodes used
- the substances present in the electrolyte and why this mixture is used
- the reactions occurring at the anode and cathode, giving half equations **[9]**

Graphite electrodes are used **[1]**.

The electrolyte is molten aluminium oxide **[1]** with the addition of cryolite **[1]**. The cryolite dissolves the aluminium oxide **[1]** and lowers the melting point of the electrolyte **[1]** to reduce the energy / amount of electricity used **[1]**.

Anode: $2O^{2-} \rightarrow O_2 + 4e^-$

(**1 mark** for correct formulae, **1 mark** for balance)

Cathode: $Al^{3+} + 3e^- \rightarrow Al$ **[1]**

Progress check

22.06 Give the name of the ore from which aluminium is extracted by electrolysis. [1]

22.07 Write a chemical equation for the reduction of zinc oxide by carbon monoxide. [1]

22.08 Why does the electrolyte in the electrolysis of aluminium oxide contain cryolite, not just aluminium oxide? [2]

22.09 Write the overall chemical equation for the electrolysis of aluminium. [2]

22.10 Explain why the anodes used in the electrolysis of aluminium oxide have to be replaced periodically. [2]

22.06 Steel alloys

The iron from the blast furnace is too brittle to be used for construction because it contains too much carbon. Steel is an alloy of iron with carbon or with carbon and other metals. Steel alloys include the following:

- Mild steel (about 0.25% carbon) is soft and malleable. So it is used for making wires and in general engineering, e.g. car bodies.
- High carbon steel (between 0.5% and 1.4% carbon) is harder. It is used to make tools such as hammers and chisels.
- Low alloy steels contain between 1–5% of other metals such as chromium, nickel and titanium. They are hard and strong and have low ductility and malleability. Nickel steels are used for bridges where strength is required. Tungsten steel is used for high speed tools because it does not change shape at high temperatures.
- Stainless steels may contain up to 20% chromium and up to 10% nickel. They are strong and resist corrosion. So they are used for the construction of reaction vessels in chemical plants, surgical instruments and cutlery.

22.07 Some uses of metals

- Aluminium: aluminium is used (as aluminium alloys) in the manufacture of aircraft because it has a relatively good strength for a low density metal. Its low density helps to reduce the weight of the aircraft. Aluminium is used in food containers because it is resistant to corrosion because of the unreactive oxide layer on its surface.
- Copper: copper is used for electrical wiring because it is an excellent electrical conductor. Copper is used in cooking utensils because it is a good conductor of heat.
- Mild steel: mild steel is used for making car bodies and some machinery because it it can be beaten into different shapes but still retains its strength.
- Stainless steel: stainless steel is used where strength and corrosion resistance are needed. So it is used to make reaction vessels in chemical plants, surgical instruments and cutlery.
- Zinc is used in **galvanising**, the process of coating a metal such as iron with zinc to prevent corrosion or rusting. It is also used in making the alloy brass, which contains 70% copper and 30% zinc. The zinc makes the alloy more corrosion resistant compared with pure copper.

TERMS

Galvanising: Coating a metal, usually iron, with a layer of zinc.

When writing about metals with low density, it is better not to use the less accurate terms light or lightweight.

22.08 Recycling metals

TERMS

Recycling: The processing of used materials into new products.

Advantages of **recycling** metals:

- raw materials such as the ores haematite and bauxite are conserved
- no energy used in transport of ores and processing of ores

- energy use is reduced in the manufacture of aluminium. Less fossil fuel is burnt to produce the high electric current required
- raw material usage, such as coal for coke in the extraction of iron, is reduced
- reduces pollution. Less carbon dioxide is emitted from both aluminium smelting works and the blast furnace for the production of iron
- less pollution due to fewer lorries and ships carrying raw materials

Disadvantages of recycling:

- more traffic on local roads to recycling centres, so **local** pollution
- time consuming to sort the materials, e.g. aluminium from car bodies
- energy consumed to collect material for recycling (although this is less than conserved by not digging up and transporting ores)

Progress check

22.11 Why are alloys used instead of pure metals? Give two reasons. [2]

22.12 Why is aluminium used for the manufacture of aircraft? [1]

22.13 Give two uses of copper. Relate each use to the relevant properties of copper. [4]

22.14 Give two advantages of recycling aluminium. [2]

Exam-style questions

Question 22.01

Iron is extracted from iron ore, which contains iron(III) oxide.

a i What is the meaning of the term ore? [1]

ii Give the name of the ore of iron containing iron(III) oxide. [1]

b Figure 23.03 shows a blast furnace.

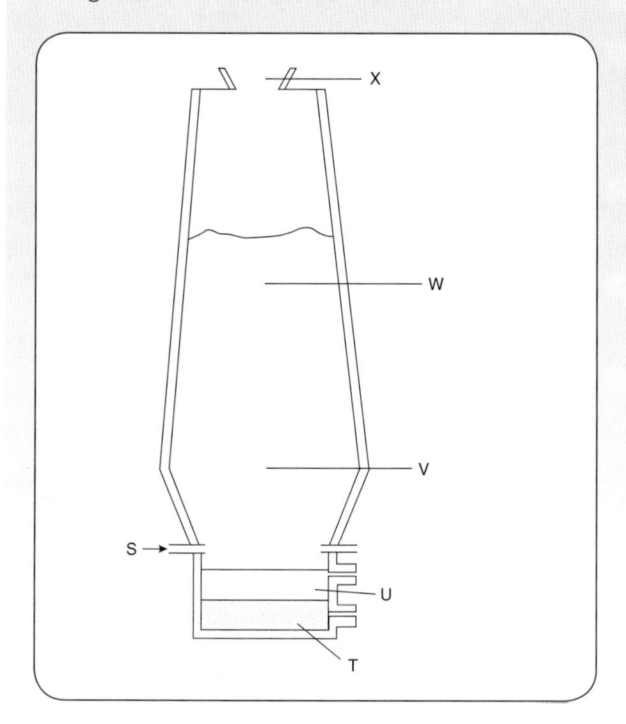

Figure 22.03

Which letter in the diagram shows:

i where the air is blown in [1]

ii where the slag is collected [1]

iii where the raw materials are put into the furnace. [1]

c One of the reactions in the blast furnace is:

$$Fe_2O_3 + ...CO \rightarrow ......Fe + ...CO_2$$

i Copy and complete this equation. [2]

ii Write a word equation for this reaction. [1]

iii Explain how this equation shows the reduction of iron(III) oxide. [1]

d In the hotter parts of the blast furnace carbon may react with iron oxide.

Copy and complete the equation for this reaction.

$$Fe_2O_3 + ...C \rightarrow 2Fe + ...CO \quad [1]$$

e Explain why limestone is added to the blast furnace. [4]

f Pure iron is not usually used in construction work. Explain why not. [1]

g Iron is converted to steel. State one use of:

i mild steel [1]

ii stainless steel [1]

Question 22.02

Zinc is a metal which is used to galvanise iron and to make the alloy brass.

a What is meant by the terms:

i galvanising? [1]

ii alloy? [1]

b Zinc sulfide, ZnS, is converted to zinc oxide, ZnO, and sulfur dioxide by reacting with oxygen in the air.

i Give the name of the sulfide ore. [1]

ii Construct a balanced equation for the reaction of oxygen with zinc sulfide. [2]

c The zinc oxide is reduced in a blast furnace. Give the names of the raw materials used. [2]

d Construct symbol equations to show how the coke (carbon) in the blast furnace is converted to carbon monoxide and how the carbon monoxide reduces thc zinc oxide. [3]

e Zinc can also be extracted by the electrolysis of molten zinc chloride.

i Suggest suitable electrodes that could be used for this reaction and give two reasons why these electrodes are used. [3]

ii Write an ionic equation for the reaction at the cathode. [1]

Revision checklist

By the end of this unit you should:

- Know that metals more reactive than carbon are generally extracted by electrolysis
- Know that metals les reactive than carbon are extracted by heating with coke
- Describe the reactions involved in the extraction of iron from haematite
- Describe how ion is converted into steel using calcium oxide and oxygen
- Describe how the properties of iron are changed by removing carbon and adding other metals
- Describe in outline, the extraction of zinc from zinc blende
- Know that aluminium is extracted from bauxite by electrolysis
- Describe the extraction of aluminium oxide by electrolysing aluminium oxide
- Be able to explain why cryolite is used in the extraction of aluminium
- Name the uses of mild steel and stainless steel
- Name the uses of copper related to its properties (electrical wiring and in cooking utensils)
- Name the uses of aluminium (aircraft manufacture and food containers)
- Be able to explain the uses of zinc for galvanising and for making brass
- Be able to discuss the advantages and disadvantages of recycling aluminium, iron and steel

Water and air

Learning outcomes

By the end of this unit you should:

- Be able to name some of the uses of water in industry and in the home
- Be able to discuss problems arising from an inadequate water supply
- Be able to describe chemical tests for water using cobalt(II) chloride and copper(II) sulfate
- Know the role of filtration and chlorination in water treatment
- State the composition of clean dry air
- Be able to describe the separation of oxygen and nitrogen from liquid air
- Know the conditions for rusting and that rusting can be prevented by paint and other coatings
- Understand how sacrificial protection and galvanising prevent rusting

23.01 Water

Uses of water

Water is used in the home for:

- drinking
- cooking
- washing and cleaning
- sanitation

Water is used in industry:

- as a solvent for many chemicals
- as a coolant to stop industrial (including chemical) processes from getting too hot
- for hydroelectric power to generate electricity
- as a raw material for some chemical processes, e.g. making ethanol
- for watering crops in agriculture

Water treatment

Water from rivers and lakes contains both insoluble and soluble impurities.

- Insoluble: soil, pieces of organic matter
- Soluble: dissolved calcium and iron compounds (and perhaps inorganic pollutants)
- Bacteria and other micro-organisms: many of these are too small to be trapped by filters

Water treatment plants remove insoluble materials and bacteria as well as removing some dissolved salts. The pH of the water is also adjusted.

Water shortages

Several problems can arise from lack or inadequate supply of water.

- Crops cannot be watered leading to poor growth or death. This leads to food shortages and famine.
- Water for washing and cleaning is not available. This can lead to the spread of disease.

There are many stages in the water treatment process, but you only need to know two of these: filtration and chlorination.

Filtration: impure water containing solids passes through a filter containing sand and gravel. This removes insoluble particles. Larger insoluble particles are trapped on the filter. The smaller water molecules and other dissolved molecules pass through the filter (Figure 23.01).

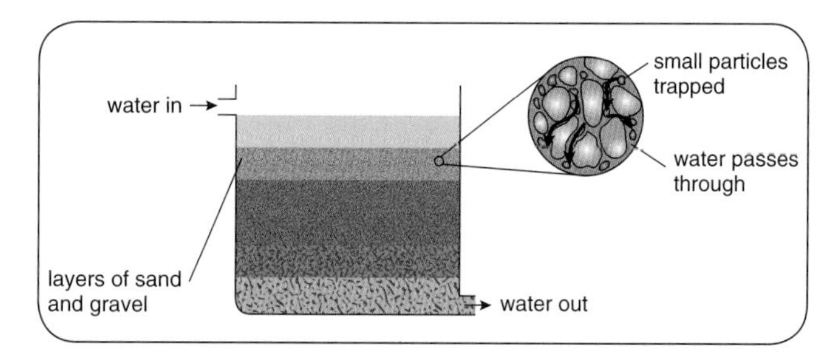

Figure 23.01 A sand and gravel filter

Chlorination: many bacteria are harmful to health. Cholera and typhoid are two of the diseases caused by bacteria in untreated water. Bacteria are not trapped on the sand filter because they are too small. Bacteria are killed by chlorination. This is the addition of chlorine to the water supply to kill bacteria and other micro-organisms.

TERMS

Chlorination: The addition of chlorine to the water supply to kill bacteria and other micro-organisms.

TIP

It is more accurate to use the term 'water treatment' for the process of producing water fit for use in the home. 'Water purification' is not very accurate because chlorine and other substances are also added.

Two tests for water

1. Anhydrous copper(II) sulfate turns from white to blue when water is added:

$$\underset{\text{white}}{CuSO_4(s)} + 5H_2O(l) \rightarrow \underset{\text{blue}}{CuSO_4.5H_2O(s)}$$

2. Anhydrous cobalt(II) chloride turns from blue to pink when water or an aqueous solution is added:

$$\underset{\text{blue}}{CoCl_2(s)} + 6H_2O(l) \rightarrow \underset{\text{pink}}{CoCl_2.6H_2O(s)}$$

Progress check

23.01 Give two uses of water in the home. [2]

23.02 Describe what you would observe when water is added to crystals of anhydrous cobalt(II) chloride. [2]

23.03 Copy and complete this equation.

$CuSO_4 + 5H_2O \rightarrow$ .... [1]

23.04 How could you prepare anhydrous nickel chloride from hydrated nickel chloride? [1]

23.05 Why are chlorination and filtration carried out during water treatment? [2]

23.02 Air

The percentage composition of clean dry air is shown in Table 23.01.

Gas present	Percentage
nitrogen	78
oxygen	21
noble gases	1
	100

Table 23.01

- The most abundant noble gas in the air is argon.
- Carbon dioxide forms 0.04% of the air but even this small percentage has a great effect on the atmosphere, being one of the gases responsible for global warming.
- In addition to these gases, damp air contains water vapour. Water vapour is one of the factors responsible for rusting.

Separating oxygen and nitrogen from the air

The gases in the air have many uses:

- Oxygen is used in steelmaking and welding and in breathing apparatus in hospitals.
- Nitrogen is used in food packaging and in silicon chip production because it is relatively unreactive.

Nitrogen and oxygen are separated by the fractional distillation of liquid air. The process is complex.

- Carbon dioxide and water vapour are removed.
- The air is cooled and undergoes a series of compressions and expansions. The temperature drops to −200 °C.
- The noble gases are still in the gas form at −200 °C. This leaves a liquid mixture of oxygen and nitrogen.
- When warmed slightly, the nitrogen, which has a lower boiling point than oxygen, vaporises first and is collected.
- Oxygen is left as a liquid.

Progress check

23.06 What is the percentage composition of
  a nitrogen in clean dry air?
  b oxygen in clean dry air? [2]

23.07a Name the process used to separate nitrogen from oxygen in liquid air. [1]
  b Explain how this process works. [3]

23.03 Corrosion and Rusting

Corrosion

Corrosion is the 'eating away' away of a metal surface by a chemical reaction. Moderately reactive metals such as iron and zinc corrode if the conditions in which they are placed are acidic. The greater the acidity of the environment in which the metal is placed, the faster is the rate of corrosion.

Rusting

Rusting is a special form of corrosion that only applies to iron and iron alloys. Rusting only occurs when both water and oxygen are present. The oxygen and water react with the iron to form hydrated iron(III) oxide (rust):

$$2Fe(s) + 1\tfrac{1}{2}O_2(g) + xH_2O(g) \rightarrow Fe_2O_3.xH_2O(s)$$

When rust flakes off, a fresh iron surface is exposed. This allows further rusting.

TERMS

Corrosion: The 'eating away' of the surface of a metal by chemical reaction.

Rusting: The reaction of iron with water and oxygen to form hydrated iron(III) oxide (rust).

TIP It is important that you distinguish between corrosion and rusting. Rusting is a reaction involving oxygen and water and the formation of hydrated iron(III) oxide. Corrosion is a general term for 'eating away' the surface of metals.

Preventing rusting

A simple way of preventing rusting is to coat iron or steel with a layer which prevents contact with oxygen and/or water. For example:

- painting (for cars, ships and bridges)
- plating with a less reactive metal (for food cans, can be plated with tin and taps can be plated with chromium)
- coating with plastic (for fence wiring)
- oiling or greasing (for moving parts of machines)

Worked example 23.01

Figure 23.02 shows an experiment to investigate rusting.

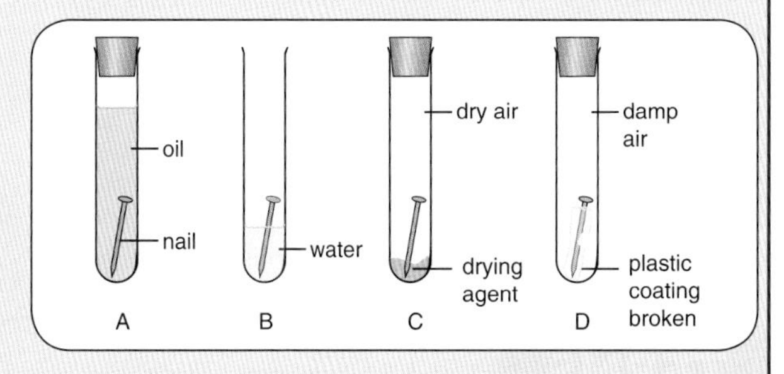

Figure 23.02

a For each tube, suggest whether the nail will rust or not. Explain your answers. **[4]**

Remember that both water and oxygen from the air are needed for rusting. If one of these is absent then rusting will not take place. Analyse each tube in turn and work out which substance if any is absent.

A does not rust because there is no oxygen (from the air) or water **[1]**.

B rusts because both oxygen from the air and water are present **[1]**.

C does not rust because there is no water present **[1]**.

D rusts because coating broken. Damp air has water and oxygen present [1].

b For each tube where the nail rusts, suggest exactly where the nail is most likely to rust. Explain your answers. [4]

You have to work out where the concentration of air / oxygen and water is the greatest and give a reason for this.

In tube B it is around the top of the water level or just below the water level [1] because the oxygen concentration in the water is highest here [1].

In tube D it is around / under the break in the plastic [1] because the rest of the nail is protected from the water and / or the air [1].

Sacrificial protection and galvanising

Galvanising is the coating of a metal with zinc. Galvanising protects the underlying iron in two ways:

1. It prevents oxygen and air coming into contact with the iron.
2. It protects by **sacrificial protection**.

Magnesium is more reactive than iron. When a strip of magnesium is placed in contact with iron, the magnesium corrodes first and protects the iron from rusting. This is an example of sacrificial protection.

TERMS

Sacrificial protection: When a more reactive metal is placed in contact with a less reactive metal, the more reactive metal corrodes first, preventing the corrosion of the less reactive metal.

Example: The sacrificial protection of iron by zinc.

- If some galvanised iron is scratched, some of the iron surface is exposed and, in the presence of water, an electrochemical cell is set up.
- Zinc is more reactive than iron, so it loses electrons more readily than iron:

$$Zn \rightarrow Zn^{2+} + 2e^-$$

- The zinc corrodes rather than iron. We say that it is a sacrificial metal.
- The iron remains protected because the electrons are accepted on its surface, keeping it in its reduced state, rather than oxidising it to Fe^{3+} ions.
- The electrons donated by the zinc react with hydrogen in the water to release hydrogen gas:

$$2H^+ + 2e^- \rightarrow H_2$$

Note that the sacrificial metal does not have to cover the surface of the iron completely for the method to work. Bars of magnesium or zinc can be attached to ships' hulls or oil pipelines to protect them from rusting.

Sample answer

A bar of magnesium is attached to an iron pipeline to prevent it from rusting. Explain using ideas of electron transfer how the magnesium prevents the steel from rusting. Include a relevant equation in your answer. **[5]**

Magnesium is more reactive than iron [1]. So magnesium loses electrons more readily than iron [1], so the iron does not become oxidised / the iron does not lose electrons [1] and the magnesium corrodes instead of the iron [1].

$$Mg \rightarrow Mg^{2+} + 2e^- \quad [1]$$

Progress check

23.08 What conditions are needed for iron to rust? [1]

23.09 Why is it incorrect to suggest that magnesium rusts? [1]

23.10 Galvanising protects iron from rusting in two ways. For each way, explain how galvanising prevents iron from rusting. [4]

Exam-style questions

Question 23.01

Water is important in everyday life.

a State two uses of water in industry. [2]

b A water treatment plant removes some impurities from the water.

 i Why is important that water is treated? [1]

 ii Explain how insoluble substances are removed from the water in a water treatment plant. [3]

 iii Explain why chlorination is used in water treatment. [1]

c Describe a chemical test for water. [3]

d Water and oxygen are needed for rusting. Give one method of preventing rusting of a bicycle chain. Explain why you chose this method. [2]

Question 23.02

Air is a mixture of gases.

a Describe the percentage composition of the three most common gases in clean, dry air. [3]

b State the name of one gas present in clean air which contributes to global warming. [1]

c Describe and explain how oxygen and nitrogen are separated from liquid air. [4]

d Oxygen is needed for rusting. Explain how galvanising prevents rusting by sacrificial protection. [3]

Revision checklist

You should be able to:

- Name some of the uses of water in industry and in the home
- Discuss the implications of an inadequate supply of water
- Describe chemical tests for water using cobalt(II) chloride and copper(II) sulfate
- Explain why filtration and chlorination are used in water treatment
- State the percentage of nitrogen, oxygen and noble gases in clean dry air
- Describe the separation of oxygen and nitrogen from liquid air
- Explain why paint and other coatings prevent rusting
- Explain how sacrificial protection and galvanising prevent rusting

Pollution

Learning outcomes

By the end of this unit you should:

- Know that carbon monoxide, sulfur dioxide, oxides of nitrogen and lead compounds pollute the air
- Know some sources of these pollutants and describe their adverse effects on buildings and health
- Be able to discuss why these pollutants are of global concern
- Know that oxides of sulfur and nitrogen contribute to acid rain
- Be able to describe the uses of lime and slaked lime in treating acidic waste (flue gas desulfurisation)
- Be able to describe and explain the presence of oxides of nitrogen in car engines and their catalytic removal
- Be able to state that carbon dioxide and methane are greenhouse gases and explain how they may contribute to climate change
- Know some sources of methane and carbon dioxide

24.01 Pollution

Table 24.01 shows some common pollutants, their sources and their adverse effects.

Pollutant	Source	Adverse effects
carbon monoxide	• incomplete combustion of coal, natural gas and petroleum hydrocarbons • incomplete combustion of other carbon-containing fuels, e.g. ethanol	• poisonous because it combines with haemoglobin in the red blood cells to prevent transport of oxygen round the body.
sulfur dioxide	• combustion of coal and natural gas • combustion of sulfide ores	acid rain leading to: • corrosion of metal structures • death of trees (especially conifers) • death of aquatic organisms due to acidification of lakes and rivers • chemical erosion of buildings made of carbonate rocks and of mortar • irritates lungs, throat and eyes
oxides of nitrogen	• reaction of nitrogen with oxygen in car engines • reaction of nitrogen with oxygen in high temperature furnaces • bacterial action in the soil	acid rain (similar effects as sulfur dioxide) and: • photochemical smog • irritates lungs, throat and eyes • breathing difficulties especially for people with asthma
lead compounds	• old water pipes and some paints • petrol in some racing cars and for old two-stroke engines	• damage to the nervous system especially to the brains of developing infants
methane	• decomposition of vegetation • waste gases from digestion of animals • rice paddy fields	• increasing global warming • explosions in coal mines (see unit 12)

Table 24.01

TERMS

Pollution: The introduction of substances into the environment which should not normally be present.

Pollution occurs when contaminating materials are introduced into the natural environment (the ground, atmosphere or water). Most pollutants have an unfavourable effect on the environment.

TIP

Remember that complete combustion of hydrocarbons (oxygen in excess) produces carbon dioxide and water. Incomplete combustion produces carbon monoxide and water and perhaps some soot (carbon).

Most of these pollutants are of global concern because:

- gaseous pollutants such as sulfur dioxide can be blown far away from where they are emitted and make their effects felt in non-industrialised areas and areas of natural beauty.
- methane and carbon dioxide can contribute to global warming and melt the ice caps. The rise in sea level will flood many low-lying areas
- global warming can also increase the average temperatures of the oceans so that food chains are disrupted
- global warming can cause desertification so that crops cannot grow
- acid rain can reduce the growth of crops and acidify lakes so that fish needed for food will be reduced in numbers

Sample answer

Lead and carbon monoxide are both atmospheric pollutants. Explain the source of these pollutants and how each of these pollutants get into the atmosphere. Give one adverse effect of each of these pollutants. **[6]**

When articles covered in old paint **[1]** are burnt **[1]** lead compounds get into the air. These compounds can cause damage to the brains of babies **[1]**. Carbon monoxide is formed when fossil fuels are burnt **[1]** and there is not enough oxygen **[1]** for them to burn completely. If you breathe in the carbon monoxide you may get poisoned and die **[1]**.

Progress check

24.01 Give the names of two pollutant gases which contribute to acid rain. [2]

24.02 State two adverse effects of acid rain on buildings. [2]

24.03 Explain why the waste gas from a domestic boiler fuelled by natural gas is dangerous if the holes letting air into the boiler are blocked. [3]

24.04 Link these pollutants with their adverse effects. [4]

Pollutants: A Lead compounds; B Methane; C Nitrogen dioxide; D Sulfur dioxide.

Effects: 1 Increases global warming; 2 Contributes to photochemical smog; 3 Causes harm to the nervous system of young children; 4 causes death of trees

24.02 Sulfur dioxide and acid rain

TERMS

Acid rain: Rain with pH below about pH 5.6 due to the reaction of acidic gases with rainwater.

- When burnt, the sulfur in fuels such as coal, petroleum and natural gas is oxidised to sulfur dioxide.
- Reactions in the atmosphere convert some of the sulfur dioxide to sulfur trioxide.
- Sulfur dioxide and sulfur trioxide react with water in the atmosphere to form acids.
- The acids dissolve in rainwater to form **acid rain.**
- Acid rain has a pH lower than about 5.6.

Reducing the adverse effects of sulfur dioxide

- Acid rain causes excess acidity in the soil and in rivers and ponds. The high acidity can lead to poor crop yields and death of aquatic organisms. Powdered limestone (which is largely calcium carbonate) is added to soils and lakes affected by acid rain to neutralise the acid:

$CaCO_3$	+	$2H^+$	→	Ca^{2+}	+	CO_2	+	H_2O
calcium carbonate		acid						

- When coal or hydrocarbons containing sulfur are burnt in power stations or furnaces, sulfur dioxide is formed. The sulfur dioxide can be removed by passing it through powdered calcium carbonate, calcium hydroxide or calcium oxide. A neutralisation reaction occurs and calcium sulfite is formed:

$$CaO(s) + SO_2(g) \rightarrow CaSO_3(s)$$

This process is called **flue gas desulfurisation.**

The calcium sulfite can be converted to harmless calcium sulfate by reacting it with air and water.

TERMS

Flue gas desulfurisation: The removal of sulfur dioxide from waste gases produced when fossil fuels containing sulfur are burned.

Worked example 24.01

Burning fossil fuels causes acid rain.

a Explain how acid rain is formed when coal is burnt in a power station. **[5]**

The 'power station' is there to make sure that you chose sulfur dioxide as the acidic gas not nitrogen dioxide. Make sure that you state every stage carefully. A common error is to suggest that sulfur dioxide dissolves in water to form sulfuric acid: so do not forget the oxidation stage!

Coal contains small amounts of sulfur **[1]**. When burnt this forms sulfur dioxide **[1]**. This gas goes into the air and is oxidised to sulfur trioxide **[1]**. Sulfur trioxide reacts with rainwater **[1]** and forms an acid **[1]**.

b Acid rain can make lakes acidic. Describe and explain how you can reduce the acidity of acidic lakes. Include a general word equation in your answer. **[5]**

'Describe' means what you have to do: for example 'add limestone'. Make sure that you chose a substance that is not too alkaline: sodium hydroxide would not be as suitable. 'Explain' in this question means that you have to write about the chemical reaction.

Add powdered limestone **[1]** to the lake.
The limestone neutralises **[1]** the acid:
carbonate + acid → salt + water + carbon dioxide
(**1 mark** for the reactants, **1 mark** for salt and **1 mark** for water + carbon dioxide).

24.03 Pollution due to nitrogen oxides

- Nitrogen oxides (NO and NO_2) are formed by the combination of nitrogen with oxygen under the high temperature and pressure of car engines.
- Nitrogen(II) oxide, NO, can be oxidised further in the atmosphere to nitrogen dioxide.
- Nitrogen dioxide dissolves in rainwater to form acid rain.

Removal of nitrogen oxides: catalytic converters

TERMS

Catalytic converter: A device fitted onto a car exhaust system to reduce the emission of oxides of nitrogen and carbon monoxide.

At very high temperatures in car engines, nitrogen combines with oxygen to form a variety of nitrogen oxides. Car exhaust fumes may also contain carbon monoxide and unburned hydrocarbons due to incomplete combustion in the engine. Cars are fitted with **catalytic converters** to their exhaust systems. Once warmed up, the following reactions take place:

- the oxidation of carbon monoxide to carbon dioxide:

$$2CO(g) + O_2(g) \rightarrow 2CO_2(g)$$

- the chemical reduction of nitrogen oxides to form harmless nitrogen gas:

$$2NO(g) \rightarrow N_2(g) + O_2(g)$$
$$2NO_2(g) \rightarrow N_2(g) + 2O_2(g)$$

- the oxidation of unburned hydrocarbons to carbon dioxide and water:

$$C_8H_{18}(g) + 12\tfrac{1}{2}\,O_2(g) \rightarrow 8CO_2(g) + 9H_2O(g)$$

Platinum and rhodium in the catalytic converter catalyse these reactions. The reactions are complex but we can write an equation for the overall reactions as:

$$2NO(g) + 2CO(g) \rightarrow N_2(g) + 2CO_2(g)$$
and $$2NO_2(g) + 4CO(g) \rightarrow N_2(g) + 4CO_2(g)$$

The overall reactions are redox reactions. The carbon monoxide is a reducing agent and the nitrogen oxides are oxidising agents.

Progress check

24.05 What is the name of the process which removes sulfur dioxide from waste gases formed by the combustion of coal and natural gas in furnaces and power stations? [1]

24.06 Describe how nitrogen oxides are formed in car engines. [3]

24.07 Write a symbol equation for the reaction (in a catalytic converter) of carbon monoxide with nitrogen(II) oxide, NO. Two non-toxic products are formed. [2]

24.04 Greenhouse gases and global warming

- When short wave radiation from the Sun hits the Earth's surface, it is absorbed.
- Most of this energy is emitted from the surface of the Earth as infrared radiation.
- Much of this infrared radiation is trapped by gases in the atmosphere (Fig 24.01).
- Gases which are good absorbers of infrared radiation include carbon dioxide and methane. Carbon dioxide is formed when hydrocarbon fuels are burnt and methane is formed as a result of bacterial action in swamps, rice paddy fields and the digestive system of animals.

TERMS

Greenhouse gas: A gas which absorbs infrared radiation emitted from the Earth's surface.

Global warming: The warming of the atmosphere due to greenhouse gases trapping infrared radiation emitted from the Earth's surface.

- A gas which absorbs infrared radiation emitted from the Earth's surface is called a **greenhouse gas**. So carbon dioxide and methane are greenhouse gases.
- The absorption of infrared radiation by greenhouse gases leads to the heating of the atmosphere. This is called **global warming**.
- An increased concentration of carbon dioxide and methane in the atmosphere due to increased burning of fossil fuels results in more infrared radiation being absorbed by the atmosphere. Global warming is increased.

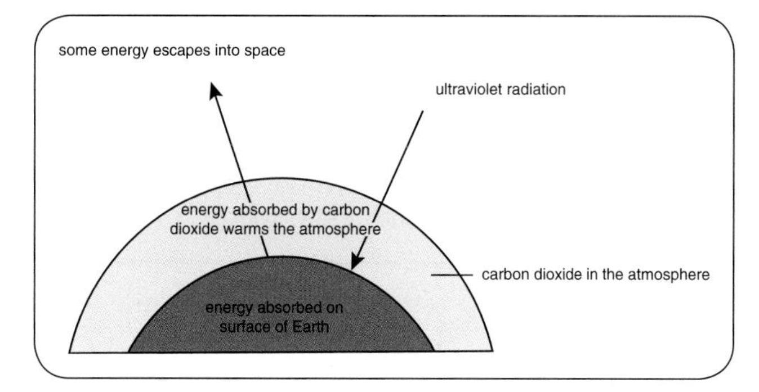

Figure 24.01

The effects of increased global warming include:

- melting of the polar ice caps, causing a rise in sea levels
- more violent and unpredictable weather
- an increase in the temperature of the Earth's surface and oceans leading to increasing desertification and the death of corals

Many substances are good greenhouse gases but the two to remember are carbon dioxide and methane.

Worked example 24.02

a When hydrocarbon fuels undergo complete combustion, carbon dioxide and water are formed. Copy and complete the equation for the complete combustion of the hydrocarbon propane, C_3H_8.

$$C_3H_8 + \ldots O_2 \rightarrow \ldots CO_2 + \ldots H_2O \quad \textbf{[3]}$$

Start by balancing the carbon: there are three C atoms on the left so there must be $3CO_2$. Then balance the hydrogen: there are eight H atoms on the left so there must be $4H_2O$ on the right ($4 \times 2 = 8$). Finally, balance the oxygen. Count up the oxygen atoms on the right $= 6 + 4 = 10$. But there are two O atoms in every oxygen molecule, so you need $5O_2$.

$C_3H_8 + \underline{5}O_2 \rightarrow \underline{3}CO_2 + \underline{4}H_2O$ (**1 mark** for each figure underlined)

b Explain why increasing the amount of hydrocarbons burnt is of global concern? **[5]**

There are two parts to this question. The first is about the effect of burning hydrocarbons. Note that the question is about hydrocarbons not coal, so the question is not about acid rain. The second is why is this of global concern? To answer the second part you need to write about an effect of global warming.

Burning hydrocarbons produces carbon dioxide **[1]**, which is a greenhouse gas **[1]** which causes global warming **[1]**. Global warming can melt the ice caps and lead to a rise in sea levels **[1]** so low-lying places will be flooded **[1]**.

Progress check

24.08 Name two gases which contribute to global warming. [2]

24.09 Explain why burning fossil fuels increases global warming. [3]

24.10 Describe two effects of global warming. [2]

Exam-style questions

Question 24.01

a Nitrogen dioxide contributes to acid rain. Describe a source of nitrogen dioxide and explain how it forms acid rain. [3]

b Flue gas desulfurisation can be used to remove sulfur dioxide from waste gases in power stations which burn fossil fuels. Explain how flue gas desulfurisation works. [4]

Question 24.02

Oxides of nitrogen are atmospheric pollutants.

a State two sources of nitrogen dioxide in the atmosphere. [2]

b Carbon monoxide is formed in car engines from the incomplete combustion of hydrocarbons.

i What is meant by the term *incomplete combustion*? [1]

ii Write a chemical equation for the incomplete combustion of the hydrocarbon decene, $C_{10}H_{22}$. [2]

c Catalytic converters remove oxides of nitrogen and carbon monoxide from the exhaust gases from car engines. Explain how catalytic converters do this. [5]

Revision checklist

You should be able to:

- Describe carbon monoxide, sulfur dioxide, oxides of nitrogen and lead compounds as atmospheric pollutants
- State the sources of these pollutants and describe their adverse effects on buildings and health
- Explain why these pollutants are of global concern
- Explain that oxides of sulfur and nitrogen contribute to acid rain
- Describe the uses of lime and slaked lime in removing acidic gases (flue gas desulfurisation)
- Explain the presence of oxides of nitrogen in car engines and their removal by catalytic converters
- State that carbon dioxide and methane are greenhouse gases and explain how they may contribute to climate change
- State some sources of methane and carbon dioxide

Carbon, nitrogen, sulfur and their compounds

Learning outcomes

By the end of this unit you should:

- Be able to state some sources of carbon dioxide in the atmosphere
- Be able to describe the carbon cycle to include the processes of combustion, respiration and photosynthesis
- Be able to describe the manufacture of lime
- Be able to name the uses of calcium carbonate in the manufacture of iron and cement
- Know the use of lime and slaked lime in neutralising acidic soils
- Know the conditions for the manufacture of ammonia and the sources of the raw materials for the Haber process
- Know the use of NPK fertilisers
- Be able to describe the displacement of ammonia from its salts
- Be able to name some sources and uses of sulfur
- Be able to name some uses of sulfur dioxide (bleach and food preservative)
- Be able to describe the manufacture of sulfuric acid by the Contact process
- Know some properties, reactions and uses of sulfuric acid

25.01 The carbon cycle

Sources of carbon dioxide in the atmosphere

- Combustion of fossil fuels, e.g. coal, methane:

$$\underset{\text{methane}}{CH_4} + 2O_2 \rightarrow CO_2 + 2H_2O$$

- **Respiration**: the production of energy in living things. The overall reaction of respiration is represented by the equation:

$$\underset{\text{glucose}}{C_6H_{12}O_6} + \underset{\text{oxygen}}{6O_2} \rightarrow \underset{\text{carbon dioxide}}{6CO_2} + \underset{\text{water}}{6H_2O}$$

- Decomposition of vegetation by bacteria and fungi.
- Thermal decomposition of limestone (see below).
- Reaction of acids with carbonates.

TERMS

Respiration: The production of energy in living things by a series of reactions involving the conversion of glucose and oxygen to carbon dioxide and water.

Removal of carbon dioxide from the atmosphere

- Photosynthesis: the process of producing glucose and oxygen from carbon dioxide and water in plants in the presence of chlorophyll and light:

$$6CO_2 + 6H_2O \rightarrow \underset{\text{glucose}}{C_6H_{12}O_6} + 6O_2$$

- Carbon dioxide dissolves in the water in seas and oceans and is removed by shellfish for making their calcium carbonate shells.

The balance of the carbon cycle

- Carbon as carbonate, carbon dioxide or organic carbon compounds is present in the sea, the air and under the Earth.
- There is a continuous cycle of these compounds between these sources. This is called the **carbon cycle** (Fig 25.01).

TERMS

Carbon cycle: The cycle of processes which move carbon compounds between the Earth, atmosphere and oceans and keeps the amount of carbon dioxide in the atmosphere fairly constant.

- There is a fairly constant amount of carbon compounds in the sea, atmosphere and under the Earth although scientists are worried that an increased amount of carbon dioxide in the atmosphere is increasing global warming (see Unit 24).

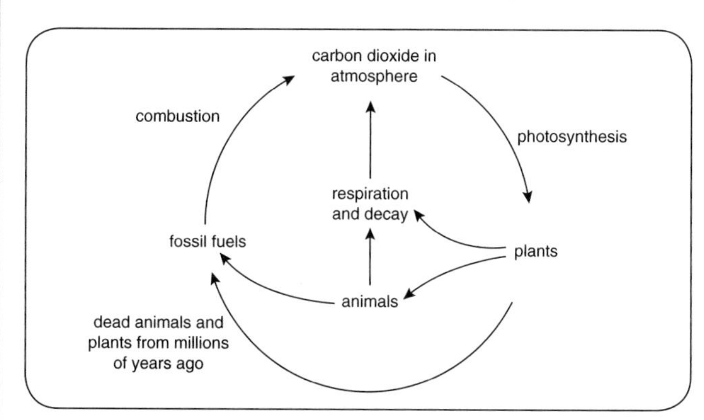

Figure 25.01 The carbon cycle

- Carbon dioxide is added to the atmosphere by respiration, combustion and carbon dioxide coming out of solution from seawater.
- Carbon dioxide is removed from the atmosphere by photosynthesis and dissolving in seawater.
- As long as these are balanced, the amount of carbon dioxide in the atmosphere remains constant.
- Scientists are worried that increasing the amounts of fossil fuels burned will increase global warming and unbalance the carbon cycle.

Progress check

25.01 Give three sources of carbon dioxide in the atmosphere. [3]

25.05 Write a word equation for the overall process of respiration. [2]

25.03 Write a symbol equation for the overall process of photosynthesis. [3]

25.04 Explain why the percentage of carbon dioxide in the atmosphere remains fairly constant. [3]

25.02 Limestone, lime and slaked lime

- Limestone is mainly calcium carbonate.
- Lime, calcium oxide, is manufactured from calcium carbonate by thermal decomposition:

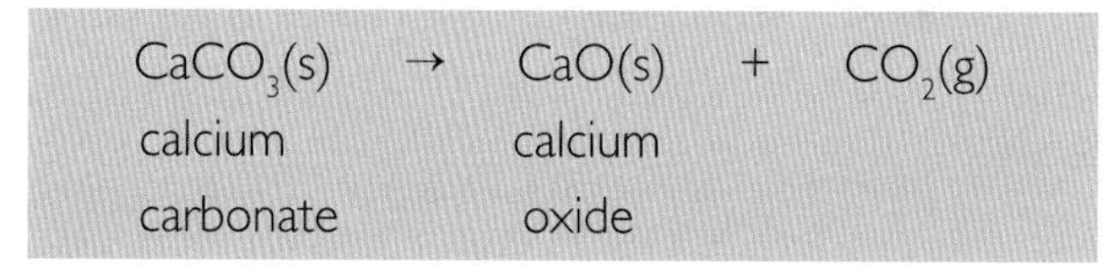

$$CaCO_3(s) \rightarrow CaO(s) + CO_2(g)$$

calcium carbonate → calcium oxide

- Slaked lime, calcium hydroxide, is made by adding a small amount of water slowly to calcium oxide.

$$CaO(s) + H_2O(l) \rightarrow Ca(OH)_2(s)$$

calcium oxide (lime) → calcium hydroxide (slaked lime)

- Limewater is a solution of calcium hydroxide in water. Limewater is alkaline.

Uses of limestone and limestone products

- Limestone (calcium carbonate) is used in the manufacture of iron and cement. It is also used in treating excess acidity in soils and lakes where it is often preferred to lime because it does not make the water or soil alkaline.
- Lime (calcium oxide) is used in lime mortar and in flue gas desulfurisation. It is also used in treating excess acidity in soils and lakes. If excess lime is used, however, the soil or water may become too alkaline.
- Slaked lime (calcium hydroxide) is used in treating acidic soils and neutralising acidic industrial waste products.

Progress check

25.05 State two uses of calcium carbonate. [2]

25.06 What is the chemical name for slaked lime and how is it formed from lime? [2]

25.07 Write a word equation for the reaction of calcium carbonate with sulfuric acid. [3]

25.03 Ammonia synthesis

TERMS

Haber process: The industrial process for making ammonia from hydrogen and nitrogen using a catalyst of iron and specific conditions of temperature and pressure.

- Ammonia is synthesised by the **Haber process**:

$$N_2 + 3H_2 \xrightleftharpoons{\text{Fe catalyst}} 2NH_3$$

- The raw materials for the synthesis ammonia are air (which contains nitrogen) and methane (which is reacted with steam to form hydrogen). Hydrogen can also be obtained from the cracking of hydrocarbons (see Unit 27).
- A catalyst of finely divided iron is used. The iron is finely divided to provide a large surface area to allow a greater number of reactant molecules to combine with the catalyst surface.
- The reaction is exothermic, so increasing the temperature shifts the position of equilibrium to the left in the direction of the reactants. So the higher the temperature, the lower is the yield of ammonia.
- An increase in pressure shifts the position of equilibrium to the right in the direction of a smaller number of gaseous molecules (see Unit 13). So the higher the pressure, the greater is the yield of ammonia.
- The optimum conditions for the reaction are 450 °C and 200 atmospheres pressure. The optimum temperature is a compromise between a higher rate of reaction at a higher temperature and a lower equilibrium yield at a higher temperature.

Worked example 25.01

Ammonia is manufactured by the Haber process. The reaction is exothermic:

$$N_2(g) + 3H_2(g) \rightleftharpoons 2NH_3(g)$$

a Table 25.01 shows the percentage yield of ammonia at various temperatures and pressures.

Temperature/°C	Pressure 200 atmospheres	Pressure 300 atmospheres	Pressure 400 atmospheres
350	43%	50%	54%
450	24%	32%	38%
550	12%	16%	18%

Table 25.01

Make sure that you answer each part of the question. 'Describe how' asks you to describe the trends in the data. Remember to put this in the comparative form (the higher the … the higher the …). 'Explain why' asks you to give the reason for this.

i Describe how and explain why the yield of ammonia changes with temperature. [4]

Yield decreases as the temperature increases [1]. For an exothermic reaction [1] the position of equilibrium shifts to the left (reactant side) as the temperature is increased [1] in the direction of the absorption of heat [1].

ii Describe how and explain why the yield of ammonia changes with pressure. [3]

Yield increases with increase in pressure [1]. There are fewer molecules of gas on the right-hand side of the equation [1]. The position of equilibrium moves to the right to oppose the increase in the number of molecules per unit volume [1].

b Finely divided iron is used in the Haber process. What is the function of the iron and explain why it is finely divided? [4]

In exam-style questions, you often have to put together information from different areas of the syllabus. This part is about rate of reaction rather than equilibrium. Remember that catalysts do not affect the position of equilibrium. The word 'explain' indicates that you have to give an answer with relevant theory.

The iron is a catalyst [1]. It increases the rate of reaction [1]. It is finely divided to increase the surface area of the catalyst [1] to allow a greater frequency of collisions of reactant molecules with the catalyst surface (or to allow more reactant molecules to come close together on the catalyst surface) [1].

25.04 NPK fertilisers

TERMS

NPK fertilisers: Fertilisers containing the elements nitrogen, phosphorus and potassium in various compounds.

NPK fertilisers contain nitrogen, phosphorus and potassium. These are all important elements that crop plants need to increase their growth. Of these elements, nitrogen is the most important because it is incorporated into proteins which are needed for growth. The crop plants absorb nitrogen from the soil. So after a few years of crop harvesting, the soil becomes depleted in nitrogen. So fertilisers are added to put nitrogen back into the soil. Most fertilisers contain ammonium salts and phosphates, e.g. ammonium phosphate, ammonium sulfate, potassium ammonium phosphate. These are made by the reaction of acids with ammonia. For example:

$2NH_3$ ammonia	+	H_2SO_4 sulfuric acid	→	$(NH_4)_2SO_4$ ammonium sulfate
$3NH_3$ ammonia	+	H_3PO_4 phosphoric acid	→	$(NH_4)_3PO_4$ ammonium phosphate

When writing equations for the reaction of ammonia with acids remember that there is no water on the right-hand side.

The displacement of ammonia from its salts

Many crop plants do not grow well if the soil is too acidic. Farmers sometimes add lime or slaked lime to the soil to neutralise the acidity. If lime or slaked lime is added to the soil too soon after a fertiliser has been added, nitrogen loss from the fertiliser may occur due to the displacement of ammonia from the ammonium salts in the fertiliser. For example:

$$2NH_4Cl + Ca(OH)_2 \rightarrow CaCl_2 + 2NH_3 + 2H_2O$$

Worked example 25.02

Ammonium sulfate is a fertiliser.

a Give the names of three elements present in most fertilisers. [2]

The question asks for the names not the symbols so the answer NPK is not enough.

Nitrogen, phosphorus and potassium (**2 marks** all three correct, **1 mark** for two correct).

b Explain why farmers put fertilisers onto the soil. **[1]**

Make sure that you mention <u>*increase*</u> *in plant growth not just 'helps plants to grow'.*

To increase plant growth / to replace the nitrogen lost from soil by harvesting previous crops **[1]**.

c Explain why adding lime to the soil at the same time as ammonium sulfate, results in a loss of nitrogen from the soil. **[4]**

You need to put four points because there are four marks. This is chemistry which comes from several sections of the syllabus. Concentrate on the chemical reaction between alkalis and ammonium salts.

Lime dissolves in soil water to form a solution of calcium hydroxide **[1]**, which is alkaline **[1]**. This reacts with ammonium sulfate **[1]** and releases ammonia **[1]**.

d Give the name of an acid and an alkali that can be used to make ammonium sulfate. **[2]**

To answer this question you need to revise naming salts from Units 7 and 15.

Sulfuric acid **[1]** ammonia **[1]**

Progress check

25.08 Name the three elements present in NPK fertilisers. [3]

25.09 Calculate the percentage by mass of nitrogen in ammonium sulfate. [2]

25.10 Calcium phosphate can be made into superphosphate fertilisers. Deduce the formula of calcium phosphate. The formula of the phosphate ion is PO_4^{3-}. [1]

25.11 In the Haber process, what is the effect on the position of equilibrium of a removing ammonia [1] b decreasing the temperature? [1]

25.05 Sulfur: sources and uses

Sulfur is found as the element underground in the USA, Mexico and Poland. It is also a by-product from the removal of sulfur from petroleum and natural gas. Sulfur can also be obtained from sulfide ores.

Uses of sulfur include:

- making sulfuric acid
- making rubber tyres more flexible (vulcanising)

25.06 Sulfur dioxide

Sulfur dioxide can be made by the direct combination of sulfur with oxygen. This is the method used in the first stage of the manufacture of sulfuric acid:

$$S + O_2 \rightarrow SO_2$$

Uses of sulfur dioxide include:

- as a bleach in the manufacture of wood pulp for paper
- as a preservative for foods and drinks by killing bacteria. Sulfites are often added to foods and these release sulfur dioxide in acidic conditions

25.07 The manufacture of sulfuric acid

TERMS

Contact process: The industrial process of making sulfuric acid using a vanadium(V) oxide catalyst and high temperature.

- Sulfuric acid is synthesised by the **Contact process**. The raw materials are sulfur and air (which contains oxygen).
- The first stage is the oxidation of sulfur:

$$S + O_2 \rightarrow SO_2$$

- The main stage is oxidation of sulfur dioxide to sulfur trioxide:

$$2SO_2 + O_2 \xrightleftharpoons{V_2O_5 \text{ catalyst}} 2SO_3$$

- A catalyst of vanadium(V) oxide, V_2O_5, is used.
- The reaction is exothermic, so increasing the temperature shifts the position of equilibrium to the left in the direction of the reactants (see Unit 13). So the higher the temperature, the lower is the yield of sulfur trioxide.
- An increase in pressure shifts the position of equilibrium to the right in the direction of a smaller number of gaseous molecules (see Unit 13). But the position of equilibrium is far to the right (the equilibrium mixture contains about 96% sulfur trioxide). So the reaction is carried out at just above atmospheric pressure because (i) it is not worth expending the extra energy or money required to produce high pressures and (ii) a higher pressure would increase the problems of dealing with the corrosive mixture of gases.
- The conditions for the reaction are 450 °C and just above atmospheric pressure. The optimum temperature is a compromise between a higher rate of reaction at a higher temperature and a lower equilibrium yield at a higher temperature.
- The sulfur trioxide is absorbed into a solution of 98% sulfuric acid to produce a thick liquid called oleum. It is not absorbed into water because a fine mist of sulfuric acid would be produced and this would be difficult to condense:

$$SO_3 + H_2SO_4 \rightarrow H_2S_2O_7$$

- Oleum is added to water to form concentrated sulfuric acid:

$$H_2S_2O_7 + H_2O \rightarrow 2H_2SO_4$$

TIP

You need to know the conditions used in both the Haber process and the Contact process and why these conditions are used in terms of the equilibrium reactions.

Sample answer

Sulfuric acid is made in the Contact process by converting sulfur dioxide to sulfur trioxide. The reaction is exothermic.

a Write a balanced equation for this reaction. **[2]**

$2SO_2 + O_2 \rightleftharpoons 2SO_3$ (1 **mark** for correct reactants and products, 1 **mark** for balance)

b The conditions in the converter are 450 °C and a pressure just above atmospheric pressure. Under these conditions, the yield of sulfur trioxide is 98%. Explain why these conditions are used and not a higher temperature and higher pressure. **[6]**

The reaction is exothermic **[1]** so increasing the temperature would decrease the yield of sulfur trioxide and move the position of equilibrium to the left **[1]**. Increasing the pressure would move the position of equilibrium to the right **[1]** in the direction of fewer moles **[1]**. But the yield of sulfur trioxide is very high **[1]** so it would be a waste of energy to increase the pressure **[1]**.

25.08 Properties and uses of sulfuric acid

- Sulfuric acid is a strong acid (see Unit 15).
- Sulfuric acid is a dibasic acid: two of its hydrogen atoms can be replaced by a metal.
- It reacts in a similar way to other acids with metal carbonates, oxides, hydroxides (and ammonia) and metals, e.g.

$$ZnO + H_2SO_4 \rightarrow ZnSO_4 + H_2O$$
$$Mg + H_2SO_4 \rightarrow MgSO_4 + H_2$$
$$Na_2CO_3 + H_2SO_4 \rightarrow Na_2SO_4 + CO_2 + H_2O$$

- If sulfuric acid is not in excess hydrogen sulfates are formed:

$$NaOH + H_2SO_4 \rightarrow NaHSO_4 + H_2O$$

Sulfuric acid has many uses:

- Concentrated sulfuric acid is used for car battery acid, as acid drain cleaners, for making phosphate fertilisers, for making soaps and detergents, and paints and dyes.
- Dilute sulfuric acid is used as a catalyst in many organic reactions and for cleaning metals.

Progress check

25.12 Give two uses of sulfur dioxide other than making sulfuric acid. [2]

25.13 Give one use of a concentrated sulfuric acid [1] b dilute sulfuric acid. [1]

25.14 Construct chemical (symbol) equations for the reactions of sulfuric acid with

a zinc [2]

b sodium hydroxide. [2]

Exam-style questions

Question 25.01

The percentage of carbon dioxide in the atmosphere is approximately 0.04%.

a Carbon dioxide is released into the atmosphere by thermal decomposition.

i What is meant by the term thermal decomposition? [1]

ii Describe two other ways by which carbon dioxide is released into the atmosphere. [2]

b i Explain why lime, CaO, is used to treat acidic soils. [3]

ii Suggest why it is better to treat acidic soils with calcium carbonate rather than with lime. [6]

c Write a generalised word equation for the reaction of a base with an acid. [2]

d Calcium sulfate is a salt which can be made by the addition of an acid to an alkali.

i Give the names of the acid and alkali used to make calcium sulfate. [2]

ii What type of chemical reaction is this? [1]

Question 25.02

a Sulfuric acid is a strong acid. Explain the meaning of the term strong acid. [1]

b Construct a symbol equation, including state symbols, for the reaction of dilute sulfuric acid with magnesium. [2]

c A 0.1 mol/dm^3 solution of sulfuric acid has a lower pH than a 0.1 mol/dm^3 solution of hydrochloric acid. Explain why. [4]

d Give two uses of concentrated sulfuric acid. [2]

Revision checklist

You should be able to:

- State sources of carbon dioxide in the atmosphere as combustion, respiration and thermal decomposition and action of acids on carbonates
- Describe the carbon cycle in terms of combustion, respiration and photosynthesis
- Describe the manufacture of lime by thermal decomposition of limestone
- Name the uses of calcium carbonate in the manufacture of iron and cement
- Describe the use of lime and slaked lime to neutralise acidic soils
- Describe the manufacture of ammonia including the sources of raw materials
- Describe the use of NPK fertilisers
- Describe the displacement of ammonia from its salts by alkalis
- State the use of sulfur to manufacture sulfuric acid
- State some uses of sulfur dioxide (bleach and food preservative)
- Describe the manufacture of sulfuric acid by the Contact process
- Describe properties, reactions and uses of sulfuric acid

Unit 26

Introduction to organic chemistry

Learning outcomes

By the end of this unit you should:

- Know that coal, petroleum and natural gas are fuels
- Be able to name methane as the main constituent of natural gas
- Be able to describe the general concept of a homologous series
- Be able to describe the general characteristics of a homologous series, including the idea of general formula
- Be able to name and draw the structures of methane, ethane, ethene, ethanol and ethanoic acid
- Be able to state the type of compound present, given a chemical name ending in -ane, -ene, -ol, or -oic acid or a molecular structure
- Be able to name and draw the structures of the unbranched alkanes, alkenes (not cis-trans), alcohols and acids containing up to four carbon atoms per molecule
- Be able to describe and identify structural isomerism

26.01 Fuels

Common fossil fuels include:

- coal: formed by the decay of vegetation in the absence of oxygen millions of years ago
- natural gas: mainly methane
- **hydrocarbons** such as methane, ethane and propane obtained by fractional distillation of petroleum

TERMS

Hydrocarbon: A compound containing only carbon and hydrogen atoms.

TIP

When defining a hydrocarbon make sure that you include the word 'only' (contains carbon and hydrogen atoms).

26.02 Organic chemistry

Organic chemistry refers to the chemistry of carbon compounds which either burn or char (go black) when heated in a flame. Inorganic compounds generally just melt on heating. All organic compounds contain carbon and most of them contain hydrogen. They may also contain other elements such as oxygen or halogens.

We may also classify organic compounds by whether they have unbranched chains, branched chains or ring structures (Figure 26.01).

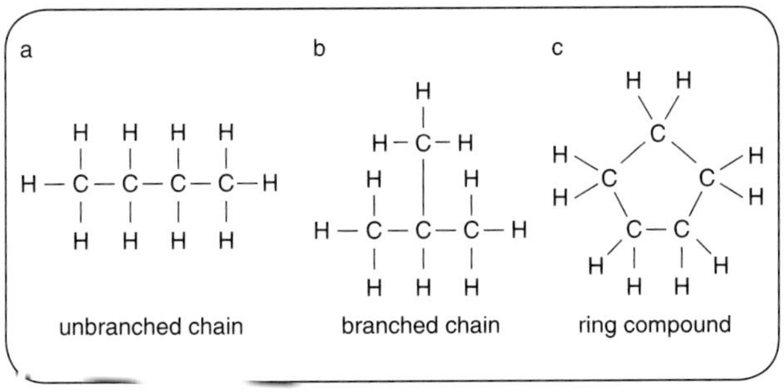

a unbranched chain

b branched chain

c ring compound

Figure 26.01

Not all compounds containing carbon are organic compounds. Carbon monoxide, carbon dioxide, carbonates and hydrogen carbonates are classified as inorganic compounds.

26.03 Different types of formulae

We can represent organic molecules using a variety of different formulae. Using butane, C_4H_{10}, and butene, C_4H_8, as examples, the molecular formula shows the number of each type of atom present in one molecule of the compound:

butane C_4H_{10}	butene C_4H_8

The structural formula shows all atoms and all bonds present in the molecule.

```
      H     H     H     H
      |     |     |     |
H  —  C  —  C  —  C  —  C  — H
      |     |     |     |
      H     H     H     H
           butane

     H     H     H     H
     |     |     |     |
H —  C  —  C  =  C  —  C  — H
     |                 |
     H                 H
           butene
```

This type of formula is also called a displayed formula. We can also draw a condensed structural formula:

$CH_3CH_2CH_2CH_3$	$CH_3CH{=}CHCH_3$
butane	butene

26.04 Homologous series

TERMS

Homologous series: A 'family' of similar compounds with similar chemical properties due to the same functional group.

Functional group: An atom or group of atoms which is responsible for the characteristic reactions of a particular homologous series.

- A **homologous series** is a group of similar compounds with similar chemical properties.
- Every member of a given homologous series has the same **functional group**.
- The functional group is responsible for the characteristic chemical properties of a particular homologous series.
- We can tell which homologous series a compound belongs to from its name. For example, the members of the carboxylic acid homologous series all end in –oic acid.

Table 26.01 shows some homologous series and functional groups.

Homologous series	Functional group	Name ending	Example
alkane	—C—H (with single bonds above and below C)	-ane	ethane, C_2H_6
alkene	>C=C<	-ene	ethene, C_2H_4
alcohol	—O—H	-ol	ethanol, C_2H_5OH
carboxylic acid	—C(=O)—O—H	-oic acid	ethanoic acid, CH_3COOH

Table 26.01

TERMS

Alkane: A hydrocarbon which contains only single bonds.

Alkene: A hydrocarbon which contains one or more double bonds.

Alcohol: A compound containing the –OH functional group, having the general formula $C_nH_{2n+1}OH$.

Carboxylic acid: A compound containing the –COOH functional group, having the general formula $C_nH_{2n+1}COOH$.

Progress check

26.01 Give the name of a fossil fuel which is

- a a solid [1]
- b a gas. [1]

26.02 What is the main constituent of natural gas? [1]

26.03 To which homologous series do the following compounds belong?

- a butane [1]
- b octanol [1]
- c hexanoic acid. [1]

26.04 Draw the full structural formula of ethanol showing all atoms and bonds. [2]

26.05 Give the molecular formula for the compound which a is an alcohol with two carbon atoms [1] b is an alkane with one carbon atom. [1]

26.06 Why is there not such a chemical as methene? [2]

Worked example 26.01

Figure 26.02 shows five organic compounds.

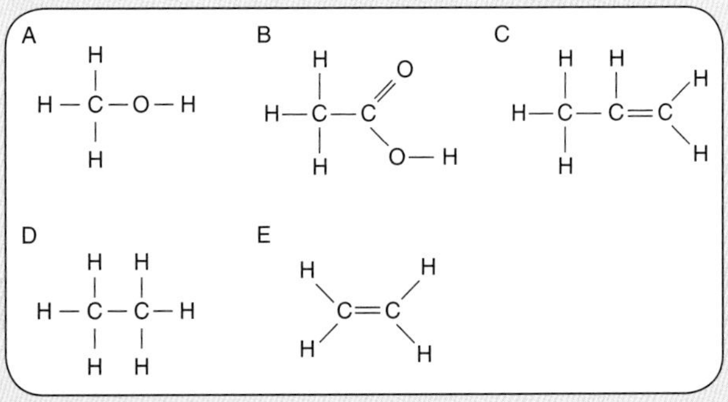

```
A                     B                        C
    H                     H      O                 H   H
    |                     |    //                  |   |    /H
H — C — O — H         H — C — C                H — C — C = C
    |                     |    \                   |        \H
    H                     H     O — H              H

D                     E
    H   H                H         H
    |   |                 \       /
H — C — C — H              C = C
    |   |                 /       \
    H   H                H         H
```

Figure 26.02

a Which two of these compounds belong to the same homologous series? Explain why. **[2]**

Look for the same functional group.

C and E **[1]** They both have a C=C double bond **[1]**

b To which homologous series does **i** compound A **ii** compound B belong? **[2]**

You should know the functional groups for alkanes, alkenes, alcohols and carboxylic acids.

i Alcohol **[1]** ii Carboxylic acid **[1]**

c Write the molecular formulae for compounds B and C. **[2]**

The molecular formula shows the number of each type of atom in a molecule of the compound.

B is $C_2H_4O_2$ **[1]** C is C_3H_6 **[1]**

d Write the full structural formula, showing all atoms and bonds, for another compound in the same homologous series as compound A. **[2]**

Do not forget to include the bond in the OH functional group:

```
    H   H
    |   |
H — C — C — O — H
    |   |
    H   H
```

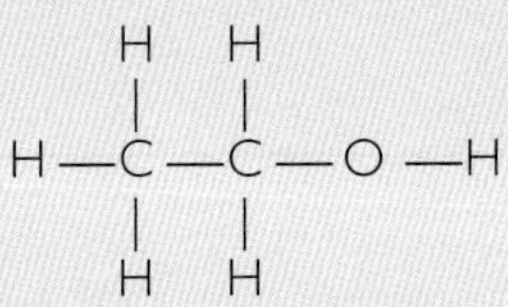

(**2 marks** if all structure correct)
(**1 mark** if OH instead of O–H)
(OR correct structure of any other alcohol)

More about homologous series

Apart from having the same functional group and similar chemical properties, each member of the same homologous series:

- can be represented by a **general formula**. For example, each member of the alkane homologous series has the general formula C_nH_{2n+2} and each member of the alkene homologous series has the general formula C_nH_{2n}, where n is the number of carbon atoms
- differs from the members immediately before or after by a CH_2 group
- shows a gradual change in physical properties as the number of carbon atoms in the compound increases

TERMS

General formula: A formula which applies to all members of a particular homologous series.

26.05 Naming simple organic compounds

- The suffix (end part of the name) of an organic compound tells us the functional group to which it belongs.
- The prefix / stem (beginning part of the name) of an organic compound tells us how many carbon atoms there are in the longest continuous carbon chain in the compound.
- The first eight prefixes are:

Number of C atoms	Prefix
1	meth-
2	eth-
3	prop-
4	but-
5	pent-
6	hex-
7	hept-
8	oct-

Table 26.02

- Some simple examples are shown in Figure 26.03.

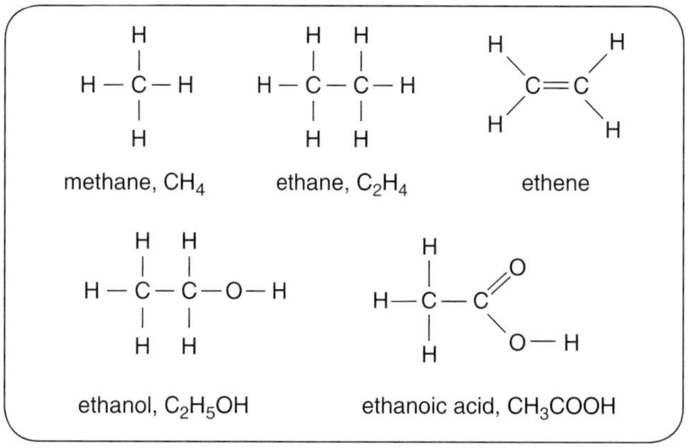

methane, CH_4 ethane, C_2H_4 ethene

ethanol, C_2H_5OH ethanoic acid, CH_3COOH

Figure 26.03

Organic compounds with three or more carbon atoms in their chains

Figure 26.04 shows some examples of unbranched organic compounds containing up to four carbon atoms in their chains.

propene but-1-ene

ethanol propan-1-ol

propanoic acid butanoic acid

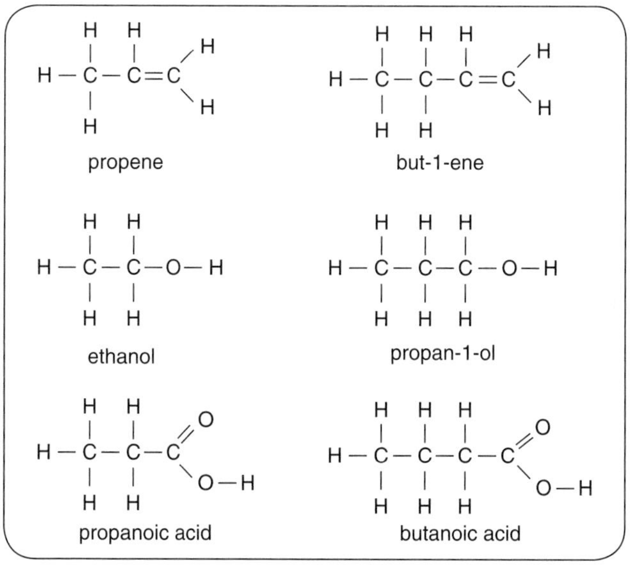

Figure 26.04

26.06 Structural isomers

TERMS

Structural isomers: Compounds with the same molecular formula but different structural formulae.

Some compounds have the same molecular formula but their atoms are arranged differently in the molecule. We call these compounds **structural isomers**. Two types of structural isomerism are:

- chain isomerism: the structure of the carbon skeleton differs. For example, butane has the same molecular formula as methylpropane, C_4H_{10}, but butane has an unbranched chain whereas methylpropane has a branched chain (Figure 26.05)

```
 H   H   H   H                    H
 |   |   |   |                    |
H—C—C—C—C—H                     H—C—H
 |   |   |   |               H    |    H
 H   H   H   H               |    |    |
                           H—C————C————C—H
                             |    |    |
                             H    H    H
   butane                    methylpropane
```

Figure 26.05

- position isomerism: the position of the functional group differs. For example, the OH functional group in butan-1-ol is different from its position in butan-2-ol (Figure 26.06)

```
  H  H  H  H               H  H  H  H
  |  |  |  |               |  |  |  |
H—C—C—C—C—H             H—C—C—C—C—H
  |  |  |  |               |  |  |  |
  H  H  H  O               H  H  O  H
            \                     \
             H                     H
   butan-1-ol              butan-2-ol
```

Figure 26.06

Note that alkyl side chains are named after the hydrocarbons, for example:

methane → methyl; ethane → ethyl;
propane → propyl; butane → butyl

TIP

When writing the formulae of different structural isomers, make sure that you are not writing the same formula twice. For example:

```
CH3 — CH2 — CH2 — CH3 is the same
      as CH3 — CH2 — CH2
                     |
                     CH3
```

TIP

You do not need to know how to name branched chain compounds, but you will comes across names with numbers in them so it is helpful to be aware that these numbers refer to the position of side chains branching off the main chain.

Sample answer

a Give the molecular formulae for the 4th and 5th members of the alcohol homologous series. **[2]**

$C_4H_{10}O$ / C_4H_9OH **[1]** $C_5H_{12}O$ / $C_5H_{11}OH$ **[1]**

b Draw the structural formula of the second member of the alcohol homologous series, showing all atoms and all bonds. **[2]**

```
   H   H
   |   |
H—C — C — O — H
   |   |
   H   H
```

(**2 marks** if all structure correct)
(**1 mark** if OH instead of O–H)

c Give the general formula of the alcohol homologous series. **[1]**

$C_nH_{2n+2}O$ / $C_nH_{2n+1}OH$ **[1]**

d i Explain why compounds A and B are not isomers.

```
                     CH3
                     |
CH3—CH2—CH2—CH—CH3          compound A

    CH3
    |
CH3—CH—CH2—CH2—CH3          compound B [2]
```

In A, counting from the right the CH_3 is at the second carbon. In B, counting from the left the CH_3 is at the second carbon **[2]**. (If **2 marks** not scored allow **1 mark** for the methyl group is in the second position.)

ii Draw an isomer of compound A. **[1]**

$$CH_3-CH_2-CH(CH_3)-CH_2-CH_3$$

(or branched chain compound with two methyl groups as side chains) **[1]**

Progress check

26.07 The members of a homologous series have the same functional group. Give two other characteristics of a homologous series. [2]

26.08 The first three members of the carboxylic acid homologous series have the formulae HCOOH, CH_3COOH and CH_3CH_2COOH. Deduce the general formula for the carboxylic acid homologous series. [1]

26.09 The molecular formulae of the first four members of the alcohol homologous series are: CH_4O, C_2H_6O, C_3H_8O and $C_4H_{10}O$. Deduce the molecular formulae of the fifth and sixth members of this homologous series. [2]

26.10 What is the meaning of the term *structural isomer*? [1]

Exam-style questions

Question 26.01

Ethene and propene belong to the same homologous series.

a What is meant by the term *homologous series*? [2]

b To which homologous series do ethene and propene belong? [1]

c Write the molecular formula and draw the full structural formula (displayed formula) of ethene. [2]

d The displayed formula of compound X is shown below.

```
    H       H
    |       |      O
H — C — C = C — C //
    |   |          \
    H   O—H         O—H
```

i Copy the formula and draw a square around the carboxylic acid functional group and draw a ring around the alkene functional group. [2]

ii State the name of the other functional group present. [1]

iii State the meaning of the term *functional group*. [1]

e Draw the full structural formula of the compound which is the main constituent of natural gas. [1]

Question 26.02

An homologous series is a group of compounds with similar chemical properties and the same functional group.

a Give three other characteristics of an homologous series. [3]

b i Give the molecular formulae for the alkenes containing 4 and 5 carbon atoms. [2]

ii Draw the structural formula of the alkene containing 3 carbon atoms showing all atoms and bonds. [1]

c Explain the meaning of the term *structural isomer*. [1]

d Bromoalkanes are compounds formed when one or more bromine atoms replace the hydrogen in an alkane. Draw two isomers of the unbranched compound with the molecular formula, C_3H_7Br. [2]

Revision checklist

You should be able to:

- State that coal, petroleum and natural gas are fuels
- State that natural gas is mainly methane
- Describe an homologous series as a 'family' of similar compounds with the same functional group
- Describe the general characteristics of a homologous series, including the idea of general formula
- Name and draw the structures of methane, ethane, ethene, ethanol and ethanoic acid
- State the type of compound present, given a chemical name ending in -ane, -ene, -ol, or -oic acid or a molecular structure
- Deduce the type of compound present from its structural formula
- Name and draw the structures of unbranched alkanes, alkenes, alcohols and carboxylic acids containing up to four carbon atoms per molecule
- Describe, identify and draw structural isomers

Petroleum fractionation and hydrocarbons

Learning outcomes

By the end of this unit you should:

- Be able to describe petroleum as a mixture of hydrocarbons and its separation into useful fractions by fractional distillation
- Be able to describe the properties of molecules within a fraction
- Know some uses of the fractions obtained by the fractional distillation of petroleum
- Be able to distinguish between saturated and unsaturated hydrocarbons in terms of bonding
- Be able to describe the properties of alkanes as being unreactive except for burning
- Be able to describe the manufacture of alkanes and hydrogen by cracking
- Be able to describe the substitution reaction of alkanes with chlorine in the presence of light
- Be able to distinguish between saturated and unsaturated hydrocarbons by reaction with aqueous bromine
- Be able to describe the addition reactions of alkenes with bromine, hydrogen and steam

27.01 Petroleum and fractional distillation

Petroleum (crude oil) is a source of hydrocarbons which is found trapped between impervious layers of rock beneath the surface of the Earth. Petroleum is a complex mixture of hydrocarbons in which some natural gas is also dissolved.

Petroleum refining involves:

- removal of impurities such as sulfur
- separation of the hydrocarbon mixture by fractional distillation into smaller groups of hydrocarbons called fractions. Each **fraction** has a limited range of relative molecular masses and number of carbon atoms. The fractions are separated because of the differences in their boiling point ranges

TERMS

Fraction: A group of molecules with a defined boiling point range which distils off at the same place during fractional distillation.

Figure 27.01 shows a fractional distillation column and the uses of these fractions.

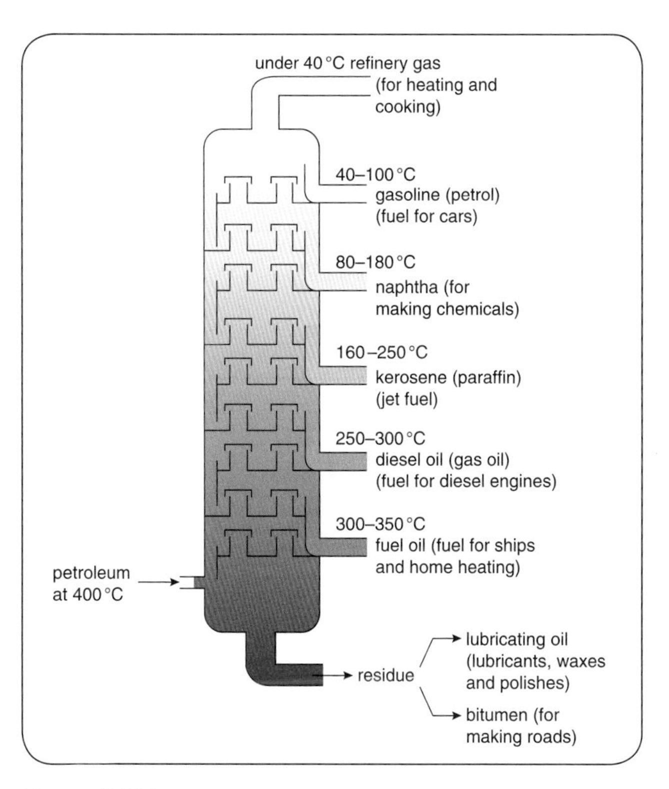

Figure 27.01

- Petroleum is heated to 400 °C.
- There is a temperature gradient in the column, higher at the bottom than at the top.
- As distillation proceeds, the more volatile hydrocarbons in the petroleum, which have lower boiling points, move further up the column.
- Hydrocarbons with lower mass and shorter chains have lower boiling points.
- As the hydrocarbon vapours move up the column, the ones with lower boiling points move further ahead of those with higher boiling points.
- At particular points in the column, the vapour containing hydrocarbons with a particular range of boiling points condense and are removed from the column.

You do not have to remember the boiling points and the range of carbon atoms in each fraction but you do need to know some uses of the fractions.

Worked example 27.01

The diagram shows a column for the separation of petroleum fractions.

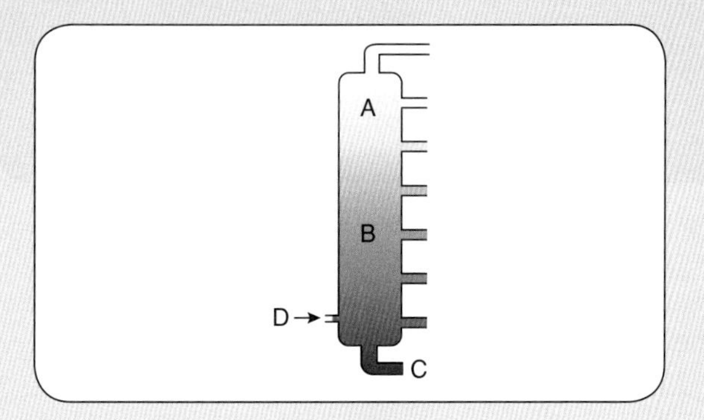

Figure 27.02

a Where in the column A, B, C or D is the temperature lowest? **[1]**

Remember that the petroleum is vaporised in a furnace near the bottom of the column. The column itself is not heated. So there is a temperature gradient in the column, hottest at the bottom.

A (at the top)

b Give the name of the process which separates petroleum into fractions and describe and explain how the separation is achieved. **[6]**

The word 'describe' indicates that you have to write something about where the different fractions come off the column. The word 'explain' indicates that you have to write about melting points and sizes of molecules.

Fractional distillation **[1]**. The column is hotter at the bottom than at the top **[1]**. The petroleum is heated so that the hydrocarbons vaporise **[1]**. The hydrocarbons rise up the distillation column until they reach the temperature at which they condense **[1]**. The distance risen depends on the boiling point of the hydrocarbons **[1]**. Larger hydrocarbon molecules have higher boiling points than smaller ones **[1]**.

27.02 Saturated and unsaturated hydrocarbons

TERMS

Saturated hydrocarbon: A hydrocarbon which has only single bonds.

Unsaturated hydrocarbon: A hydrocarbon which has one or more double (or triple) bonds.

- Hydrocarbons are compounds containing carbon and hydrogen only.
- Hydrocarbons with only single bonds in their molecules are called **saturated hydrocarbons**. Examples are ethane and propane (see Figure 27.03).
- Hydrocarbons containing one or more double or triple carbon- carbon bonds are called **unsaturated hydrocarbons**. Examples are ethene and propene (see Figure 27.03).

- We can distinguish saturated and unsaturated hydrocarbons by reaction with aqueous bromine (see under Alkenes below).

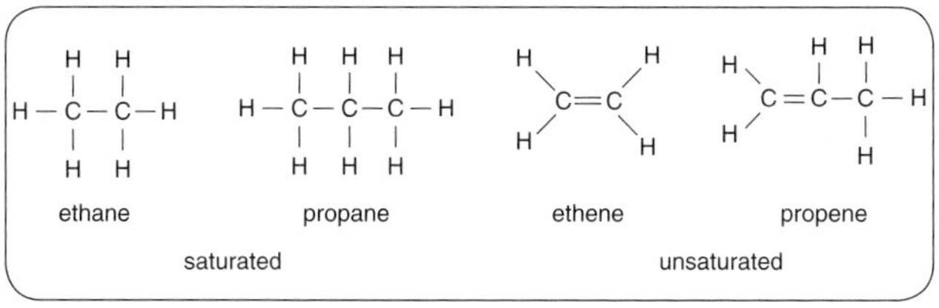

ethane | propane | ethene | propene

saturated | unsaturated

Figure 27.03

When defining a hydrocarbon you must indicate that there are no other elements present other than carbon and hydrogen. The inclusion of the word 'only' is useful here.

Progress check

27.01 Give a use for each of the following petroleum fractions:

a bitumen [1]

b refinery gas [1]

c gasoline. [1]

27.02 What is meant by the term *hydrocarbon*? [1]

27.03 How do saturated compounds differ from unsaturated compounds in terms of bonding? [2]

27.03 Alkanes

- Alkanes are saturated hydrocarbons. They only have single covalent bonds. The structures of the first four unbranched alkanes are shown in Figure 27.04.

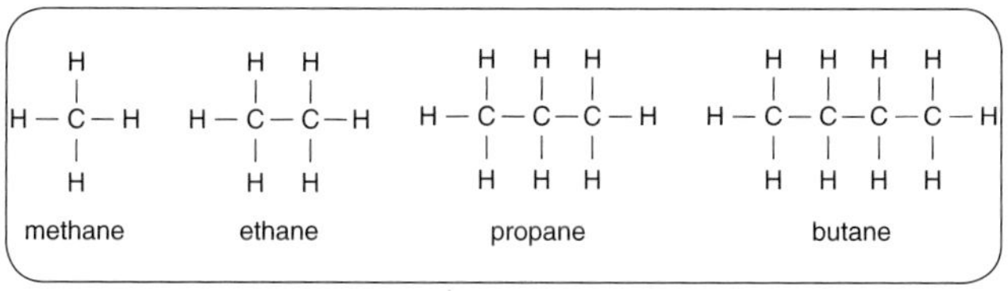

methane | ethane | propane | butane

Figure 27.04

- The alkanes are colourless compounds showing a gradual change in physical properties as the number of carbon atoms in the unbranched chain increases.
- The alkanes are generally unreactive except in terms of combustion (burning), cracking and reaction with halogens in the presence of light.

Combustion (burning) of alkanes

In the presence of excess oxygen, alkanes burn with a blue flame which does not appear sooty. We say that the alkane undergoes complete combustion. The products are carbon dioxide and water. For example:

$$CH_4(g) + 2O_2(g) \rightarrow CO_2(g) + 2H_2O(l)$$

$$C_3H_8(g) + 5O_2(g) \rightarrow 3CO_2(g) + 4H_2O(l)$$

If oxygen is not in excess, alkanes undergo incomplete combustion. The products of incomplete combustion are carbon monoxide and water. Some carbon may also remain unreacted. This may cause the flame to be yellow and sooty.

$$2C_2H_6(g) + 5O_2(g) \rightarrow 4CO(g) + 6H_2O(l)$$

When balancing equations for combustion reactions balance the carbon, then the hydrogen. Leave the balancing of the oxygen until last.

Cracking alkanes

Some fractions from the distillation of petroleum are more useful than others. We use more gasoline (petrol) than can be supplied by the fractional distillation of petroleum. We use a process called **cracking** to convert fractions containing larger, less useful hydrocarbons into smaller, more useful hydrocarbons. Kerosene and diesel oil are often cracked to make more gasoline (petrol), more alkenes or hydrogen. Cracking is a form of thermal decomposition.

TERMS

Cracking: The process by which large, less useful hydrocarbon molecules are broken down into smaller, more useful alkanes and alkenes.

In catalytic cracking gaseous kerosene or diesel oil fractions are passed through a mixed catalyst of silicon(IV) oxide and aluminium oxide at 400–500 °C. Catalytic cracking produces a relatively high proportion of hydrocarbons for use in gasoline (petrol), for example:

$C_{10}H_{22}(g)$	→	$C_8H_{18}(g)$	+	$C_2H_4(g)$
decane		octane		ethene
(long chain alkane)		(shorter chain alkane)		(alkene)

Hydrogen can also be made by cracking at high pressure and at temperatures above 700 °C. A higher proportion of alkenes are formed:

$C_{10}H_{22}(g)$	→	$C_7H_{14}(g)$	+	$C_3H_6(g)$	+	$H_2(g)$
decane		heptene		propene		
(long chain alkane)		(shorter chain alkene)		(alkene)		

Reaction of alkanes with halogens

- Alkanes do not react with halogens in the dark.
- In the presence of sunlight or ultraviolet light a reaction does take place.
- When one mole of chlorine reacts with one mole of methane, a chlorine atom replaces a hydrogen atom in the methane.

$$CH_4 + Cl_2 \xrightarrow{\text{light}} CH_3Cl + HCl$$

- This is an example of a **substitution reaction**.

TERMS

Substitution reaction: A reaction in which one atom or group of atoms in a compound is replaced by another.

- In the presence of excess chlorine, the hydrogen atoms are substituted one by one until there are none left.

$$CH_3Cl + Cl_2 \rightarrow CH_2Cl_2 + HCl$$
$$CH_2Cl_2 + Cl_2 \rightarrow CHCl_3 + HCl$$
$$CHCl_3 + Cl_2 \rightarrow CCl_4 + HCl$$

Progress check

27.04 Name these compounds:

a C_2H_4 [1]

b CH_4 [1]

c C_2H_6. [1]

27.05 Write the structural formulae for:

a an unbranched alkane with seven carbon atoms [1]

b a branched alkane with five carbon atoms. [1]

27.06 a What is meant by the term *substitution reaction*? [1]

b What condition is needed for chlorine to react with methane? [1]

27.04 Alkenes

Alkenes are unsaturated hydrocarbons. The structures of some alkenes are shown in Figure 27.05.

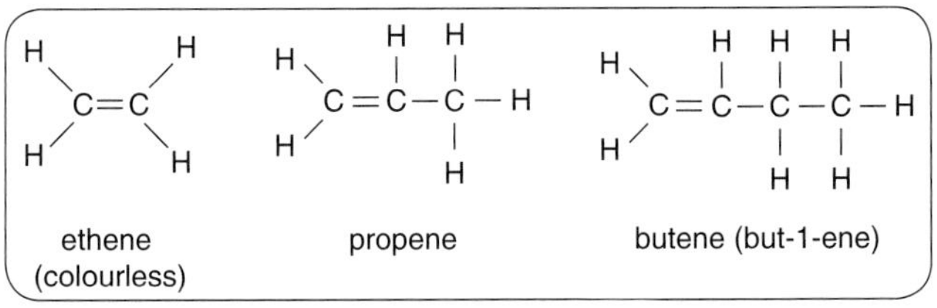

ethene (colourless) | propene | butene (but-1-ene)

Figure 27.05

Combustion of alkenes

The complete combustion of an alkene in oxygen produces carbon dioxide and water:

$$C_2H_4(g) + 3O_2(g) \rightarrow 2CO_2(g) + 2H_2O(l)$$

Reaction of alkenes with bromine

- Most of the reactions of alkenes are **addition reactions**.

> **TERMS**
>
> **Addition reaction**: In organic chemistry, a reaction where two molecules combine to give a single product.

- Alkenes react with halogens to form dihalogenoalkanes. The halogen adds across the double bond and no other product is formed. Bromine reacts with ethene to form 1,2-dibromoethane:

$$H_2C{=}CH_2 + Br_2 \longrightarrow H{-}CHBr{-}CHBr{-}H$$

ethene			1,2-dibromoethane
(colourless)	(red-brown)		(colourless)

Figure 27.06

- Bromine is red-brown in colour but 1,2-dibromoethane is colourless. So when we add a drop of bromine to excess alkene, the bromine becomes decolourised.

Testing with aqueous bromine (bromine water)

Liquid bromine is too hazardous for use in schools. So we use aqueous bromine (bromine water) to distinguish between alkanes and alkenes and between saturated and unsaturated compounds.

- Bromine water is red-brown when concentrated and orange when dilute.
- When bromine water is added to an alkane in the dark no reaction occurs. The bromine water remains red-brown (or orange). No C=C double bond is present. The hydrocarbon is saturated.
- When bromine water is added to an alkene in the dark there is a reaction. The bromine water is decolourised (goes colourless). A C=C double bond is present. The hydrocarbon is unsaturated.

When describing the bromine water test do not write that 'the bromine water goes clear'. In chemistry, the word clear does not mean colourless.

Worked example 27.02

The structure of two compounds A and B are shown below:

Compound A: $H{-}CH_2{-}CH_2{-}CH_2{-}H$ (H–C–C–C–H, each carbon with two H)

Compound B: $H{-}CH_2{-}CH{=}CH{-}H$ (H–C–C=C–H)

Compound A   Compound B

a Describe the similarities and differences between compounds A and B in terms of structure, bonding and in their reaction with aqueous bromine. **[6]**

Make sure that you write about each compound and each of (1) structure: what type of compounds are they? (2) bonding: type of bonding and details (3) details of colour changes in the bromine test for <u>both compounds</u>.

In marking this six of the possible nine points are allowed for maximum marks.

They are both simple molecules **[1]**. They are both hydrocarbons / they both contain hydrogen and carbon only **[1]**. Both have covalent bonds **[1]**. B has a double bond / is unsaturated **[1]**. A has no double bonds / is saturated **[1]**. B has fewer hydrogen atoms than A **[1]**. They both have three carbon atoms **[1]**. A will not react with aqueous bromine / aqueous bromine stays orange **[1]**. Compound B decolourises aqueous bromine **[1]**.

b Give the molecular formula of compound A and compound B. **[2]**

You have to count the number of each type of atom in each of A and B. Do not draw the formula showing the bonds.

A: C_3H_8 **[1]** B: C_3H_6 **[1]**

Reaction of alkenes with hydrogen

The addition of hydrogen to an alkene to form an alkane is an example of a hydrogenation reaction. Hydrogen gas is reacted with the alkene at 150 °C using a nickel catalyst:

$$CH_3CH_2CH{=}CH_2 + H_2 \xrightarrow{\text{Ni, 150 °C}} CH_3CH_2CH_2CH_3$$

butene → butane

Hydrogenation reactions are used to change vegetable oils into margarine.

Reaction of alkenes with steam

This is another example of an addition reaction. It is also called a hydration reaction because water is added. The reaction is important for the industrial preparation of alcohols (see Unit 28). For example:

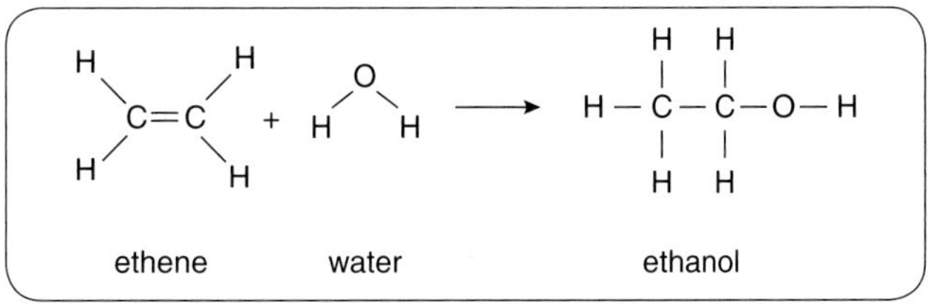

Figure 27.07

The conditions for the reaction are:

- catalyst of concentrated phosphoric acid
- temperature 330 °C
- pressure 60–70 atmospheres

Make sure that you know the conditions (temperature, pressure and name of catalyst) for organic reactions. These can be written above the arrow in the equation.

Sample answer

a What does the term *addition reaction mean?* **[2]**

An addition reaction is where two molecules combine [1] to give a single product [1].

b Describe the addition reactions of propene with **i** bromine and **ii** steam. Write symbol equations for these reactions and give any essential conditions. **[9]**

i Correct formula for propene, C_3H_6 or structural formula [1]. Propene decolourises (aqueous) bromine [1]: $C_3H_6 + Br_2 \rightarrow C_3H_6Br_2$ [1]

ii Forms propanol with steam [1]: $C_3H_6 + H_2O \rightarrow C_3H_7OH$ [1] Conditions are catalyst [1] of concentrated phosphoric acid [1] high temperature / 330 °C [1] high pressure / 60–70 atmospheres [1].

Progress check

27.07 Describe a test for an unsaturated compound, giving the colour changes. [2]

27.08 Draw the full structural formula of ethene showing all atoms and bonds. [1]

27.09 Name these compounds:

a C_4H_8 [1]

b C_5H_{10} [1]

27.10 Deduce the general formula for an alkene. [1]

27.11 State the conditions required for the addition reaction of hydrogen with propene and explain why it is an addition reaction. [3]

Exam-style questions

Question 27.01

a What is the meaning of the term *cracking*? Explain why cracking is used in an oil refinery. [4]

b Describe the conditions for cracking. [3]

c Copy and complete the equation for the cracking of dodecane, $C_{12}H_{26}$, to form ethene and one other alkane. $C_{12}H_{26}$ → ……… + ……… [2]

d i Describe the difference between unsaturated and saturated hydrocarbons in terms of molecular structure. [2]

ii Describe a test to distinguish between saturated and unsaturated compounds. [3]

e Copy and complete the symbol equation for the complete combustion of ethane.

$2C_2H_6 + 7……$ → $……CO_2 + ……H_2O$ [3]

Question 27.02

Butane and butene are hydrocarbons.

a Write a symbol equation for the complete combustion of butane, C_4H_{10}. [2]

b Draw the structural formulae for two structural isomers of butane and two structural isomers of butene. [4]

c Ethane is an alkane. Ethane reacts with halogens by a substitution reaction.

Describe, with an equation, the substitution reaction when one mole of ethane reacts with one mole of chlorine. What condition is required for this reaction to take place? [3]

d Draw the full structural formulae, showing all atoms and bonds, of the products of the reaction of propene with i bromine ii hydrogen. [2]

e Construct a symbol equation to show the cracking of the hydrocarbon $C_{16}H_{34}$ to form butene, one other hydrocarbon and hydrogen. [3]

Revision checklist

You should be able to:

- Describe petroleum as a mixture of hydrocarbons and know how it is separated into useful fractions by fractional distillation
- Describe the properties of molecules within a fraction (range of C atoms and boiling points)
- State some uses of the fractions obtained by the fractional distillation of petroleum
- Distinguish saturated from unsaturated hydrocarbons in terms of bonding
- Describe the properties of alkanes as being generally unreactive, except in terms of burning
- Describe the manufacture of alkanes by cracking
- Describe the substitution reaction of alkanes with chlorine in the presence of light
- Distinguish between saturated and unsaturated hydrocarbons by reaction with aqueous bromine
- Describe the properties of alkenes in terms of addition reactions with bromine, hydrogen and steam

Alcohols and carboxylic acids

Learning outcomes

By the end of this unit you should:

- Be able to name the uses of ethanol as a solvent and as a fuel
- Be able to describe the properties of ethanol in terms of burning
- Be able to describe the formation of ethanoic acid by the oxidation of ethanol fermentation and by oxidation with acidified.The burning is core aqueous potassium manganate(VII)
- Be able to describe the manufacture of ethanol by fermentation and the catalytic addition of steam to ethene
- Be able to describe the fermentation of simple sugars to produce ethanol
- Be able to outline the advantages and disadvantages of the manufacture of ethanol by fermentation and hydration of ethane
- Be able to describe the properties of aqueous ethanoic acid
- Be able to describe ethanoic acid as a typical weak acid
- Be able to describe the reaction of carboxylic acids with alcohols in the presence of a catalyst to give an ester.
- Be able to name and draw the structures of esters which can be made from unbranched alcohols and carboxylic acids containing up to four carbon atoms

28.01 Alcohols

Alcohols contain the –OH functional group (hydroxyl group). The structures of three alcohols are shown in Figure 28.01.

```
     H             H   H              H   H   H   H
     |             |   |              |   |   |   |
 H — C — O — H  H — C — C — O — H  H — C — C — C — C — O — H
     |             |   |              |   |   |   |
     H             H   H              H   H   H   H

   methanol        ethanol               butanol
                                       (butan-1-ol)
```

Figure 28.01

Ethanol is used as a solvent and as a fuel. It can be used as a fuel for cars either on its own or mixed with petrol.

28.02 Reactions of alcohols

Combustion

Alcohols burn in excess air. Carbon dioxide and water are formed. For example:

$$CH_3CH_2OH + 3O_2 \rightarrow 2CO_2 + 3H_2O$$

ethanol

TIP

- When writing equations for the combustion of alcohols, do not forget the oxygen in the alcohol.
- You need not worry about adding state symbols to equations involving organic reactions because organic solvents are often involved in the reaction.

Worked example 28.01

a Ethene reacts with steam to form ethanol. Complete the symbol equation for this reaction .... + ... → C_2H_5OH **[2]**

Steam is water, H_2O. Even if you cannot remember the formula for ethene, you can work it out because *there are equal numbers of each type of atom on each side of the equation.*

C_2H_4 **[1]** H_2O **[1]**

b Write a word equation for the complete combustion of ethanol. **[2]**

Combustion is burning in oxygen. Ethanol contains carbon, hydrogen and oxygen, so the oxides of carbon and hydrogen are the products.

ethanol + oxygen → carbon dioxide + water

(**2 marks** if all correct, **1 mark** if one error or omission)

c Methanol is an alcohol. Describe how you could compare the energy given out by methanol and ethanol when they burn. **[4]**

You should be prepared to answer questions involving practical aspects of chemistry at any point during the course. See Units 2 and 11.

Compare same amount of each alcohol burned **[1]** to heat up the same amount of water **[1]**. Compare the temperature rise **[1]** keeping everything else constant **[1]**.

d Describe the structure and bonding in ethanol. **[3]**

You should write about both the structure (molecular, giant ionic or giant covalent) as well as the bonding (covalent or ionic). With organic compounds you can also write about saturated or unsaturated.

Structure is simple molecular **[1]** Bonding is covalent **[1]** C–C bonds are unsaturated / only contains single carbon-carbon bonds **[1]**.

Oxidation of ethanol

Oxidation with potassium manganate(VII)

- Alcohols are oxidised to carboxylic acids by heating with acidified potassium manganate(VII).
- A few drops of concentrated sulfuric acid catalyst are needed to get the best conditions for oxidation.
- The heating is done under reflux. This involves heating the reactants with a condenser attached vertically to the reaction flask. This prevents the volatile alcohol from escaping.

The oxidation of ethanol to ethanoic acid using potassium manganate(VII), $KMnO_4$, can be represented by the equation:

$$CH_3CH_2OH + 2[O] \xrightarrow{H^+} CH_3COOH + H_2O$$

The actual equation is complex so we simplify it by writing [O] to represent the oxidising agent, $KMnO_4$ and H^+ to represent the acid catalyst.

In this reaction, the purple manganate(VII) ions are converted to colourless Mn^{2+} ions by the reducing agent, ethanol.

Oxidation by fermentation

Ethanol can be oxidised to ethanoic acid by atmospheric oxygen. This is speeded up by fermentation. During fermentation, enzymes in bacteria or yeasts catalyse the oxidation of organic substances.

Esterification

Alcohols react with carboxylic acids to produce esters (see below).

28.03 The manufacture of ethanol

Ethanol can be manufactured by fermentation or by the **hydration** of ethene (the reaction of ethene with steam).

TERMS

Hydration: An addition reaction in which water is combined with another compound.

Manufacture by fermentation

TERMS

Fermentation: The breakdown of organic substances by microorganisms with effervescence and release of energy. It commonly refers to the breakdown of sugars by enzymes in yeast in the absence of oxygen to form ethanol.

- Yeasts produce enzymes which catalyse **fermentation** reactions in organic materials.
- Most plant materials can be fermented but fermentation is most rapid when simple sugars such as glucose or sucrose are added.
- Fermentation is used to make alcoholic drinks. The main alcohol in these drinks is ethanol.
- The overall reaction for the fermentation of glucose is:

$$\underset{\text{glucose}}{C_6H_{12}O_6} \xrightarrow{\text{enzymes in yeast}} \underset{\text{ethanol}}{2C_2H_5OH} + \underset{\text{carbon dioxide}}{2CO_2}$$

The conditions needed for fermentation are:

- Temperatures of between about 15 and 35 °C. Too low a temperature will slow down the rate of the enzyme-catalysed reaction too much. Too high a temperature will denature the enzymes.
- Absence of oxygen. The yeasts responsible for alcoholic fermentation are anaerobic: they respire in the absence of oxygen. If oxygen is allowed into the mixture, bacteria may grow and spoil the alcoholic fermentation by producing acids.
- Reaction mixture with a pH value near pH 7. Too acidic or alkaline pH values slow down the rate of the enzyme-catalysed reaction.
- Presence of water. Yeast is a living organism, so water is needed for it to survive.

Fermentation to produce ethanol occurs in the absence of oxygen. It is a common error to suggest that oxygen is needed for this type of fermentation.

Manufacture by hydration

Pure ethanol can be produced by reacting steam with ethene:

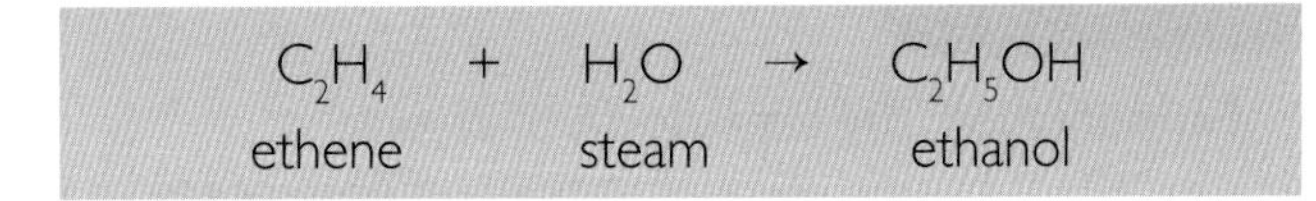

$$\underset{\text{ethene}}{C_2H_4} + \underset{\text{steam}}{H_2O} \rightarrow \underset{\text{ethanol}}{C_2H_5OH}$$

This is an addition reaction. It is also called a hydration reaction because water is added.

The conditions for the reaction are:

- catalyst of concentrated phosphoric acid
- temperature 330 °C
- pressure 60–70 atmospheres

Comparing manufacturing methods

Both hydration and fermentation have advantages and disadvantages. Table 28.01 shows these.

Fermentation	Hydration (addition of steam to ethene)
simple equipment required	complex pressure vessels required
uses renewable reactants (sugar from plants)	uses non-renewable reactant (ethene from cracking petroleum fractions)
works at room temperature and atmospheric pressure so energy requirement is low	requires high temperature and pressure so energy requirement is high
catalyst does not have to be manufactured (enzymes in yeast) and only damaged if ethanol concentration is too high	catalyst has to be produced and has to be 'cleaned'
uses a batch process: you have to start again from the beginning after the fermentation mixture is removed	uses a continuous process: ethanol can be removed continuously and ethene and steam fed into the reactor continuously
impure ethanol produced: the ethanol has to be purified by distillation	pure ethanol produced
reaction is slow	fast reaction
reaction has low atom economy: carbon dioxide and side products produced as well as ethanol	reaction has high atom economy: only ethanol produced; there are no waste products which have to be dumped
carbon dioxide produced is a greenhouse gas	no greenhouse gas produced (although pollutants are produced as a result of the burning of fossil fuels to maintain high temperature and pressure)

Table 28.01

Sample answer

Ethanol can be made by the hydration of ethene.

a Write an equation for this reaction showing the structural formulae with all atoms and bonds of each reactant and product. **[3]**

(**1 mark** for structural formula of each reactant and product)

H H C=C H H + H O H → H H H–C–C–O–H H H

ethene water ethanol

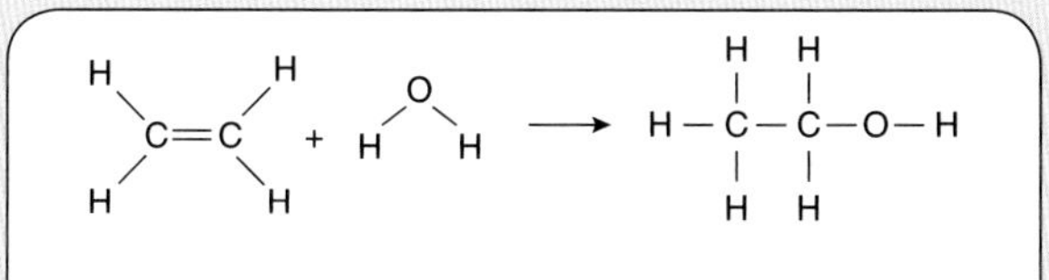

Figure 28.02

b Give three advantages of manufacturing ethanol by hydration that makes this process better than the manufacture of ethanol by fermentation. Give a reason for each choice you make. **[6]**

For each choice: one mark for advantage and one mark for further comment up to a maximum of six marks.

It is a continuous process **[1]** because you do not have to shut down the reactor **[1]**.

Pure ethanol is produced **[1]** so energy is not wasted purifying it **[1]**.

The reaction is faster **[1]** so can keep up with demand for ethanol **[1]**.

It has a high atom economy **[1]** so that no potentially harmful products are produced and no waste products are made **[1]**.

No carbon dioxide produced **[1]** so there are fewer greenhouse gas emissions **[1]**.

Progress check

28.01 Give two uses of ethanol. [2]

28.02 Copy and complete the symbol equation for the complete combustion of ethanol.
$CH_3CH_2OH + \ldots O_2 \rightarrow \ldots CO_2 + \ldots H_2O$ [3]

28.03 Write a word equation for the fermentation of glucose, $C_6H_{12}O_6$. [2]

28.04 What are the catalysts used in the fermentation of glucose? [1]

28.05 Describe three advantages of producing ethanol by fermentation rather than by the hydration of ethene.

28.06 Draw the structural formula of propanol showing all atoms and bonds. [2]

28.04 Carboxylic acids

Carboxylic acids contain the –COOH functional group (carboxylic acid group). The structures of three carboxylic acids are shown in Figure 28.03.

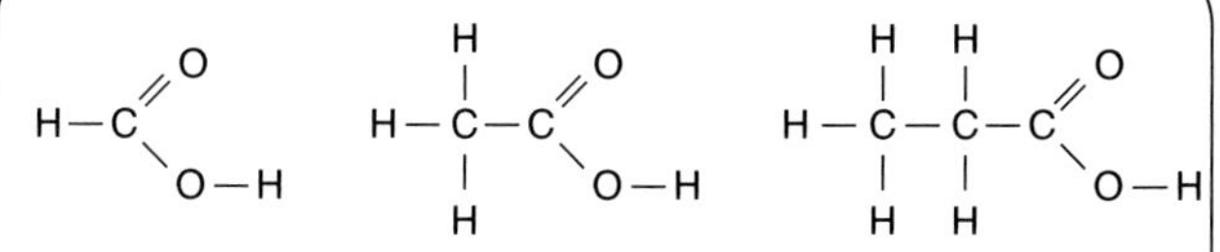

H–C(=O)–O–H methanic acid; H H–C–C(=O)–O–H H ethanoic acid; H H H–C–C–C(=O)–O–H H H propanoic acid

methanoic acid ethanoic acid propanoic acid

Figure 28.03

Carboxylic acids are colourless liquids which have typical acidic properties. For example, they turn blue litmus paper red and form salts called ethanoates.

When writing formulae and naming carboxylic acids, remember to count the carbon atom in the COOH group. For example CH_3CH_2COOH is propanoic acid and not ethanoic acid.

Worked example 28.02

Ethanoic acid has typical acidic properties.

a Draw the structure of ethanoic acid showing all atoms and bonds. **[2]**

The question asks for all atoms and all bonds. The word 'all' is important here if you are to get full marks.

H–C(H)(H)–C(=O)–O–H

ethanoic acid

b Ethanoic acid reacts with zinc carbonate to form a salt called zinc ethanoate and two other compounds. Name these two other compounds. **[2]**

Although you have not studied this, you have been told that ethanoic acid is a typical acid. So you have to refer back to the general reaction of an acid with a carbonate.

carbon dioxide **[1]** water **[1]**

c Convert this symbol equation into a word equation. CH_3COONa is sodium ethanoate.

$NaOH + CH_3COOH \rightarrow CH_3COONa + H_2O$ **[2]**

You should be able to recognise the formula of ethanoic acid as either CH_3COOH or CH_3CO_2H. The rest of the question is working out the names from basic principles (Unit 7).

sodium hydroxide + ethanoic acid → sodium ethanoate + water

(**2 marks** if all correct, **1 mark** if one error)

Reaction of ethanoic acid with metals

An aqueous solution of ethanoic acid reacts with metals to form a salt and hydrogen:

$2CH_3COOH + Mg \rightarrow (CH_3COO)_2Mg + H_2$

ethanoic acid — magnesium ethanoate

Reaction of ethanoic acid with hydroxides

An aqueous solution of ethanoic acid reacts with hydroxides of reactive metals. A salt and water are formed. This is a neutralisation reaction:

$KOH + CH_3COOH \rightarrow CH_3COOK + H_2O$

ethanoic acid — potassium ethanoate

Reaction of ethanoic acid with carbonates

An aqueous solution of ethanoic acid reacts with carbonates. A salt, water and carbon dioxide are formed. For example:

$K_2CO_3 + 2CH_3COOH \rightarrow 2CH_3COOK + H_2O + CO_2$

ethanoic acid — potassium ethanoate

Ethanoic acid as a weak acid

Ethanoic acid, CH_3COOH, is a typical weak acid. It only partially ionises (partly dissociates) in solution. There are many more molecules of unionised acid present than there are ions. There is an equilibrium between molecules of unionised acid and their ions.

$CH_3COOH \rightleftharpoons H^+ + CH_3COO^-$

ethanoic acid — ethanoate ions

When the concentration of acids are equal:

- Ethanoic acid has a higher pH than hydrochloric acid.
- Ethanoic acid will react more slowly with metals, metal carbonates and metal oxides (at the same temperature) than hydrochloric acid because ethanoic acid has a lower concentration of H^+ ions.
- Ethanoic acid does not conduct electricity as well as hydrochloric acid because ethanoic acid has a lower concentration of H^+ ions.

28.05 Esters

TERMS

Ester: A substance with the general formula RCOOR', formed by the reaction between a carboxylic acid and an alcohol.

- **Esters** are made by the reaction of alcohols with carboxylic acids.
- The functional group in esters is:

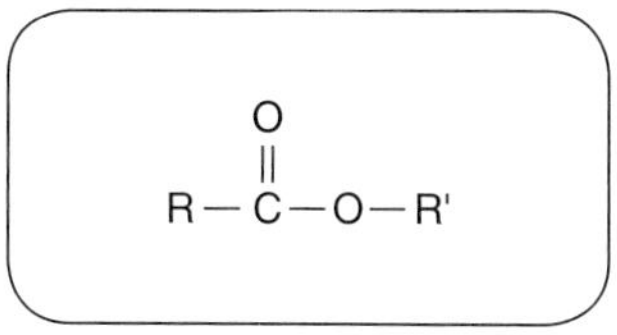

$$\mathrm{R-\overset{\overset{\Large O}{||}}{C}-O-R'}$$

Figure 28.04

- R can be either hydrogen or an alkyl group (methyl, ethyl, etc). R' can be an alkyl group but not hydrogen.
- Esters are named after the acid from which they are made. The -oate part of the name comes last and the name of the alcohol prefix comes first.

Example 1:

$C_3H_7COOCH_3$ is methyl butanoate

C_3H_7COOH butanoic acid $HOCH_3$ methanol

Example 2:

$HCOOC_3H_7$ is propyl methanoate

$HCOOH$ methanoic acid HOC_3H_7 propanol

Example 3:

$CH_3COOC_2H_5$ is ethyl ethanoate

CH_3COOH ethanoic acid HOC_2H_5 ethanol

- Figure 28.05 shows the structural formulae of some esters.

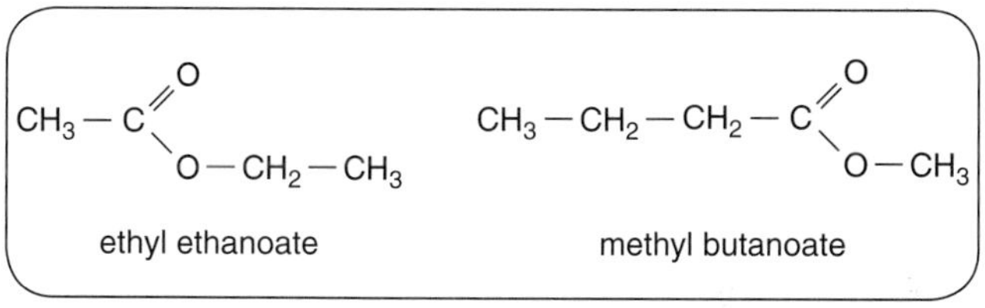

Figure 28.05

Esterification: making esters

TERMS

Esterification: The reaction between a carboxylic acid and an alcohol to produce an ester and water.

- Esters are synthesised by warming the alcohol with the carboxylic acid.
- This is an **esterification** reaction.
- A few drops of concentrated sulfuric acid is added to catalyse the reaction.
- An example is:

$$CH_3COOH + CH_3OH \xrightleftharpoons{H^+ \text{ catalyst}} CH_3COOCH_3 + H_2O$$

ethanoic acid, methanol, methyl ethanoate

- If larger amounts of ester are required, the mixture is refluxed gently to prevent loss of volatile compounds.
- After warming, the reaction mixture is poured into a dilute solution of sodium carbonate. This reacts with excess acid so that we can smell the typical sweet smell of the ester.

Progress check

28.07 Give two properties of ethanoic acid. [2]

28.08 Give the name and formula of the functional group present in propanoic acid. [2]

28.09 Write the structural formula for the ester methyl methanoate. [1]

28.10 Construct the symbol equation for the reaction of butanoic acid with sodium hydroxide. [2]

28.11 Name these esters:

a $CH_3CH_2COOC_2H_5$

b $HCOOC_3H_7$.

Exam-style questions

Question 28.01

The molecular formula of ethanol is C_2H_6O.

a Write the full structural formula of ethanol showing all atoms and bonds. [2]

b Ethanol can be manufactured by fermentation. Give three conditions needed for fermentation. [3]

c Ethanol can also be produced from ethene. State the name of the other reactant needed and give the conditions for the reaction. [4]

d Ethanol can be oxidised to ethanoic acid. Ethanoic acid has typical acidic properties. Its salts are called ethanoates.

i Describe how you could show that ethanoic acid has acidic properties. [2]

ii To which homologous series does ethanoic acid belong? [1]

iii Write a word equation for the reaction of sodium with ethanoic acid. [2]

iv Copy and complete the symbol equation for the reaction of ethanoic acid with sodium carbonate.

K_2CO_3 + 2........... →
......CH_3COOK + ... + ... [4]

Question 28.02

Ethanoic acid is a typical weak acid.

a Explain why a 0.1 mol/dm^3 solution of ethanoic acid reacts more slowly with magnesium than a 0.1 mol/dm^3 solution of hydrochloric acid. [5]

b Write a balanced equation for the reaction of ethanoic acid with magnesium carbonate. The formula for one of the products, magnesium ethanoate, is $(CH_3COO)_2Mg$. [3]

c Ethanoic acid reacts with butanol to form an ester.

i Give the name of the ester produced. [1]

ii Draw the structural formula for this ester showing all atoms and all bonds. [2]

iii Describe the conditions needed for esterification. [2]

d Ethanoic acid can be made from ethanol. Describe a method other than a fermentation method for producing ethanoic acid from ethanol. Give details of reactants and conditions. [3]

Revision checklist

You should be able to:

- Name some uses of ethanol (solvent, fuel)
- Describe the combustion of ethanol
- Describe the oxidation of ethanol to ethanoic acid by oxidation with acidified aqueous potassium manganate(VII)
- Describe the manufacture of ethanol by fermentation and by the catalytic addition of steam to ethene
- Describe the fermentation of sugars to produce ethanol
- Outline the advantages and disadvantages of the manufacture of ethanol by fermentation and hydration of ethene
- Describe the properties of aqueous ethanoic acid
- Describe acid as a typical weak acid
- Describe the formation of esters by reacting alcohols with carboxylic acids
- Name and draw the structures of esters which can be made from unbranched alcohols and carboxylic acids containing up to four carbon atoms

Polymers

Learning outcomes

By the end of this unit you should:

- Be able to define polymers as large molecules built up from smaller units called monomers
- Understand the formation of addition polymers from monomers with reference to the polymerisation of ethene
- Be able to explain the difference between addition polymerisation and condensation polymerisation
- Understand that different polymers have different units or linkages
- Be able to deduce the structure of a polymer product from a given alkene monomer and and vice versa
- Be able to describe the formation of nylon and terylene by condensation polymerisation
- Know some uses of plastics and man-made fibres
- Be able to describe the pollution problems caused by nonbiodegradable plastics

29.01 Polymers

Polymers are large molecules (macromolecules) built up by linking at least 50 small molecules called **monomers**. Figure 29.01 shows the **polymerisation** of ethene monomers to make the polymer poly(ethene).

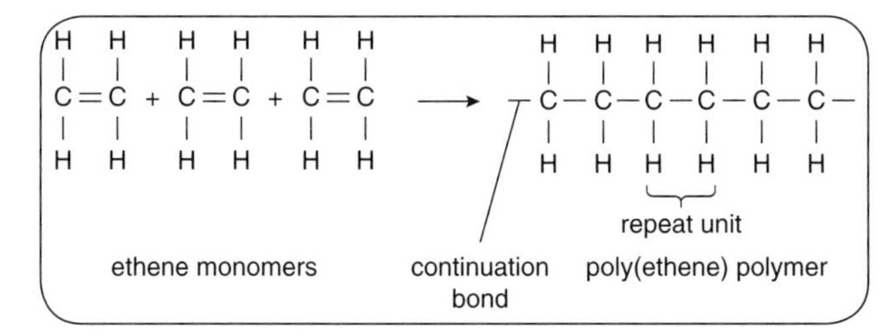

Figure 29.01

TERMS

Monomer: A small reactive molecule that reacts and joins together with itself or another molecule to form the repeating units of a polymer.

Polymer: A large molecule built up from smaller units called monomers.

Polymerisation: The chemical reaction where monomers react to form a polymer.

- The repeat units in polymers are connected by covalent bonds.
- The continuation of the polymer chain is shown by continuation bonds.

There are two types of polymerisation:

- Addition polymerisation: the monomers have a C=C double bond. They combine to form the polymer and no other compound is formed.
- Condensation polymerisation: the monomers are linked to form the polymer and a small molecule is eliminated (see below).

29.02 Addition polymerisation

TERMS

Addition polymerisation: A reaction in which monomers containing C=C double bonds react to form polymers without any other substance being formed.

- In **addition polymerisation**, one of the bonds in the C=C double bond of each alkene monomer breaks and forms a bond with an adjacent monomer.

- The polymer formed has single bonds.
- Many polymers are synthesised from alkene monomers.
- Other polymers are based on alkenes where one or more hydrogen atoms are substituted by groups such as –Cl or –OH.
- The name of the polymer is formed by putting the name of the monomer in brackets and adding poly- as a prefix. For example, propene (monomer) → poly(propene).

Progress check

29.01 Give the name of the polymer whose repeating unit is $-CH_2-CH_2-$ [1]

29.02 The equation shows polymerisation:

$nC_2H_4 \rightarrow -[-CH_2-CH_2-]-_n$ (where n is a large number)

Name the monomer in this reaction. [1]

29.03 What type of polymer is poly(ethene)? [1]

29.04 Explain the meaning of the term *addition polymerisation*. [2]

Deducing the structure of an addition polymer

When the monomer contains alkyl groups or other functional groups such as –Cl or –OH, we can represent the equation in a relatively simple way. Figure 29.02 represents the polymerisation of propene monomers, $CH_3CH=CH_2$, to form poly(propene).

```
CH3 H    CH3 H    CH3 H         CH3 H  CH3 H  CH3 H
 |   |    |   |    |   |          |  |   |  |   |  |
 C = C  +  C = C  +  C = C  →   – C– C – C– C – C– C –
 |   |    |   |    |   |          |  |   |  |   |  |
 H   H    H   H    H   H          H  H   H  H   H  H
```

Figure 29.02

To draw the polymer we:

1. draw the monomer so that the side chains are arranged vertically AND
2. change the C=C double bond to a single bond
3. write down the number of formula units required by the question
4. draw single bonds between the repeat units
5. put 'continuation bonds' at both ends of the chain

Figure 29.03 shows the process.

```
                  CH3 H       CH3 H
                   |   |       |   |
CH3CH=CH2  →  C = C   →   C – C   →  join units and add continuation bonds
                   |   |       |   |
                   H   H       H   H
```

Figure 29.03

Another way of drawing the polymer is shown in Figure 29.04. This shows the polymer poly(chloroethene).

```
   Cl  H         [ Cl  H ]
   |   |         [ |   | ]
 n C = C   →   –[ C – C ]–
   |   |         [ |   | ]
   H   H         [ H   H ]n
```

Figure 29.04

1. draw the structure of the monomer AND
2. change the C=C double bond to a single bond
3. put 'continuation bonds' at both ends of the repeat unit
4. put square brackets through the continuation bonds
5. put an 'n' at the bottom right to show a large number of repeat units

NOTE: The 'n' in front of the monomer in the diagram shows that there are a large number of them.

When drawing an addition polymer do not forget (i) the double bond changes to a single bond, (ii) 'continuation bonds'.

Deducing the monomer from the addition polymer

The procedure is as follows:

1. identify the repeat unit in the polymer
2. remove the 'continuation bonds'
3. make the C-C single bond into a C=C double bond

Figure 29.05 shows the procedure for poly(ethenol).

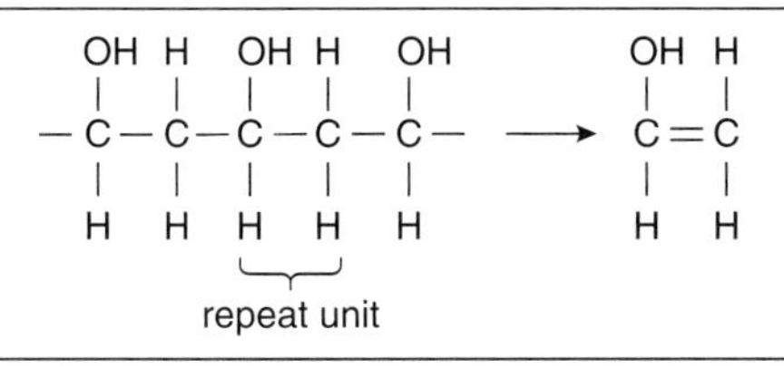

Figure 29.05

Worked example 29.01

a The diagram shows the structure of molecule X.

```
    CH3  OH
    |    |
H — C =  C — H  (molecule X)
```

i Explain why X does not undergo condensation polymerisation. **[1]**

Do not be tempted to refer to addition polymerisation. You should answer this question by referring to a condensation reaction which requires two functional groups to react.

It has not got two functional groups which react with each other **[1]**.

ii X forms an addition polymer. Draw a section of this addition polymer to show three repeat units. **[2]**

First draw the structure of one repeating unit by rotating the H atoms so that they are at right angles to the C=C. Then change the double bond into a single bond. Then draw another two units in line with the one that you have drawn and join them by single bonds. Lastly do not forget to add on the continuation bonds.

```
  CH3  OH   CH3  OH   CH3  OH
  |    |    |    |    |    |
— C  — C  — C  — C  — C  — C —
  |    |    |    |    |    |
  H    H    H    H    H    H
```

(**1 mark** for correct structure, **1 mark** for continuation bonds)

b The structure of polymer Z is shown below.

```
  Cl  H   Cl  H   Cl  H
  |   |   |   |   |   |
— C — C — C — C — C — C —
  |   |   |   |   |   |
  Cl  H   Cl  H   Cl  H
```

Draw the structure of the monomer used to make this polymer. **[2]**

First identify the repeating unit. Then draw this and change the single bond into a double bond. Lastly make sure that you have removed the continuation bonds.

```
Cl    H
|     |
C  =  C
|     |
Cl    H
```

(**2 marks** if all correct, **1 mark** if single C–C bond)

29.03 Condensation polymerisation

TERMS

Condensation reaction: A reaction where two organic molecules join together with the elimination of a small molecule such as water.

Condensation polymerisation: A reaction where two organic molecules join together in a condensation reaction to form the repeating units of a polymer.

- In **condensation polymerisation**, monomer molecules are linked together and a small molecule such as water or hydrogen chloride is eliminated.
- An example of a **condensation reaction** is shown in Figure 29.06 where an amide linkage is formed by the reaction of a carboxylic acid with a chemical called an amine.

```
    O                                O
    ||                               ||
CH3C—O—H + H—N—CH3   →   CH3 — C — N—CH3 + H2O
             |                     |
             H                     H
                           amide        water
                           linkage      eliminated
```

Figure 29.06

- Condensation polymerisation may involve two different monomers.
- Each monomer has at least two functional groups. For example:

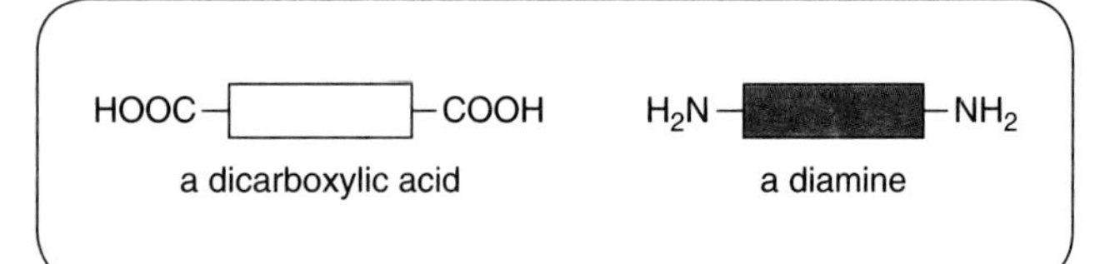

Figure 29.07

In this diagram the boxes represent the rest of the molecule.

Polyamides

TERMS

Polyamide: A polymer whose monomers are bonded to each other by the amide link, –CONH–.

- Nylon is a polyamide.
- The monomers for polyamides are dicarboxylic acids and diamines.
- Each –COOH group reacts with an $-NH_2$ group to form an amide linkage and a molecule of water is eliminated (Figure 29.08).

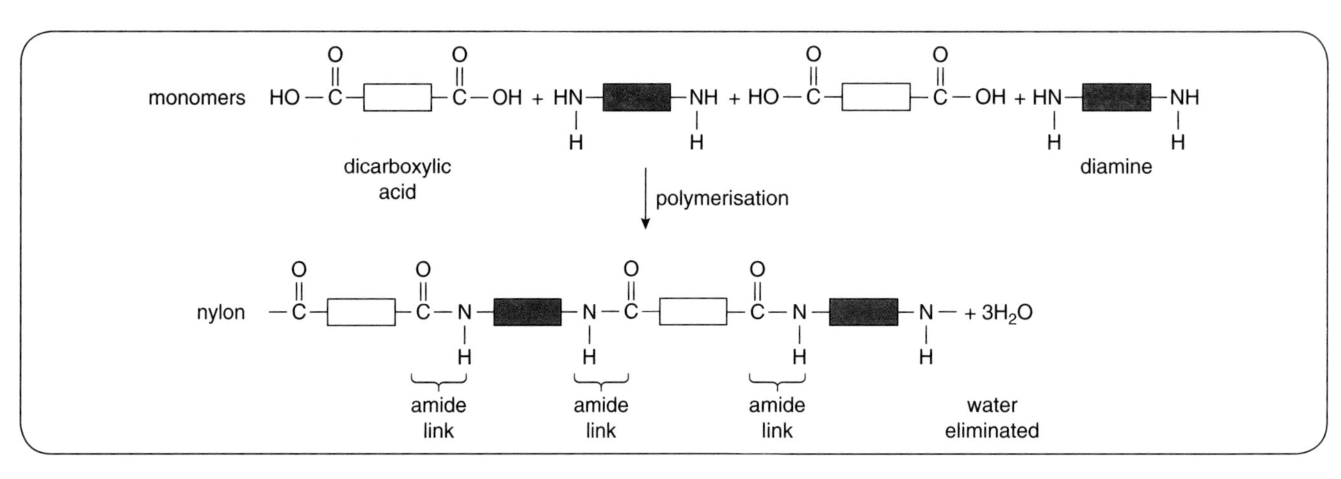

Figure 29.08

TIP

You should be able to draw block diagrams as shown in Figure 29.08 showing the displayed formula of the linkage.

Polyesters

TERMS

Polyester: A polymer whose monomers are bonded to each other by the ester link, –COO–.

- Terylene is a polyester.
- The monomers for polyesters are dicarboxylic acids and diols. Diols have two alcohol functional groups.
- When these molecules react, each –COOH group reacts with an –OH group to form an ester linkage and a molecule of water is eliminated (Figure 29.09).

When writing formulae for condensation polymers, make sure that the atoms in the linkage group are in the correct order.

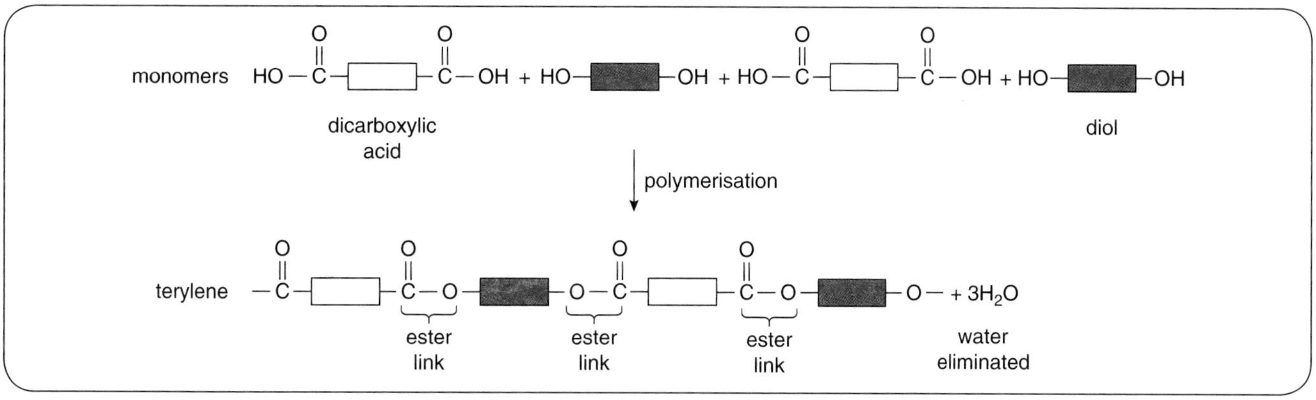

monomers HO—C(=O)—□—C(=O)—OH + HO—■—OH + HO—C(=O)—□—C(=O)—OH + HO—■—OH

dicarboxylic acid   diol

polymerisation

terylene —C(=O)—□—C(=O)—O—■—O—C(=O)—□—C(=O)—O—■—O— + $3H_2O$

ester link   ester link   ester link   water eliminated

Figure 29.09

Sample answer

Terylene is a synthetic polymer which is used to make clothing.

a Terylene is made by condensation polymerisation. Explain the meaning of the term condensation polymerisation. **[3]**

A reaction where two organic molecules join together **[1]** with the elimination of a small molecule **[1]** to form the repeating units of a polymer **[1]**.

b Write the formulae for the two monomers used to make terylene. Show only the functional groups joined to a square representing the rest of the molecule. **[2]**

HOOC–□–COOH **[1]** HO–□–OH **[1]**

c A form of terylene containing an ester linkage can be made from a monomer with the following structure:

HO–X–COCl where X is the rest of the molecule.

i Draw three units of the polymer formed from this monomer and put a ring around an ester linkage. **[4]**

Three units drawn **[1]** units and bonding correct **[1]** continuation bonds present **[1]** ester link correctly ringed **[1]**.

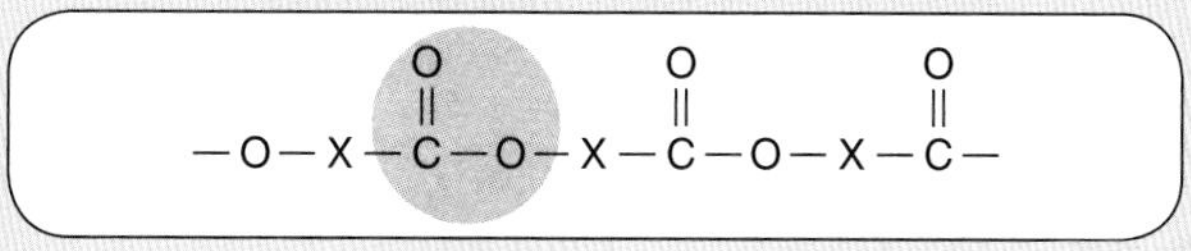

—O—X—C(=O)—O—X—C(=O)—O—X—C(=O)—

Figure 29.10

ii Give the name of the molecule eliminated in this reaction. **[1]**

hydrogen chloride **[1]**

Progress check

29.05 Draw two units of the polymer chain formed by the polymerisation of the monomer with the formula $C_6H_5CH{=}CH_2$. [2]

29.06 Draw the structure of the polymer formed from the monomer $CH_3CH{=}CHCl$, showing one repeat unit and using the letter 'n' to indicate the repetition of this unit. [2]

29.07 Draw a an amide linkage [1] b an ester linkage [1].

29.08 Draw the structure of the monomers of the polymer with the following structure: [2]

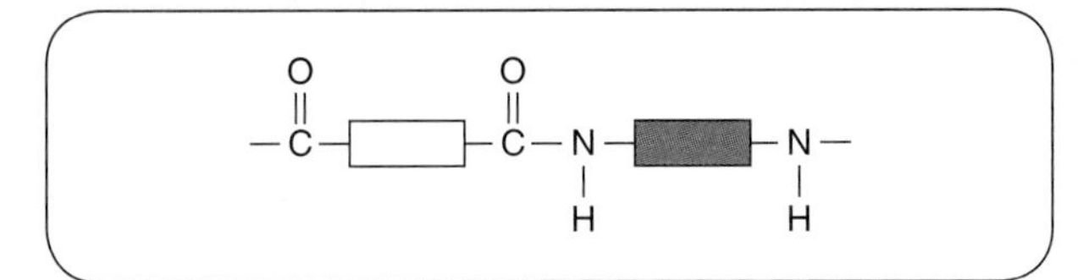

Figure 29.11

29.04 Uses of polymers

- Poly(ethene): plastic bottles, plastic bags, cling film wrap
- Poly(propene): crates for bottles, ropes
- Poly(chloroethene): water pipes, gutters, electrical cable insulation
- Poly(tetrachloroethene): non-stick pans
- Nylon: fishing lines, clothing, carpets, ropes
- Terylene: bedsheets, clothing, towels

29.05 Pollution by plastics

Some plastics are biodegradable. They can be broken down in the environment by bacteria and fungi. Others are **non-biodegradable**. They cannot be broken down in this way. The main environmental problems with non-biodegradable plastics are as follows:

TERMS

Non-biodegradable: Unable to be broken down in the environment by microorganisms or other living things.

- Blocking drains and causing flooding as a result.
- Small pieces getting trapped in the airways or gullets of animals (especially sea birds and fish) leading to death.
- Birds and fish getting trapped in discarded plastic netting. This can cause injury and eventual death.
- When burned they may produce toxic gases. Some of these contribute to acid rain, and others, e.g. carbon dioxide, contribute to global warming. Poisons called dioxins may also be produced on burning.
- Toxic additives to the plastic may leach out of plastic dumped on the soil or in the water and poison organisms in these environments.

Biodegradable plastics can also be harmful to the environment:

- Toxic additives leak out and have the same harmful effects as non-biodegradable plastics.
- They may break down into microscopic particles which are harmful to aquatic life.

Progress check

29.09 Give two harmful effects of burning plastics. [1]

29.10 Give one use of terylene and one use of nylon. [1]

Exam-style questions

Question 29.01

Poly(ethene) is an addition polymer.

a Describe the structure and bonding in poly(ethene). [2]

b i Draw the structural formula showing all atoms and bonds of the monomer used to make poly(ethene). [1]

ii What feature of this molecule makes it possible to undergo addition polymerisation? [1]

c Poly(ethene) is a non-biodegradable plastic. Describe three pollution problems caused by poly(ethene). [3]

Question 29.02

Nylon is a synthetic polymer which is used to make ropes and clothing.

a Nylon is made by condensation polymerisation. Explain the meaning of the term condensation polymerisation. [3]

b Write the formulae for the two monomers used to make nylon. Show only the functional groups joined to a square representing the rest of the molecule. [2]

c A form of nylon can be made from a monomer with the following structure:

NH_2–X–COOH, where X is the rest of the molecule.

i Draw three units of the polymer formed from this monomer and put a ring around the amide linkage. [4]

ii Give the name of the molecule eliminated in this reaction. [1]

Question 29.03

a Describe two differences between addition polymerisation and condensation polymerisation. [2]

b The diagram shows the structure of molecule X.

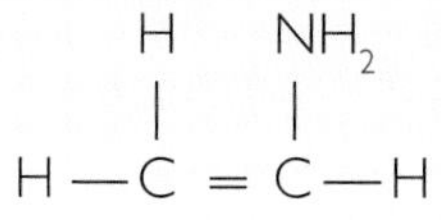

```
   H    NH2
   |    |
H—C  =  C—H   (molecule X)
```

i Explain why X does not undergo condensation polymerisation. [1]

ii X forms an addition polymer. Draw a section of this addition polymer to show three repeat units. [2]

Revision checklist

You should be able to:

- Define polymers as macromolecules built up from smaller units called monomers
- Describe the polymerisation of ethene to form poly(ethene)
- Explain the differences between condensation and addition polymerisation
- Understand that different polymers have different units and/or different linkages (ester, amide)
- Deduce the structure of a polymer from a given alkene monomer
- Deduce the structure of an alkene monomer from the structure of a polymer
- Describe the formation of nylon and terylene by condensation polymerisation
- State some uses of plastics and man-made fibres
- Describe some pollution problems caused by non-biodegradable plastics

Natural macromolecules

Learning outcomes

By the end of this unit you should:

- Know that proteins and carbohydrates are constituents of food
- Be able to describe proteins as possessing an amide link between amino acids
- Be able to describe the simplified structure of proteins using different blocks between the amide links
- Be able to describe complex carbohydrates in terms of a large number of sugar units joined through condensation polymerisation
- Be able to describe the hydrolysis of proteins to amino acids
- Describe the hydrolysis of complex carbohydrates, by acids or enzymes to give simple sugars
- Be able to describe, in outline, the usefulness of chromatography in separating and identifying the products of hydrolysis of carbohydrates and proteins

30.01 Proteins and carbohydrates

TERMS

Carbohydrates: The general name for simple and complex sugars having the general formula $C_x(H_2O)_y$.

Proteins: Condensation polymers formed from amino acids which are joined by peptide (amide) groups.

- Two of the most important components of food are **carbohydrates** and **proteins**.
- Carbohydrates such as glucose provide us with energy. They are converted into carbon dioxide and water in the body by a complex series of reactions (respiration). Energy is released in respiration to make muscles work and drive reactions in the body.
- Proteins are needed for growth and other essential functions in the body. The enzyme catalysts in our bodies are all proteins.

30.02 The structure of proteins

TERMS

Amino acids: Small molecules containing one or more NH_2 and COOH functional groups.

- Proteins are polymers made up from **amino acid** monomers.
- Amino acids are small molecules containing one or more NH_2 and COOH functional groups.
- There are 20 different amino acids that are present in most proteins. These have different side chains. Two of these amino acids are shown in Figure 30.01.

alanine	serine
$H_2N-CH(CH_3)-COOH$	$H_2N-CH(CH_2OH)-COOH$
side chain: CH_3	side chain: CH_2OH

Figure 30.01

- Proteins are condensation polymers formed from amino acids.
- Water is the small molecule eliminated.
- The amino acids form amide linkages as shown in Figure 30.02.

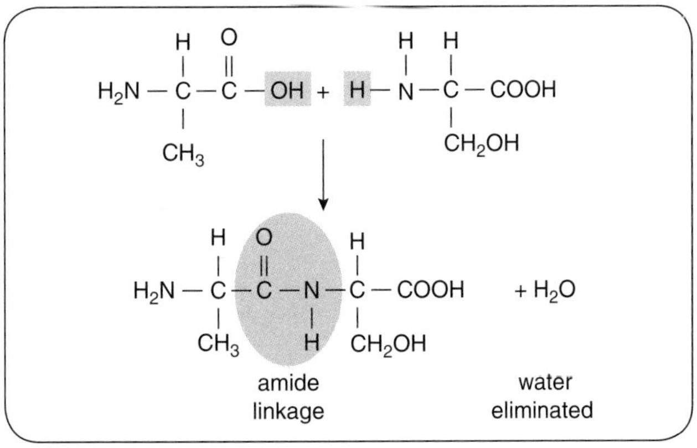

$H_2N-CH(CH_3)-CO-OH + H-NH-CH(CH_2OH)-COOH \rightarrow H_2N-CH(CH_3)-CO-NH-CH(CH_2OH)-COOH + H_2O$

amide linkage

water eliminated

Figure 30.02

- We can describe the structure of proteins using a 'block diagram' like the one shown in Figure 30.03 which does not distinguish the side chains of the amino acids.

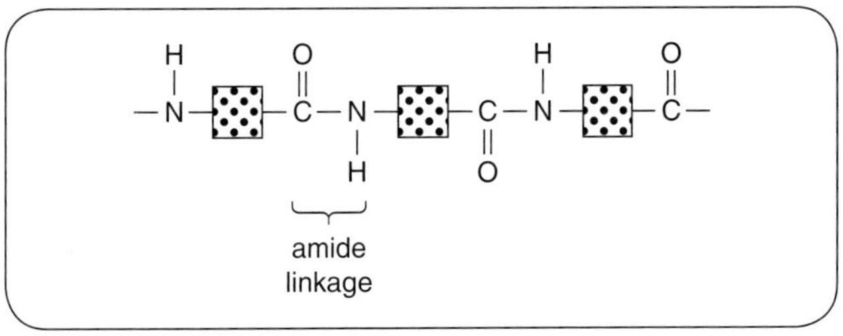

$-NH-\square-CO-NH-\square-CO-NH-\square-CO-$

amide linkage

Figure 30.03

TIP

You do not have to know the detailed structure of amino acids or proteins but you should recognise the repeating units and the amide link.

Worked example 30.01

Describe the polymerisation of amino acids to make proteins. In your answer describe the simplified structure of amino acids and explain the polymerisation reaction in simple terms. **[8]**

Remember that the polymerisation is similar to that of amines with carboxylic acids. Make sure that you name the groups and give the formulae of the functional groups which react. In answering this type of question, you should also write about the type of reaction and mention the products formed.

Amino acids have an amine group **[1]** NH_2 **[1]** and a carboxylic acid group **[1]** COOH **[1]** in the same molecule. The NH_2 of one amino acid reacts with the COOH group of another amino acid **[1]**. The type of polymerisation reaction is condensation polymerisation **[1]**. Water **[1]** is eliminated **[1]**.

Progress check

30.01 State the function of proteins and carbohydrates in the body. [2]

30.02 Name the linkage between the monomer units in proteins. [1]

30.03 A monomer has the simplified structure H_2N–□–COOH. Give the name of this monomer. [1]

30.04 The formation of proteins from amino acids is a condensation reaction. Give the meaning of the term *condensation reaction*. [2]

30.03 The structure of carbohydrates

- Carbohydrates are compounds of carbon, hydrogen and oxygen having the general formula $C_x(H_2O)_y$.
- Simple carbohydrates (simple sugars or monosaccharides) include glucose, $C_6H_{12}O_6$, and fructose.
- Complex carbohydrates (polysaccharides) such as starch and cellulose are polymers formed from simple sugar monomers.
- Complex carbohydrates are usually made up of the same monomers (unlike proteins which have 20 different monomers).
- Water is the small molecule eliminated when simple sugars polymerise:

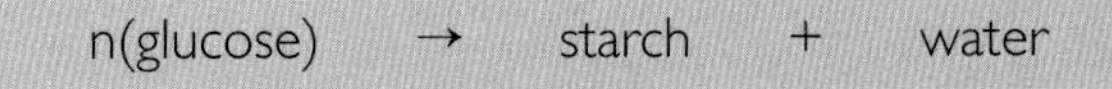

- The simple sugars form –O– linkages as shown in Figure 30.04. Note how we represent the simple sugar molecule.

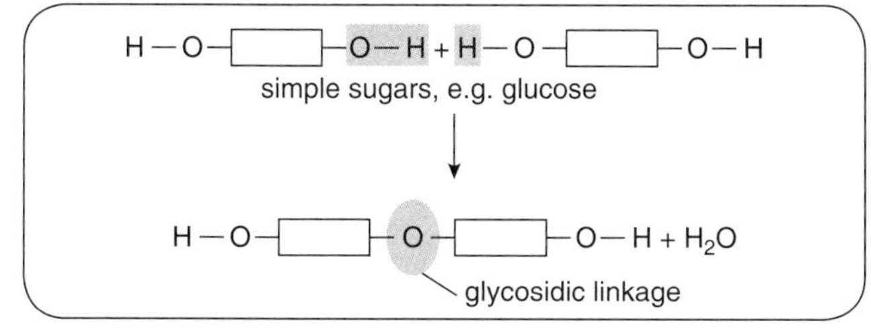

Figure 30.04

- We can describe the structure of complex carbohydrates using a 'block diagram' like the one shown in Figure 30.05 which does not show the structure of the sugars.

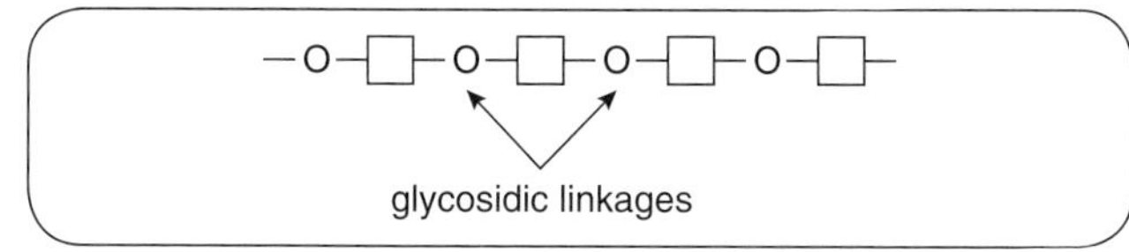

Figure 30.05

30.04 The hydrolysis of proteins and carbohydrates

TERMS

Hydrolysis: The breakdown of a compound by water (often catalysed by acid or alkali).

- **Hydrolysis** is the breakdown of a compound by water.
- Hydrolysis can be speeded up by the addition of an acid or alkali.
- Proteins are hydrolysed to amino acids by refluxing with 6 mol/dm^3 hydrochloric acid for several hours. The condenser is in the reflux position (vertical) to condense the hydrochloric acid vapour which would otherwise escape from the flask (Figure 30.06).

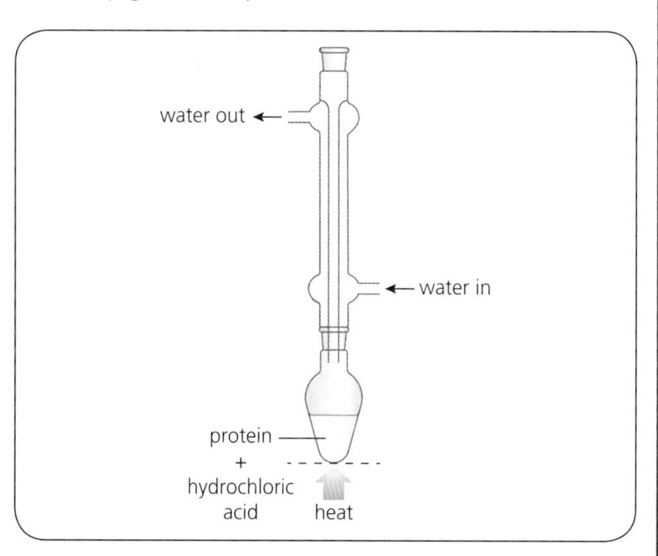

Figure 30.06

- After the reaction has finished the excess acid is neutralised with aqueous ammonia.
- The reaction can be represented as: protein + water → amino acids or as shown in Figure 30.07.

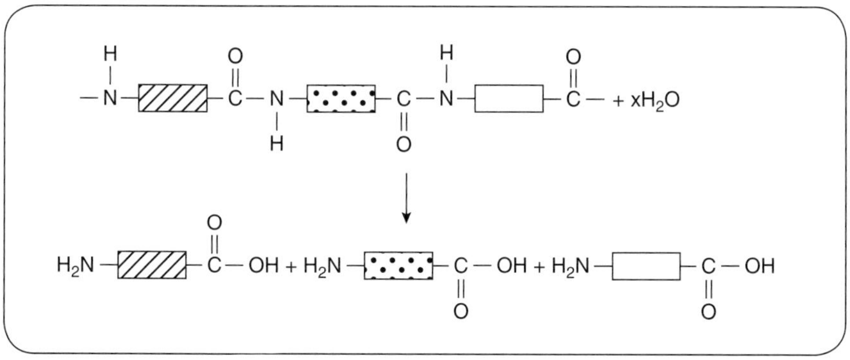

Figure 30.07

TERMS

Complex carbohydrates (polysaccharides): Polymers formed from simple carbohydrate monomers, e.g. starch.

- **Complex carbohydrates** such as starch are also hydrolysed to simple sugars such as glucose by refluxing with moderately concentrated hydrochloric acid:

starch + water → glucose

Figure 30.08 shows a block diagram of this process.

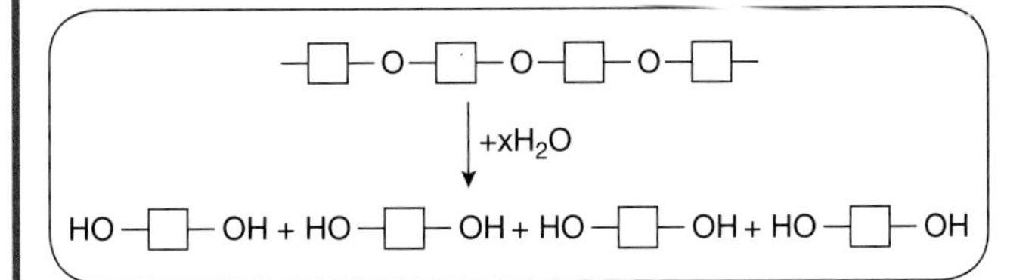

Figure 30.08

- Enzymes can also be used to hydrolyse proteins and complex carbohydrates. These reactions occur at room temperature and occur naturally in living organisms.

For the optimum conditions for the use of enzymes see Unit 12.

30.05 Identifying hydrolysis products

- Paper chromatography is used to identify the hydrolysis products of proteins or complex carbohydrates.
- Two-dimensional paper chromatography is used. This is because some amino acids and some simple carbohydrates have similar R_f values.
- The chromatography is carried out in one direction, the paper is turned round 90° and chromatography carried out again using a different solvent (see Figure 30.09).
- The chromatogram is allowed to dry and then sprayed with a locating agent (see Unit 3).
- Coloured spots are developed after warming gently.
- The R_f value in each solvent is characteristic of each amino acid or simple sugar.

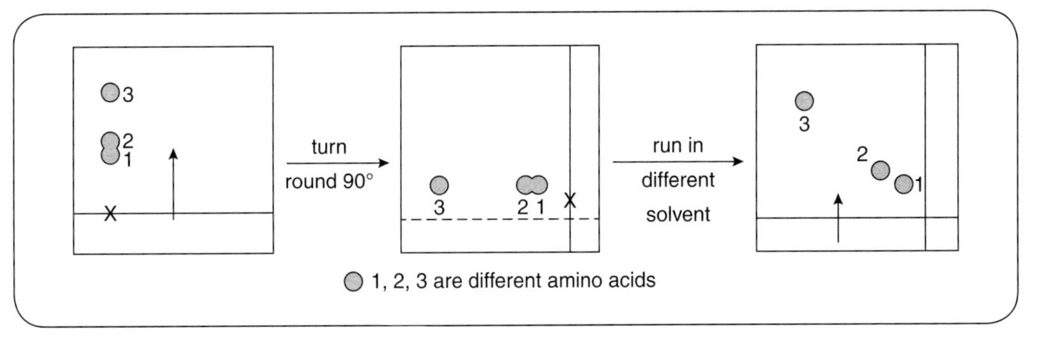

Figure 30.09

TIP

You do not have to know the names of specific locating agents used to show the presence of amino acids or simple sugars.

Sample answer

Starch is a polymer of glucose.

a Write the formula of the functional group present in glucose. **[1]**

−OH **[1]**

b Draw a simplified diagram of starch. **[2]**

–□–O–□–O–□–O–□–O– (**1 mark** for O link, **1 mark** for at least three units with continuation bonds)

c Specific enzymes hydrolyse starch to glucose.

i What is meant by the term *hydrolyse*? **[1]**

Breaking down a compound by water / acid / alkali. **[1]**

ii Write a word equation for the hydrolysis of starch. **[1]**

starch → glucose + water **[1]**

iii Describe the reagents and conditions needed for the hydrolysis of starch using inorganic reagents. **[3]**

Heat / reflux **[1]** with moderately concentrated **[1]** hydrochloric acid **[1]**.

Progress check

30.05 Write a word equation for the polymerisation of glucose (a simple sugar). [3]

30.06 Write a word equation for the hydrolysis of proteins. [2]

30.07 A polymer has the structure:
–□–O–□–O–□–O–□–O–.
Is this polymer a protein, starch or terylene? Explain your answer. [2]

30.08 Give the meaning of the term *locating agent.* [2]

Exam-style questions

Question 30.01

Proteins are polymers of amino acids.

a Write the formulae of the two functional groups present in every amino acid. [2]

b State the name and draw the structure of the linkage between the amino acids in a protein. [2]

c Proteases are enzymes which hydrolyse proteins to amino acids.

i What is meant by the term *hydrolyse*? [1]

ii Describe the conditions needed for the hydrolysis of proteins using inorganic reagents. [2]

iii Aqueous ammonia is added to the mixture obtained by hydrolysis. Explain why. [1]

d The mixture of amino acids is separated by paper chromatography. First in one direction, then in another using a different solvent. Explain why a different solvent is used. [2]

e A locating agent is sprayed on the chromatogram. Explain why a locating agent is used. [3]

Question 30.02

Starch is a complex carbohydrate. The diagram represents a molecule of starch.

–□–O–□–O–□–O–□–O–

a Draw the structure of the monomer used to make starch. [1]

b Starch can be hydrolysed by enzymes.

i To which class of food substances do enzymes belong? [1]

ii Explain why enzymes speed up a reaction. [2]

c Hydrolysis can be carried out by refluxing with hydrochloric acid. The condenser is attached to the flask in a vertical position. Explain why the hydrolysis is carried out in this way. [2]

d The monomers can be identified by paper chromatography. Explain how they can be identified. [4]

Revision checklist

You should be able to:

- Name proteins and carbohydrates as constituents of food
- Describe proteins as having an amide link between amino acids
- Describe the structure of proteins
- Describe complex carbohydrates in terms of a large number of sugar units joined together by condensation polymerisation
- Describe the hydrolysis of proteins to amino acids
- Describe the hydrolysis of complex carbohydrates by acids or enzymes to give simple sugars
- Describe, in outline, the usefulness of chromatography in separating and identifying the products of hydrolysis of carbohydrates and proteins

Appendix: The Periodic Table

alkali metals | halogens | noble gases

Group I	Group II											Group III	Group IV	Group V	Group VI	Group VII	Group VIII
							1 **H** Hydrogen 1					←		non metals		→	2 **He** Helium 4
3 **Li** Lithium 7	4 **Be** Beryllium 9											5 **B** Boron 11	6 **C** Carbon 12	7 **N** Nitrogen 14	8 **O** Oxygen 16	9 **F** Fluorine 19	10 **Ne** Neon 20
11 **Na** Sodium 23	12 **Mg** Magnesium 24											13 ***Al*** Aluminium 27	14 **Si** Silicon 28	15 **P** Phosphorus 31	16 **S** Sulfur 32	17 ***Cl*** Chlorine 35.5	18 **Ar** Argon 40
19 **K** Potassium 39	20 **Ca** Calcium 40	21 **Sc** Scandium 45	22 **Ti** Titanium 48	23 **V** Vanadium 51	24 **Cr** Chromium 52	25 **Mn** Manganese 55	26 **Fe** Iron 56	27 **Co** Cobalt 59	28 **Ni** Nickel 59	29 **Cu** Copper 64	30 **Zn** Zinc 65	31 **Ga** Gallium 70	32 **Ge** Germanium 73	33 **As** Arsenic 75	34 **Se** Selenium 79	35 **Br** Bromine 80	36 **Kr** Krypton 84
37 **Rb** Rubidium 85	38 **Sr** Strontium 88	39 **Y** Yttrium 89	40 **Zr** Zirconium 91	41 **Nb** Niobium 93	42 **Mo** Molybdenum 96	43 **Tc** Technetium –	44 **Ru** Ruthenium 101	45 **Rh** Rhodium 103	46 **Pd** Palladium 106	47 **Ag** Silver 108	48 **Cd** Cadmium 112	49 **In** Indium 115	50 **Sn** Tin 119	51 **Sb** Antimony 122	52 **Te** Tellurium 128	53 **I** Iodine 127	54 **Xe** Xenon 131
55 **Cs** Caesium 133	56 **Ba** Barium 137	57 * **La** Lanthanum 139	72 **Hf** Hafnium 179	73 **Ta** Tantalum 181	74 **W** Tungsten 184	75 **Re** Rhenium 186	76 **Os** Osmium 190	77 **Ir** Indium 192	78 **Pt** Platinum 195	79 **Au** Gold 197	80 **Hg** Mercury 201	81 ***Tl*** Thallium 204	82 **Pb** Lead 207	83 **Bi** Bismuth 209	84 **Po** Polonium 209	85 **At** Astatine 210	86 **Rn** Radon 222
87 **Fr** Francium 223	88 **Ra** Radium 226	89 † **Ac** Actinium 227															

metalloids

← metals →

*58–71 Lanthanoid series
†90–103 Actinoid series

58 **Ce** Cerium 140	59 **Pr** Praseodymium 141	60 **Nd** Neodymium 144	61 **Pm** Promethium 145	62 **Sm** Samarium 150	63 **Eu** Europium 152	64 **Gd** Gadolinium 157	65 **Tb** Terbium 159	66 **Dy** Dysprosium 163	67 **Ho** Holmium 165	68 **Er** Erbium 167	69 **Tm** Thulium 169	70 **Yb** Ytterbium 173	71 **Lu** Lutetium 175
90 **Th** Thorium 232	91 **Pa** Protactinium 231	92 **U** Uranium 238	93 **Np** Neptunium 237	94 **Pu** Plutonium 244	95 **Am** Americium 243	96 **Cm** Curium 247	97 **Bk** Berkelium 247	98 **Cf** Californium 251	99 **Es** Einsteinium 252	100 **Fm** Fermium 257	101 **Md** Mendelevium 258	102 **No** Nobelium 259	103 **Lr** Lawrencium 262

Key: a **X** b

a = atomic number
X = atomic symbol
b = relative atomic mass

Answers to Progress Check questions

Unit 1

1.01 Solids keep their own shape **(1)** and keep the same volume **(1)**.

Liquids have no fixed shape / they take the shape of their container **(1)**. They have a fixed volume **(1)**.

Gases have no fixed shape **(1)** and no fixed volume **(1)** (they expand to fit the shape of any closed container in which they are placed).

1.02 Solids: particles close together / touching **(1)** particles only vibrate **(1)**. Liquids: particles close together / touching **(1)**, motion is sliding over each other / moving slowly **(1)**.

1.03 **a** Vapour: particles are far apart **(1)** and irregularly / randomly arranged **(1)**

b Solid: particles are close together **(1)** and regularly arranged **(1)**

1.04 There is little / no space between the particles **(1)** particles are imagined as hard spheres **(1)** (in the simple kinetic theory).

1.05 **a** condensation **(1)** **b** evaporation **(1)** **c** sublimation **(1)**

1.06 increases **(1)**; boiling **(1)**; energy **(1)**; attractive **(1)**; molecules **(1)**; constant **(1)**

1.07 Brownian motion: **(1)** random movement **(1)** of suspensions of visible particles **(1)**.

OR

Diffusion **(1)** random movement of particles **(1)** in liquids / gases **(1)**.

1.08 Diffusion is the random movement **(1)** of molecules in liquids and gases from higher to lower concentrations **(1)** leading to mixing and spreading of the particles throughout the mixture **(1)** because they continually collide with each other **(1)**.

1.09 H_2 has a lower relative molecular mass than H_2S **(1)**. Molecules with lower relative molecular masses move faster than those with higher relative molecular masses **(1)**.

Unit 2

2.01 840 g **(1)**

2.02 0.023 kg **(1)**

2.03 0.036 dm^3 **(1)**

2.04 450 cm^3 **(1)**

2.05 Burette **(1)**

2.06 Gas syringe **(1)** upturned measuring cylinder or burette filled with water **(1)**.

2.07 Measuring cylinder is not accurate enough **(1)** because it does not measure to nearest 0.1 cm^3 **(1)**.

2.08 An experiment where the independent variable affects the dependent variable / experiment where one of the variables is given particular values and another is dependent on these values **(1)**. All other variables are controlled **(1)**.

2.09 Volume at 30 s **(1)**. The values are generally increasing by 5 cm^3 every 10 s **(1)** apart from the volume at 30 s which does not fit the pattern of the rest of the data **(1)**.

Unit 3

3.01 Impure; because it does not have a sharp melting point **(1)**.

3.02 Impurities may harm skin **(1)**.

3.03 Impurities lower the melting point **(1)**. Salt makes the water impure / mixture of ice and salt on the road **(1)** so water does not freeze / water less likely to freeze **(1)**.

3.04 So the ink does not run up the chromatography paper / so the ink does not interfere with the spots / you might not be able to see some of the spots **(1)**.

3.05 (See middle diagram of Figure 3.01)

Chromatography paper correctly positioned in beaker **(1)**.

Bottom of paper dipping into solvent **(1)** Solvent below the spot **(1)**.

3.06 $18 \div 45 = 0.4$ **(1)**

3.07 Simple distillation **(1)**

3.08 Fractional distillation **(1)**

3.09 Filtration **(1)**

3.10 Chromatography **(1)**

Unit 4

4.01 Relative masses: proton = 1, neutron = 1, electron = 0.00054 or 1/1836

Relative charges: proton = +1, neutron = uncharged, electron = −1 (**1 mark** each)

4.02 Atoms of the same element which have the same proton number but a different nucleon number **(1)**.

4.03 Nucleon number − proton number = 54 − 26 = 28 **(1)**.

4.04 S has 32 − 16 = 16 neutrons **(1)** so in one molecule there are 8 × 16 = 128 neutrons **(1)**.

4.05 **a** 2,8,4 **b** 2,8,8,2 **c** 2,7 (**1 mark** each)

4.06 8 electrons

4.07 **a** Al^{3+} is 2,8 (3 electrons lost from 2,8,3 structure) **(1)**

O^{2-} is 2,8 (2 electrons gained to 2,6 structure) **(1)**

b 8 electrons in outer shell **(1)** stable noble gas structure **(1)**

Unit 5

5.01 B and C (**1 mark** each)

5.02 An element is a substance containing only one type of atom. **(1)**

5.03 Any two of (**1 mark** each for the compound):

compound: fixed composition / cannot be separated by physical means / formed by chemical change / heat changes on formation / physical properties are different from elements forming them.

Any two of (**1 mark** each for the mixture):

mixture: no fixed composition / can be separated by physical means / no chemical change / not usually heat changes on formation / physical properties are the average of the elements forming them.

5.04 Salt and water can be separated by physical means **(1)** AND by distillation **(1)** the water evaporates and condenses (in the condenser) **(1)**. The salt remains in the distillation flask **(1)**.

OR

By heating the solution **(1)** the water evaporates **(1)** leaving the salt in the evaporating basin **(1)**.

5.05 Any two of (**1 mark** each): does not conduct heat (or electricity) / dull surface / brittle / not sonorous ALLOW: low melting or boiling point / low density.

5.06 Any two of: lithium, sodium, potassium, rubidium, caesium, gallium, lead (**1 mark** each).

5.07 Carbon as graphite **(1)**.

5.08 It contains metals / non-metals other than iron / it is a mixture of iron with other elements **(1)**.

Unit 6

6.01 **a** Bromide ion formed by gain of electron **(1)** potassium ion formed by loss of electron **(1)**.

b $Li \rightarrow Li^{+} + e^{-}$

(**1 mark** for Li and **1 mark** for + e^{-} (or + e))

6.02 **a**

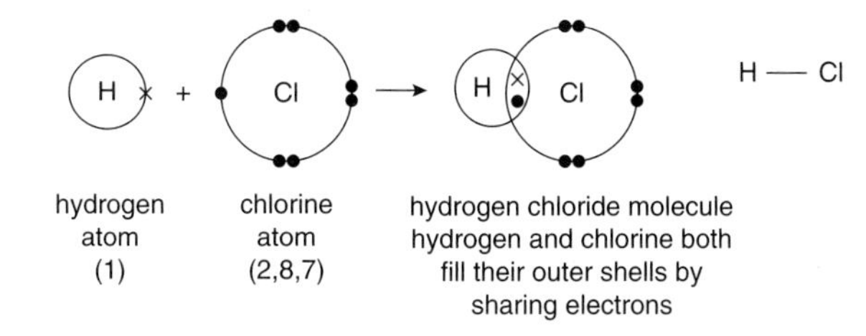

2 marks if all the structure is correct. If two marks are not scored **1 mark** if the bonding electron pair between the H and Cl is shown.

b

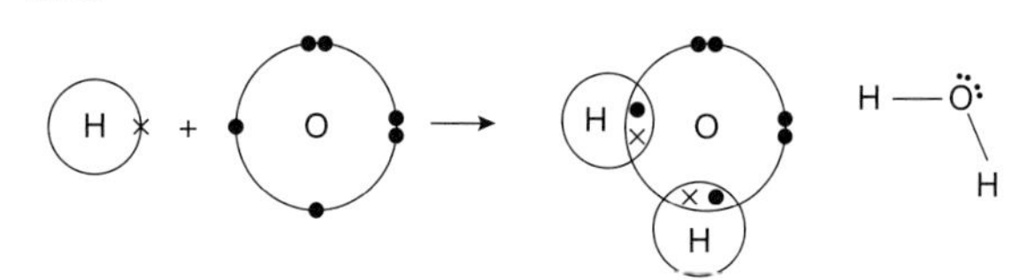

3 marks if the structure is totally correct

2 marks if the lone pairs are missing

1 mark if the lone pairs of electrons on the oxygen atoms are missing but there is a pair of electrons between one of the H and O atoms.

6.03 (Giant) ionic **(1)**

6.04 Volatility is high / low boiling point **(1)** does not conduct **(1)** insoluble in water / soluble in organic solvents **(1)**.

6.05 Graphite **(1)** giant **(1)** covalent **(1)**

6.06 Graphite has layers **(1)**, the layers can slide over each other **(1)**.

Diamond: all bonds are strong / lots of strong bonds **(1)**, it takes a lot of energy to break these bonds / it takes a high temperature to break these bonds **(1)**.

6.07 Magnesium oxide is ionic **(1)**. The attractive forces between the ions are strong. ALLOW: The bonding between the ions is strong **(1)**. It takes a lot of energy / high temperature to overcome the attractive forces between the ions **(1)**.

Sulfur dioxide has a simple molecular structure / is a molecule **(1)**. The forces between the molecules are weak / intermolecular forces are weak **(1)**. So it does not take much energy / only needs a low temperature to overcome these forces **(1)**.

6.08 Metallic bonding is (due to) the attractive forces **(1)** between positive metal ions **(1)** and the delocalised electrons **(1)** in a metallic lattice. Metals conduct because the delocalised electrons **(1)** are free to move throughout the whole structure **(1)**.

Unit 7

7.01 a Al_2O_3 **(1)**

b H_2S **(1)**

c $Mg(NO_3)_2$ **(1)**

d $SiCl_4$ **(1)**

7.02 a K_2SO_4 **(1)**

b $MgCO_3$ **(1)**

c $Al(NO_3)_3$ **(1)**

7.03 a C_2H_4 **(1)**

b $C_2H_2ClF_3$ (allow any order) **(1)**

c KI **(1)**

7.04 a $(NH_4)_2SO_4$ **(1)**

b $Ce(NO_3)_3$ **(1)**

c $Fe(NO_3)_2$ **(1)**

d NH_4NO_3 **(1)**

7.05 a copper(II) nitrate + magnesium → copper + magnesium nitrate **(1)**

b calcium oxide + sulfuric acid → calcium sulfate + water **(1)**

c chlorine + potassium bromide → bromine + potassium chloride **(1)**

d ammonia + sulfuric acid → ammonium sulfate **(1)**

7.06 a $2Mg + O_2 \rightarrow 2MgO$ **(1)**

b $4P + 3O_2 \rightarrow 2P_2O_3$ **(1)**

c $2Na + 2H_2O \rightarrow 2NaOH + H_2$ **(1)**

d $4Na + O_2 \rightarrow 2Na_2O$ **(1)**

e $2SO_2 + O_2 \rightarrow 2SO_3$ **(1)**

7.07 a $Ca(s) + 2HCl(aq) \rightarrow CaCl_2(aq) + H_2(g)$

(Correct reactants and products **(1)**, correct balance **(1)**, correct state symbols **(1)**)

b $2K(s) + Br_2(g) \rightarrow 2KBr(s)$

(Correct reactants and products **(1)**, correct balance **(1)**, correct state symbols **(1)**)

c $4Al(s) + 3O_2(g) \rightarrow 2Al_2O_3(s)$

(Correct reactants and products **(1)**, correct balance **(1)**, correct state symbols **(1)**) d $H_2SO_4(aq) + 2NH_3(aq) \rightarrow (NH_4)_2SO_4(aq)$

(Correct reactants and products (**1**), correct balance (**1**), correct state symbols (**1**))

e $MgCO_3(s) + 2HCl(aq) \rightarrow MgCl_2(aq) + CO_2(g) + H_2O(l)$

(Correct reactants and products (**1**), correct balance (**1**), correct state symbols (**1**))

7.08 a $Mg + 2H^+ \rightarrow Mg^{2+} + H_2$

(Correct reactants and products (**1**), correct balance (**1**))

b $Cl_2 + 2Br^- \rightarrow Br_2 + 2Cl^-$

(Correct reactants and products (**1**), correct balance (**1**))

c $Cr^{3+} + 3OH^- \rightarrow Cr(OH)_3$

(Correct reactants and products (**1**), correct balance (**1**))

Unit 8

8.01 a 110 **b** 44 **c** 100 **d** 148 **e** 342 (**1 mark** each)

8.02 a 0.4 **b** 2 **c** 0.05 **d** 0.1 **e** 0.3 (**1 mark** each)

8.03 a 30 g **b** 14 g **c** 34.8 g **d** 1920 g **e** 52.8 g (**1 mark** each)

8.04 $2Na \rightarrow Na_2O_2$

$2 \times 23 \quad 78$ (**1**)

$6.9 \times \frac{78}{46} = 11.7\,g$ (**1**)

8.05 $SnO_2 + 2C$

$151 \quad 2 \times 12$ (**1**)

$28 \times \frac{2 \times 12}{151} = 4.45\,g$ (**1**)

8.06 mol methane $= \frac{60}{16} = 3.75\,mol$ (**1**)

two mols O_2 react with methane $= 2 \times 3.75 = 7.5\,mol$ (**1**)

mol oxygen $\times 24 = 180\,dm^3$ (**1**)

8.07 $Sn = \frac{45.6}{119} = 0.383$, $Cl = \frac{54.4}{35.5} = 1.53$ (**1**)

Divide by the lowest value, 0.383 (**1**)

gives empirical formula $SnCl_4$ (**1**)

8.08 a $CaCO_3 = \frac{7}{100}\,mol = 0.07\,mol$ (**1**)

So $HCl = 2 \times 0.07 = 0.14\,mol$ (**1**)

b 2 mol $HCl \rightarrow 1$ mol CO_2 (**1**)

0.25 mol $\rightarrow$ 0.125 mol CO_2 (**1**)

$0.125 \times 24 = 3\,dm^3\ CO_2$ (**1**)

8.09 a mol, Fe $= \frac{6.4}{56} = 0.114\,mol$ (**1**)

mol $H_2SO_4 = 0.10$ (**1**) so H_2SO_4 is limiting reactant

b 0.1 mol $H_2SO_4 \rightarrow$ 0.1 mol H_2 (**1**)

$0.1 \times 24 = 2.4\,dm^3$ (**2**)

Unit 9

9.01 a $\frac{4}{40} \times \frac{1000}{250} = 0.4\,mol/dm^3$ (**1**)

b $\frac{22.2}{111} \times \frac{1000}{360} = 0.56\,mol/dm^3$ (**1**)

c $\frac{6}{60} \times \frac{1000}{125} = 0.8\,mol/dm^3$ (**1**)

9.02 a $0.5 \times 2 = 1\,mol\ (\times 58.5) = 58.5\,g$ (**1**)

b $0.1 \times \frac{50}{1000} = 0.005\,mol\ (\times 160) = 0.8\,g$ (**1**)

c $1.5 \times \frac{100}{1000} = 0.15\,mol\ (\times 148) = 22.2\,g$ (**1**)

9.03 a $\frac{2}{0.5} = 4\,dm^3 / 4000\,cm^3$ (**1**)

b $\frac{0.05}{0.8} = 0.0625\,dm^3 / 62.5\,cm^3$ (**1**)

9.04 a aqeous sodium chloride **b** water **c** sodium chloride

(**2 marks** if all three correct, **1 mark** if one correct)

9.05 (point at which) an indicator (**1**) changes colour (**1**)

9.06 2nd, 3rd and 4th (**1**)

These are within $0.1\,cm^3$ of each other / very close to each other / the 1st and 5th are not as close to the others (**1**).

Rough titration is (usually) inaccurate (**1**).

9.07 a mol KOH $= 0.120 \times \frac{25}{1000} = 0.003\,mol$ (**1**)

b 0.003 mol (because the mole ratio in the equation is 1:1)

c concentration of $HCl = 0.003 \times \frac{1000}{42.5} = 0.071\,mol/dm^3$ (**1**)

Unit 10

10.01 Aluminium is a good conductor of electricity (**1**) and has low density (**1**). The steel core gives strength (**1**) (so that the cable does not break under its own weight).

10.02 Insulator: A substance which does not conduct electricity (**1**).

Electrolysis: The decomposition / breakdown of an ionic compound (**1**), molten or in aqueous solution, by an electric current (**1**).

Anode: The positive electrode (**1**).

Electrolyte: A compound containing ions which conducts electricity (**1**).

10.03 **a** anode: oxygen (**1**), cathode: hydrogen (**1**)

b anode: chlorine (**1**), cathode: hydrogen (**1**)

c anode: bromine (**1**), cathode: lithium (**1**)

d anode: chlorine (**1**), cathode: hydrogen (**1**)

10.04 Hydrogen produced at the cathode from H^+ ions in the acid (**1**) since these are the only positive ions present (**1**).

Oxygen produced at the anode (**1**) since hydroxide ions form from the ionisation of water (**1**) lower in the discharge series / reactivity than sulfate ions (**1**).

Hydroxide ions from self-ionisation of water used in the electrolysis (**1**), leaving sulfate ions and hydrogen ions in solution (**1**).

10.05 In each of these questions there is **1 mark** for the correct formulae of reactants and products and **1 mark** for the correct balance.

a anode: $2Cl^- \rightarrow Cl_2 + 2e^-$ cathode: $2H^+ + 2e^- \rightarrow H_2$

b anode: $4OH^- \rightarrow O_2 + 2H_2O + 4e^-$ cathode: $2H^+ + 2e^- \rightarrow H_2$

c anode: $2Cl^- \rightarrow Cl_2 + 2e^-$ cathode: $Mg^{2+} + 2e^- \rightarrow Mg$

d anode: $2O^{2-} \rightarrow O_2 + 4e^-$ cathode: $Al^{3+} + 3e^- \rightarrow Al$

10.06 Hydrogen produced at the cathode (**1**) since hydrogen ions from ionisation of water (**1**) lower in the discharge series / reactivity than sodium ions (**1**).

Oxygen produced at the anode (**1**), since hydroxide ions from ionisation of water (**1**) lower in the discharge series / reactivity (**1**) than sulfate ions.

10.07 Any two of: nickel cathode (**1**); copper anode (**1**); electrolyte soluble copper salt in aqueous solution (**1**); idea of complete circuit with power pack / batteries (**1**).

10.08 Any two of (**1 mark** each): prevent corrosion / prevent scratching or harden surface / make object look nice or shiny.

10.09 Magnesium and copper (**1**) greatest difference in reactivity (**1**).

10.10 Electrochemical cell produces electricity and electrolysis needs electricity (**1**).

Electrochemical cell: negative electrode releases electrons to external circuit and electrolysis: negative electrons accepts electrons from external circuit (**1**).

Unit 11

11.01 **a** Endothermic (**1**) heat energy needed (**1**) (to change the reactant into products).

b Exothermic (**1**) paraffin is a fuel (**1**) fuels release energy when burnt (**1**)

11.02 release; surroundings; endothermic; heat; combustion (or burning); decomposition (**1 mark** each)

11.03 $3 \times 805 = 2415$ kJ (**1**)

11.04 Energy absorbed in bond breaking (**1**) energy released on bond making (**1**) more energy is released than absorbed (**1**).

11.05 Uranium-235 (**1**)

11.06 Non-polluting / does not produce nitrogen oxides / small amount of material produces a large amount of energy (**1**).

11.07 Oxygen and hydrogen react in the cell (**1**) to form water (**1**). The electrons flow round the external circuit from the negative to the positive electrode (**1**) to produce electrical power.

11.08 Not polluting / does not produce carbon dioxide (**1**) gives off more energy per gram of fuel (**1**).

Unit 12

12.01 Change in concentration or mass of reactant or product **(1)** divided by time **(1)**.

12.02 Any two of the following pairs:

Measure increasing volume of gas **(1)** using gas syringe **(1)**.

Measure increasing volume of gas **(1)** using upturned measuring cylinder full of water **(1)**.

Measure decrease in total mass **(1)** of the reactants **(1)**.

12.03 Initial rate of reaction is faster **(1)**, then rate of reaction decreases **(1)**.

12.04 Particles of flour dust are very small **(1)** so they have a relatively large surface area per unit volume **(1)** rate of reaction fast enough / high enough to be explosive **(1)**.

12.05 **a** decreases reaction rate **(1)** **b** increases reaction rate **(1)** **c** increases reaction rate **(1)**

12.06 Reaction rate faster with powder **(1)**. The powder has a large surface area / volume ratio **(1)**.

12.07 **a** The higher the temperature, the faster the particles move / particles have more energy **(1)**. At a higher temperature more particles have energy greater **(1)** than the activation energy **(1)**.

b The particles are closer together at higher pressure **(1)** there are more particles per unit volume **(1)** so the frequency of collisions (rate of collisions) is greater **(1)** leading to a faster reaction.

12.08 **a** Measuring a decrease in acidity **(1)** since hydrochloric acid is acidic but the product water is not acidic and magnesium chloride is not very acidic **(1)**.

b Measuring an increase in the volume of oxygen produced **(1)** because oxygen is a gas **(1)**.

12.09 Powder has a larger surface area volume ratio **(1)**. More particles exposed on the surface **(1)** so there are more frequent collisions **(1)**.

12.10 Increase light intensity **(1)** increase concentration of carbon dioxide **(1)**.

12.11 Reaction catalysed by light / reaction is photochemical **(1)**.

Silver ions converted to silver (atoms) **(1)**.

Silver is grey (when particles small) **(1)**.

Unit 13

13.01 **a** salts containing water of crystallisation **(1)**

b A reaction which can go in the forward and in the backward direction **(1)**.

13.02 $NiSO_4.7H_2O \rightarrow NiSO_4 + 7H_2O$ (**2 marks** if all correct; **1 mark** if one error)

13.03 **1 mark** each for any three of: it is dynamic / the forward and backward reaction occur at the same rate / the concentration of reactants and products remain constant / only occurs in a closed system.

13.04 For an exothermic reaction, an increase in temperature shifts the position of equilibrium to the left **(1)**.

13.05 **a** Position of equilibrium moves to the right / more methanol formed **(1)**.

b Position of equilibrium moves to the left (decrease in pressure shifts reaction in direction of larger volume) **(1)**.

c Position of equilibrium moves to the right (to oppose the removal of methanol) **(1)**.

13.06 Increase the concentration of reactants / decrease concentration of products **(1)**; increase the pressure **(1)**; decrease the temperature **(1)**.

Unit 14

14.01 **a** loss of oxygen / ALLOW: decrease in oxidation state / gain of electrons **(1)**

b A reaction in which oxidation and reduction occur together **(1)**.

14.02 **a** reduced because oxygen has been removed / oxidation number of copper decreased

b reduced because oxidation number of oxygen decreases

c oxidised because oxidation number of hydrogen increases

d oxidised because carbon gains oxygen / oxidation number of carbon increases (**1 mark** each)

14.03 a Br^- **b** Zn **c** Fe^{2+} (**1 mark** each)

14.04 a reduction **b** reduction **c** oxidation **d** oxidation (**1 mark** each)

Unit 15

15.01 a weakly acidic **b** strongly basic **c** neutral **d** strongly acidic (**1 mark** each).

15.02 a magnesium sulfate + water

b sodium chloride + carbon dioxide + water

c zinc nitrate + water

d potassium chloride + water

(**1 mark** for each correct product in each question.)

15.03 a $2NH_3 + H_2SO_4 \rightarrow \underline{(NH_4)_2}\,\underline{SO_4}$ **(1)**

b $\underline{H_2SO_4} + 2\underline{KOH} \rightarrow K_2SO_4 + 2H_2O$ **(2)**

c $MgCO_3 + 2HCl \rightarrow \underline{MgCl_2} + \underline{CO_2} + \underline{H_2O}$ **(3)**

d $NH_4Cl + NaOH \rightarrow NH_3 + \underline{NaCl} + \underline{H_2O}$ **(2)**

e $Ba + \underline{H_2SO_4} \rightarrow BaSO_4 + \underline{H_2}$

15.04 Add a drop of solution to universal indicator **(1)** compare the colour with a pH colour chart for universal indicator **(1)**.

15.05 Add an indicator / litmus / methyl orange to the alkali **(1)** then add acid slowly / drop by drop **(1)** until the indicator changes colour **(1)**.

15.06 Calcium carbonate is not acidic or alkaline **(1)** so adding excess will not make the soil greater than pH 7 **(1)**. Calcium oxide reacts with water to form a strongly alkaline solution **(1)** so adding excess calcium oxide makes the soil alkaline **(1)** (which is not a good condition for plant growth).

15.07 a basic **b** acidic **c** acidic **d** basic (**1 mark** each)

15.08 a $HClO_2 \rightarrow H^+ + ClO_2^-$ **(1)**

b $2HClO_2 + Na_2O \rightarrow 2NaClO_2 + H_2O$ (**1 mark** for correct formulae, **1 mark** for correct balance)

c $2HClO_2 + Zn \rightarrow Zn(ClO_2)_2 + H_2$ (**1 mark** for correct formulae, **1 mark** for correct balance)

15.09 a $Al_2O_3 + \underline{2KOH} \rightarrow 2KAlO_2 + \underline{H_2O}$ **(2)**

b $Al_2O_3(s) + 6HCl(l) \rightarrow 2AlCl_3(aq) + 3H_2O(l)$ (**1 mark** for correct formulae, **1 mark** for correct balance)

15.10 a A strong acid completely ionises in water **(1)**. A weak acid partially ionises in water **(1)**.

b Take the same concentration of the strong and weak acid **(1)**. React with magnesium / reactive metal / calcium carbonate **(1)** more bubbles with the strong acid / faster reaction with the strong acid **(1)**.

Unit 16

16.01 Sodium hydroxide, sodium carbonate, sodium oxide (sodium is too reactive to use) (**1 mark** each).

16.02 So that all the (sulfuric) acid is used up **(1)**.

16.03 The reaction is too violent / too dangerous **(1)**.

16.04 a copper(II) oxide + sulfuric acid → copper(II) sulfate + water **(1 mark for correct reactants, 1 mark for correct products)**

b nickel oxide + hydrochloric acid → nickel chloride + water (**1 mark** for correct reactants, **1 mark** for correct products)

c zinc oxide + nitric acid → zinc nitrate + water (**1 mark** for correct reactants, **1 mark** for correct products)

16.05 a iron(II) carbonate + sulfuric acid → iron(II) sulfate + water + carbon dioxide

(**1 mark** for correct reactants, **1 mark** for correct products)

b magnesium carbonate + hydrochloric acid → magnesium chloride + water + carbon dioxide

(**1 mark** for correct reactants, **1 mark** for correct products)

16.06 a titration of ammonia **(1)** with sulfuric acid **(1)**

b add zinc / zinc oxide / zinc hydroxide / zinc carbonate **(1)** with hydrochloric acid

16.07 a lead nitrate **(1)** and any soluble chloride, e.g. sodium chloride or hydrochloric acid **(1)**

b barium chloride / barium nitrate **(1)** and any soluble sulfate, e.g. sodium sulfate or sulfuric acid **(1)**

c silver nitrate **(1)** and any soluble iodide, e.g. potassium iodide **(1)**

16.08 a $Pb^{2+}(aq) + 2Cl^{-}(aq) \rightarrow PbCl_2(s)$ (**1 mark** for formulae, **1 mark** for state symbols)

b $Ba^{2+}(aq) + SO_4^{2-}(aq) \rightarrow BaSO_4(s)$ (**1 mark** for formulae, **1 mark** for state symbols.)

c $Ag^{+}(aq) + I^{-}(aq) \rightarrow AgI(s)$ (**1 mark** for formulae, 1 mark for state symbols)

16.09 **a** insoluble **b** insoluble **c** soluble **d** soluble **e** insoluble (**1 mark** each)

Unit 17

17.01 a flame test **(1)** yellow flame **(1)**

b White precipitate on addition of sodium hydroxide **(1)**, no precipitate on addition of aqueous ammonia **(1)**.

c flame test **(1)** red flame **(1)**

d White precipitate with aqueous ammonia **(1)** which does not dissolve in excess **(1)**.

17.02 a light blue **b** red-brown **c** white (**1 mark** each)

17.03 a light blue **(1)** precipitate **(1)**; in excess ammonia the precipitate dissolves **(1)** to form a dark blue solution **(1)**.

b flame coloured blue-green **(1)**

17.04 a Fe^{3+} ions **(1)** hydrogen gas **(1)**

b iron + hydrochloric acid → iron(II) chloride + hydrogen **(2)** (**1 mark** if one error)

17.05 **a** calcium **(1)** **b** sulfur dioxide **(1)** **c** (calcium) sulfite **(1)** **d** white **(1)** precipitate **(1)**

Unit 18

18.01 Proton number / atomic number **(1)**

18.02 Arsenic / As **(1)**

18.03 Group VIII / noble gases **(1)**

18.04 Number of electrons in outer shell is the same as the Group number **(1)**.

18.05 Elements on the left conduct **(1)** but elements on the right do not **(1)**.

18.06 Basic oxides are metallic oxides so found on the left **(1)**. Non-metal oxides are found on the right and are acidic oxides **(1)**.

18.07 Between metals and non-metals / some elements in Groups III to V / between elements having basic oxides and acidic oxides **(1)**.

18.08 In Groups I to III electrons are lost **(1)** (to form positive ions). The charge on the ions is the same as the Group number **(1)**. Group IV elements do not form ions readily **(1)** (apart from Sn and Pb). In Groups V to VII electrons are gained **(1)** (to form negative ions). The charges on the ions are: Group V −3, Group VI −2 and Group VII −1 **(1)**.

18.09 Ratio of *Cl* to element increases then decreases across the Period **(1)**. For Groups I to IV: $NaCl$, $MgCl_2$, $AlCl_3$, $SiCl_4$ **(1)**. For Groups V to VII PCl_3, SCl_2, $ClCl$ **(1)**.

Unit 19

19.01 caesium + water → caesium hydroxide + hydrogen (**1 mark** for caesium hydroxide, **1 mark** for hydrogen)

19.02 **a** 2,8,8,1 **(1)** **b** 2,8,8 **(1)**

19.03 Down the group: Melting point decreases **(1)** density increases **(1)** hardness decreases / gets softer **(1)**.

19.04 **1 mark** each for any three of: bubbles / metal disappears / metal goes into a ball / moves (about) on the surface of the water.

19.05 $2Rb + 2H_2O \rightarrow 2RbOH + H_2$ (**1 mark** for correct formulae, **1 mark** for correct balance)

19.06 **1 mark** each for any two of: are good catalysts / do not react with cold water / forms ions with different charges.

19.07 **1 mark** each for any three of: high melting points / boiling points / high densities / hard / strong / forms coloured compounds.

19.08 X **(1)** (because it is a coloured compound and a good catalyst).

19.09 **1 mark** each for any two of: coloured compounds / catalytic activity / form ions with different charges.

19.10 A mixture **(1)** of a metal with another metal **(1)** or non-metal **(1)**.

19.11 **1 mark** each for any two of: harder / (more) resistant to corrosion / stronger.

Unit 20

20.01 **a** green **b** red-brown / brown / orange **c** purple (**1 mark** each)

20.02 **a** They are more reactive up the group (**1**).
b They are more dense lower in the group (**1**).

20.03 **a** 2,8,7 (**1**) **b** 2,8 (**1**)

20.04 potassium bromide + chlorine → potassium chloride + bromine (**1 mark** for each of the products)

20.05 **a** $Cl_2 + 2KBr \rightarrow 2KCl + Br_2$ (**1 mark** for correct symbols, **1 mark** for correct balance)

b $Cl_2 + 2Br^- \rightarrow 2Cl^- + Br_2$ (**1 mark** for correct symbols, **1 mark** for correct balance)

20.06 Chlorine has diatomic (**1**) molecules (**1**). Argon is monatomic / has single atoms (**1**).

20.07 It is unreactive / inert. Does not allow other gases to interfere with the process (**1**).

20.08 Outer electron shell is full (**1**).

Unit 21

21.01 Iron with steam gives iron oxide (**1**) and hydrogen (**1**); calcium with water gives calcium hydroxide (**1**) and hydrogen (**1**).

21.02 Calcium and magnesium (since they are above carbon in the reactivity series) (**1 mark** for both correct).

21.03 Calcium (**1**) ALLOW: calcium + magnesium

21.04 Copper and silver (**1 mark** for both correct).

21.05 $Zn + 2HCl \rightarrow ZnCl_2 + H_2$ (**1 mark** for H_2, **1 mark** for 2(HCl)).

21.06 Calcium is more reactive than carbon (**1**) iron is less reactive than carbon (**1**).

21.07 Mix magnesium powder with the iron oxide (**1**) heat strongly (**1**).

$3Mg + Fe_2O_3 \rightarrow 3MgO + 2Fe$ (**1 mark** for correct formulae, **1 mark** for correct balance)

21.08 Magnesium ionises more readily than zinc (**1**) since it is higher in the reactivity series (**1**). Magnesium loses electrons more readily than zinc / magnesium is a better reducing agent than zinc / zinc accepts electrons more readily than magnesium (**1**).

21.09 $2NaNO_3 \rightarrow 2NaNO_2 + O_2$ (**1 mark** for correct formulae, **1 mark** for correct balance)

21.10 Reactivity of metals increases down the group (**1**). The more reactive the metal, the more difficult it is to decompose the carbonate (**1**). So more difficult to decompose the carbonate as you go down the group (**1**).

21.11 $2Ca(NO_3)_2 \rightarrow 2CaO + 4NO_2 + O_2$ (**1 mark** for correct formulae, **1 mark** for correct balance)

Unit 22

22.01 Air, coke, limestone, haematite (iron ore) (**1** each)

22.02 To make carbon dioxide / to heat the furnace (**1**).

22.03 iron(III) oxide + carbon monoxide → iron + carbon dioxide (**1**).

22.04 Oxygen oxidises impurities (**1**) calcium oxide reacts with impurities to form slag (**1**).

22.05 Zinc, lead, iron (**1**)

22.06 Bauxite

22.07 $ZnO + CO \rightarrow Zn + CO_2$ (**1**)

22.08 Cryolite dissolves the aluminium oxide (**1**) and lowers the melting point of the electrolyte (**1**).

22.09 $2Al_2O_3 \rightarrow 4Al + 3O_2$ (**1 mark** for correct formulae, **1 mark** for correct balance)

22.10 The carbon electrodes react with oxygen (**1**) to form carbon dioxide (**1**), 'burning away' the electrode.

22.11 **1 mark** each for any two of: The alloys are stronger / harder / more resistant to corrosion.

22.12 It has low density (**1**) (combined with strength).

22.13 Electrical wiring (**1**) because it is a good conductor of electricity (**1**); in cooking utensils (**1**) because it is a good conductor of heat (**1**).

22.14 Any two of: raw materials conserved / less energy used / reduces pollution or less carbon dioxide emitted / less pollution from transport of raw materials. (**2**)

Unit 23

23.01 **1 mark** each for any two of: drinking / cooking / washing / cleaning / sanitation / watering (house plants).

23.02 Cobalt chloride turns from blue **(1)** to pink **(1)**.

23.03 $CuSO_4 + 5H_2O \rightarrow CuSO_4.5H_2O$ **(1)**

23.04 Heat gently **(1)**

23.05 Chlorination: for killing bacteria **(1)**; Filtration: for removing insoluble substances / for removing large particles **(1)**.

23.06 **a** nitrogen 78% **(1)** **b** oxygen 21% **(1)**

23.07 **a** fractional distillation **(1)**

b Nitrogen and oxygen have different boiling points **(1)** raise the temperature of the column gradually **(1)** the more volatile fraction / the fraction with the lower boiling point / the nitrogen evaporates first **(1)**.

23.08 Air and water **(1)** (both needed for the mark).

23.09 Rust only refers to iron(III) oxide / only iron rusts **(1)**.

23.10 Oxygen and water cause rusting **(1)** idea that these cannot reach the surface **(1)**.

Sacrificial protection: A more reactive metal placed in contact with the iron **(1)** reacts instead of the iron **(1)**.

Unit 24

24.01 Sulfur dioxide **(1)** nitrogen dioxide **(1)**

24.02 **1 mark** each for any two of: erosion of mortar / erosion of limestone building material / corrosion of metalwork.

24.03 Incomplete combustion **(1)** produces carbon monoxide **(1)** which is toxic / poisonous / kills you **(1)**.

24.04 A Lead compounds: 3 affect nervous system of young children **(1)**

B Methane: 1 increases global warming **(1)**

C Nitrogen dioxide: 2 photochemical smog / irritant to eyes and respiratory system **(1)**

D Sulfur dioxide: 4 irritant to eyes and respiratory system **(1)**

24.05 Flue gas desulfurisation **(1)**

24.06 Nitrogen and oxygen **(1)** combine under the high temperature **(1)** and high pressure **(1)** (in the car engine).

24.07 $2NO + 2CO \rightarrow N_2 + 2CO_2$ (**1 mark** for correct formulae, **1 mark** for correct balance)

24.08 Methane **(1)** carbon dioxide **(1)**

24.09 Carbon dioxide is produced **(1)** which is a greenhouse gas **(1)** carbon dioxide absorbs heat energy **(1)**.

24.10 Any two effects of global warming (**1 mark** each): desertification / melting ice caps / flooding / increase in temperature of seas / more storms or more extreme weather.

Unit 25

25.01 **1 mark** each for any three sources of carbon dioxide: respiration / combustion (of carbon compounds) / reaction of acids with carbonates / thermal decomposition of carbonates.

25.02 glucose + oxygen → carbon dioxide + water

(**1 mark** for correct reactants, **1 mark** for correct products)

25.03 $6CO_2 + 6H_2O \rightarrow C_6H_{12}O_6 + 6O_2$

(**1 mark** for correct formulae for reactants, **1 mark** for correct formulae for products, **1 mark** for correct balance)

25.04 Respiration removes oxygen from the atmosphere and gives off carbon dioxide **(1)**.

Photosynthesis removes carbon dioxide from the atmosphere and gives off oxygen **(1)**.

The release of carbon dioxide in respiration is balanced by the absorption of the carbon dioxide in photosynthesis **(1)**.

25.05 **1 mark** for each of any two of: building stone / removing impurities in the extraction of iron / making lime mortar / making cement.

25.06 Calcium hydroxide **(1)**; calcium hydroxide is formed by the action of water on calcium oxide **(1)**.

25.07 calcium carbonate + sulfuric acid →
calcium sulfate + carbon dioxide + water

(**1 mark** for each of the correct products)

25.08 Nitrogen **(1)** phosphorus **(1)** potassium **(1)**

25.09 $(NH_4)_2SO_4$: $2N = 2 \times 14 = 28$ $M_r = 132$;

$$\%N = 100 \times \frac{28}{132} = 21.2\%$$

(**2 marks** for 21.2% or 21%. If two not scored, allow **1 mark** for correct molar mass of ammonium sulfate if error carried forward from wrong molar mass or incorrect mass of nitrogen)

25.10 $Ca_3(PO_4)_2$ **(1)**

25.11 a position of equilibrium to the right **(1)**

b position of equilibrium to the right **(1)**

25.12 a $Zn + H_2SO_4 \rightarrow ZnSO_4 + H_2$

(**1 mark** for products and **1 mark** for reactants)

b $2NaOH + H_2SO_4 \rightarrow Na_2SO_4 + 2H_2O$

(**1 mark** for products and **1 mark** for reactants)

ALLOW: balanced equation with formation of $NaHSO_4$

Unit 26

26.01 a coal **(1)** **b** methane **(1)**

26.02 Methane **(1)**

26.03 a alkane **(1)** **b** alcohol **(1)** **c** carboxylic acid **(1)**

26.04

```
     H    H
     |    |
H —  C —  C — O — H
     |    |
     H    H
```

all correct = **2 marks**
if OH instead of O–H = **1 mark**

26.05 a C_2H_6O / C_2H_5OH **(1)** b CH_4 **(1)**

26.06 (Alk)enes have a double bond between carbon atoms **(1)**.

So the first alkene in the alkene homologous series must have two carbon atoms / compounds starting with meth- only have one carbon atom **(1)**.

26.07 One mark each for any two of: have the same general formula / differ by CH_2 group / have similar chemical properties / show a trend in physical properties.

26.08 $C_nH_{2n}O_2$ / $C_nH_{2n+1}COOH$ **(1)**

26.09 $C_5H_{12}O$ / $C_5H_{11}OH$ (1) and $C_6H_{14}O$ / $C_6H_{13}OH$ **(1)**

26.10 Compounds with the same molecular formula but different structural formulae. **(1)**

Unit 27

27.01 a Tar for road surfaces / roofing **(1)** **b** fuel for cooking / heating **(1)** **c** fuel for cars **(1)**

27.02 Compound of carbon and hydrogen only **(1)**.

27.03 Saturated compounds have no carbon–carbon double bonds **(1)** unsaturated have carbon–carbon double bonds **(1)**.

27.04 a ethene **(1)** **b** methane **(1)** **c** ethane **(1)**

27.05 a $CH_3CH_2CH_2CH_2CH_2CH_2CH_3$ **(1)**

b

```
        CH3                 CH3
        |                   |
 CH3CH2CHCH2     OR     CH3CCH3
                            |
                            CH3
```

(1)

27.06 a Reaction in which one type of atom replaces another type of atom in a compound **(1)**.

b light **(1)**

27.07 Bromine water / aqueous bromine added to the unsaturated compound **(1)** turns from orange to colourless **(1)**.

27.08

```
H     H
|     |
C  =  C
|     |
H     H
```

(1)

27.09 a butene **(1)** **b** pentene **(1)**

27.10 C_nH_{2n} **(1)**

27.11 Nickel catalyst **(1)**, heat / 150 °C **(1)**; only one product is formed from two reactants **(1)**.

Unit 28

28.01 **1 mark** each for any two of: solvent / fuel / drinks /cleaning fluid / medical wipes / other suitable use.

28.02 $CH_3CH_2OH + \underline{3}O_2 \rightarrow \underline{2}CO_2 + \underline{3}H_2O$

(**1 mark** for each of the underlined numbers.)

28.03 glucose → ethanol + carbon dioxide

(**1 mark** for ethanol, **1 mark** for carbon dioxide)

28.04 Enzymes **(1)**

28.05 **1 mark** each for any three of: simple equipment required / uses renewable reactants / works at room temperature and atmospheric pressure so energy requirement is low / catalyst does not have to be manufactured (enzymes in yeast).

28.06

```
     H    H
     |    |
H —  C —  C — O — H
     |    |
     H    H
```

(**2 marks** if all correct; **1 mark** if OH instead of O–H)

28.07 **1 mark** each for any two of: liquid (at room temperature and pressure)/ colourless / low pH or acidic or turns blue litmus red / any other suitable chemical property.

28.08 Carboxylic acid **(1)**, –COOH **(1)**

28.09 $HCOOCH_3$ **(1)**

28.10 $NaOH + C_3H_7COOH \rightarrow C_3H_7COONa + H_2O$

(**2 marks** if all correct; **1 mark** if one error)

28.11 **a** ethyl propanoate **(1)** **b** propyl methanoate **(1)**

Unit 29

29.01 Poly(ethene) **(1)**

29.02 Ethene **(1)**

29.03 Addition **(1)**

29.04 Addition means that the molecules add together and only one product is formed **(1)**. Polymerisation is the addition of monomers to make a polymer **(1)**.

29.05

```
    C6H5   H    C6H5   H
     |     |     |     |
—    C   — C —   C   — C —
     |     |     |     |
     H     H     H     H
```

(**2 marks** if all correct; **1 mark** if continuation bonds missing)

29.06

```
  ┌ CH3 Cl ┐
  │  |   | │
 ─┼─ C — C ┼─
  │  |   | │
  └  H   H ┘n
```

(2)

29.07 **a**

```
— C — N —
  ||  |
  O   H
```

b

```
— C — O —
  ||
  O
```

29.08 HOOC–□–COOH **(1)** H_2N–■–NH_2 **(1)**

29.09 **1 mark** each for any two of: may produce toxic gases / global warming / acid rain.

29.10 Terylene–one of: sheet / clothing / towels **(1)**.

Nylon–one of: fishing lines or nets / clothing / carpets / ropes **(1)**.

Unit 30

30.01 Proteins for growth **(1)**, carbohydrates for energy **(1)**

30.02 Amide linkage / peptide link **(1)**

30.03 Amino acid **(1)**

30.04 A reaction where two organic molecules join together **(1)** with the elimination of a small molecule **(1)**.

30.05 n (glucose) → starch + water

(**1 mark** for indication of many glucose molecules, **1 mark** for starch, **1 mark** for water)

30.06 protein + water → amino acids (**1 mark** for protein + water, **1 mark** for amino acids)

30.07 Starch **(1)** because the links are –O– / glycosidic link **(1)**.

30.08 A substance that reacts with colourless spots **(1)** on a chromatogram to make them visible **(1)** (as coloured spots).

Answers to Exam-style questions

Unit 1

Question 1.01

a B melting, D is sublimation. E is boiling (or evaporation if it takes place below the boiling point) (**2 marks** if all correct, **1 mark** if one or two correct).

b Particles irregularly arranged in a liquid **(1)** but become regularly arranged in a solid **(1)**. Particles sliding over each other in liquid **(1)** but only vibrating (or no movement) in solid **(1)**.

c i B: energy absorbed **(1)** (to overcome attractive forces between particles in a solid).

ii F: energy released **(1)** (kinetic energy is reduced when particles come closer together. This energy given out to raise the temperature of the surroundings).

Question 1.02

a Any three of: diffusion / particles are in random movement / particles spread everywhere / particles mix with air particles / idea of bulk movement of particles from where they are in high concentration to where they are in low concentration **(3)**.

b Took longer to turn red **(1)** HBr has higher M_r than HC*l* **(1)** so HBr moves more slowly **(1)**.

c Particles in a liquid have attractive forces keeping them together **(1)**. As temperature rises, attractive forces weaken **(1)** and some molecules have enough energy to overcome these attractive forces so escape from the surface of the liquid **(1)**.

Question 1.03

Particles of smoke move by Brownian motion **(1)**.

Molecules of gases in the air are moving randomly / in a zig-zag way **(1)**.

The particles in the air are hitting the smoke particles randomly **(1)**.

(At any one time) there is a greater force on one side of the smoke particles **(1)**.

Particles move in direction of greater force **(1)**.

Unit 2

Question 2.01

a Gas bubbled through water **(1)**: gas is soluble in water so will not get to gas jar **(1)**. Gas jar is wrong way up **(1)**: chlorine is denser than air **(1)**.

b beneath the flask **(1)**

c Chlorine is poisonous **(1)**.

d Sulfuric acid is corrosive / burns the skin **(1)**:

e wear gloves / war lab coat / wear eye protection **(1)**

Question 2.02

a To make sure all the water is all the same temperature / so there are no hot spots **(1)**.

b Any two answers with corresponding reasons, for example:

Same amount of water **(1)**: temperature rise will not reflect the correct amount of energy / idea of energy spread out more if more water (or reverse argument) **(1)**.

Same distance of flame from can **(1)**: the further the flame from the can, the more energy lost to the surroundings (or reverse argument) **(1)**.

Same copper can **(1)**: different cans may lose different amounts of heat to surroundings **(1)**.

c So that an average could be taken **(1)** to get more accurate results / to confirm the results were correct **(1)**.

d Measuring cylinder **(1)**: large amount of water used so percentage error not great **(1)** OR burette **(1)** to get the maximum accuracy possible **(1)**.

e The accuracy / inaccuracy of the thermometer limits the accuracy of the whole experiment **(1)**.

Unit 3

Question 3.01

a (See Figure 3.5)

Fractional distillation **(1)**: diagram of distillation flask connected to vertical tube **(1)** condenser connected to the tube and labelled **(1)** heat applied to the flask **(1)** B distils over first / goes into the flask first **(1)** A remains in the flask / distils over second **(1)**.

b Pentane because it has a lower boiling point **(1)**.

c Measure the boiling point **(1)** (boiling point) is sharp if B is pure / if (boiling point) is not sharp B is impure **(1)**.

d Warm gently / heat gently **(1)** until crystallisation point / to form a saturated solution **(1)** leave to crystallise then filter off the crystals **(1)** dry crystals between filter paper **(1)**.

Question 3.02

a Spray with a chemical which reacts with carbohydrates / spray with locating agent **(1)** which makes the spots coloured **(1)**.

b Three spots **(1)**

c One of the spots appears larger / longer / more spread out **(1)**.

Two or more carbohydrates may have the same / similar R_f values **(1)**.

d $R_f = \dfrac{\text{distance from base line to centre of spot}}{\text{distance of solvent front from base line}}$ **(1)**

= ALLOW: 0.74 to 0.76 **(1)**

e C because the spot had not moved much / spot carried up the paper least by solvent **(1)**.

Unit 4

Question 4.01

a D **(1)** as one electron in the outer shell **(1)**.

b B **(1)** has the lowest number of electrons hence lowest number of protons **(1)**.

c D **(1)** has four shells containing electrons **(1)**.

d C **(1)** has full outer shell of electrons / has 8 electrons in outer shell **(1)**.

e i Atoms of the same element which have the same proton number but a different nucleon number **(1)**.

ii electrons = 2 + 8 + 4 = 14 **(1)**
protons = 14 (protons = number of electrons) **(1)**
neutrons = 30 − 14 = 16 **(1)**

Question 4.02

a Br-79 neutrons = 79 − 35 = 44 and Br-81 neutrons = 81 − 35 = 46 **(1)**

b i seven electrons in outer shell **(1)**

ii 2 + 8 + 18 + 7 = 35 electrons **(1)**

Number of electrons = number of protons, so number of protons = 35 **(1)**.

c i 2,8,18,8 (one extra electron in forming ion) **(1)**

ii 2,8 (two electrons lost from Mg atom) **(1)**

Unit 5

Question 5.01

a i Any three of (**1 mark** each): conducts heat (or electricity) / lustrous / malleable / ductile

ii soft **(1)** low melting point **(1)**

b Low boiling point / liquid **(1)** not lustrous **(1)**
IGNORE: references to colour.

c Sodium bromide has high melting point **(1)** whereas sodium and bromine have lower melting points **(1)**.

Sodium bromide is a solid **(1)**, whereas bromine is a liquid **(1)**.

ALLOW (if no marks scored): Has different properties from sodium or bromine **(1)**.

Question 5.02

a A compound is a substance made up of two or more different atoms (or ions) bonded together **(1)**.

b Any three of (**1 mark** each): does not conduct heat (or electricity) / dull / not malleable / brittle (or not ductile) / dull sound when hit / low melting or boiling point / low density.

c Iron sulfide is black / not a mixture of colours **(1)**, mixture is yellow with silvery bits/ silvery with yellow bits / greyish **(1)**.

Iron sulfide gives off a bad-smelling gas with hydrochloric acid (1); mixture would not give off bad-smelling gas (1).

Unit 6

Question 6.01

a Graphite has a layered structure but diamond has a tetrahedral / pyramidal structure (1). Each carbon atom in graphite joined to three others but in diamond each carbon atom is joined to four others (1).

b In graphite (some of) the electrons can move along the layers (1). In diamond there are no electrons which move (across the structure) (1).

c High melting point because there are a lot of covalent bonds / there are a lot of strong bonds (1). It takes a high temperature / lots of energy to break these bonds (1).

d i Correct dot-and-cross diagram for methane

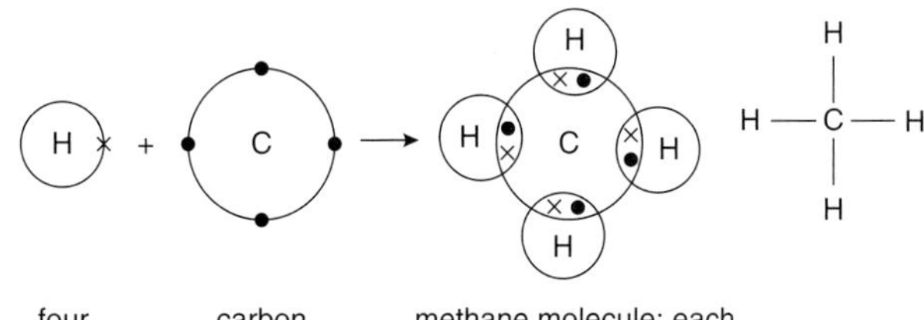

(1)

ii Volatility is high / low boiling point (1). Solubility: insoluble in water / soluble in organic solvent (1). Conductivity: does not conduct (1).

Question 6.02

a Carbon dioxide has simple molecular structure (1) because has weak intermolecular forces / weak forces between molecules (1). It does not take much energy / only needs low temperature to overcome these forces (1).

Silicon dioxide is a covalent giant structure (1) with strong bonding (1). It takes a lot of energy / needs high temperature to overcome these forces (1).

b

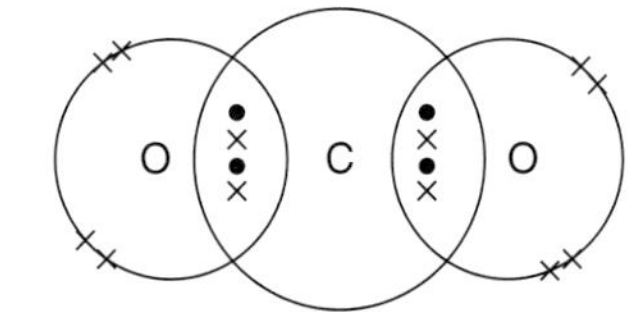

O=C=O

Carbon dioxide molecule

Four bonded pairs of electrons around the carbon atom (1), two lone pairs of electrons on each oxygen (1).

c Tetrahedral / pyramidal structure in both (1), Si has four atoms around it, so has C (1).

d No free electrons / no delocalised electrons (1).

e i Delocalised electrons (1) are fee to move (1).

ii There are attractive forces between ions and delocalised electrons (1). The layers move (1) because when a force is applied the forces are overcome (1). Bonds are reformed (when force removed) (1).

Unit 7

Question 7.01

a $MgCO_3 + 2HCl \rightarrow MgCl_2 + CO_2 + H_2O$ (1 **mark** for CO_2 and 1 **mark** for 2(HC*l*))

b magnesium + water → magnesium oxide + hydrogen (1 **mark** for correct reactants, 1 **mark** for correct products)

c Magnesium sulfate is $MgSO_4$ (1) magnesium sulfide is MgS (1).

Question 7.02

a nitrogen + hydrogen → ammonia (1)

(One condition over the arrow, for example catalyst / heat / temperature (1))

b $3H_2$ (1 **mark** for H_2, 1 mark for balance.)

c i $2NH_3 + H_2SO_4 \rightarrow (NH_4)_2SO_4$ (1)

ii ammonium sulfate (1)

Question 7.03

a $2K(s) + 2H_2O(l) \rightarrow 2KOH(aq) + H_2(g)$

(Correct reactants and products (1), correct balance (1), correct state symbols (1))

b i $MgBr_2 + 2AgNO_3 \rightarrow Mg(NO_3)_2 + 2AgBr$

(Correct reactants and products (1), correct balance (1))

ii $Br^-(aq) + Ag^+(aq) \rightarrow AgBr(s)$

(Correct reactants and products (1), correct state symbols (1))

iii magnesium and nitrate (1)

Question 7.04

a $Fe_2O_3 + 2Al \rightarrow Al_2O_3 + 2Fe$

(Correct reactants and products **(1)**, correct balance **(1)**)

b i (g) is gas **(1)** ii (aq) is aqueous solution / dissolved in water **(1)**

c $Fe(s) + 2H^+(aq) \rightarrow Fe^{2+}(aq) + H_2(g)$

(Correct reactants and products **(1)**, correct state symbols **(1)**)

d $Fe_2(SO_4)_3$ **(1)**

Unit 8

Question 8.01

a C H Cl

$\frac{1.92}{12} = 0.16 \quad \frac{0.32}{1} = 1 \quad \frac{5.78}{35.5} = 0.163$ **(1)**

Divide by lowest: $\frac{0.16}{0.16} = 1 \quad \frac{0.32}{0.16} = 0.2 \quad \frac{0.163}{0.16} = 1$

Empirical formula: CH_2Cl **(1)**

b empirical formula mass = 49.5 **(1)**

$\frac{99}{49.5} = 2$ so molecular formula = $C_2H_4Cl_2$ **(1)**

c $C_2H_2 + 2Cl_2$

Moles of ethyne and chlorine $\frac{13}{26} = 0.5 \quad \frac{69}{71} = 0.972$ **(1)**

Use of 1:2 mole ratio, e.g. 0.5 moles C_2H_2 requires 1.0 mol Cl_2

So ethyne in excess **(1)**

Question 8.02

a Moles $CO_2 = \frac{2000}{1000} = 2.0\,dm^3 \quad \frac{2.0}{24} = 0.0833\,mol$ **(1)**

If $NaHCO_3$ is pure we would need 2×0.083 mol = 0.166 mol **(1)**

mass of $NaHCO_3 = 0.166 \times 84 = 13.94$ g **(1)**

$\frac{\text{mass of pure substance}}{\text{mass of impure substance}} \times 100 = \frac{13.94}{20.4} \times 100$
$= 68.3\%$ **(1)**

b $2NaHCO_3 \rightarrow Na_2CO_3$

Correct molar masses 84 → 106 **(1)**

Mole ratio 2×84 106 **(1)**

26.1 → $26.1 \times \frac{106}{2 \times 84}$
$= 16.47$ g **(1)**

At 72% yield $= 16.47 \times \frac{72}{100} = 11.9$ g **(1)**

Question 8.03

Moles of Ca $= \frac{0.8}{40} = 0.02$ mol **(1)**

Moles of ethanoic acid $= \frac{3}{60} = 0.05$ mol **(1)**

$Ca + 2CH_3COOH \rightarrow (CH_3COO)_2Ca + H_2$

1 mol + 2 mol → 1 mol + 1 mol

Therefore 2 moles of ethanoic acid react with 1 mole Ca, so

$2 \times 0.02 = 0.04$ (**1 mark**). So Ca is the limiting reactant and ethanoic acid is in excess (**1 mark**).

Excess ethanoic acid = 0.01 mol (**1 mark**)

So mass in excess $= 0.01 \times 60 = 0.6$ g (**1 mark**)

Unit 9

Question 9.01

a $0.2 \times \frac{20}{1000} = 0.004$ mol **(1)**

b 2×0.004 mol = 0.008 mol **(1)**

c $0.008 \times \frac{1000}{32.5} = 0.246$ (or 0.25) mol/dm^3 **(1)**

Question 9.02

a $1.71 \times \frac{25}{1000} = 0.04275$ (or 0.043) g/dm^3 **(1)**

$\frac{1.71}{171} = 0.01$ mol so concentration $= 0.01 \times \frac{25}{1000}$
= 0.00025 mol/dm^3 **(1)**

b 0.00025 → 2×0.00025 mol $HCl = 0.00050$ mol **(1)**

c $0.0005 \times \frac{1000}{22.5} = 0.022$ mol/dm^3 **(1)**

Question 9.03

a $0.1 \times \frac{25}{1000} = 0.0025$ mol **(1)**

b 0.0025 → $\frac{0.0025}{2}$ mol sulfuric acid = 0.00125 mol **(1)**

c $0.00125 \times \frac{1000}{22.4} = 0.56$mol/dm^3 **(1)**

Unit 10

Question 10.01

a conducts electricity **(1)** inert electrode **(1)**

b anode: chlorine **(1)** cathode: zinc **(1)**

c Completed circuit to anode and cathode with power pack / battery **(1)**.

Two electrodes dipping into electrolyte **(1)**. Electrolyte labelled molten zinc chloride **(1)**.

d anode: chlorine **(1)** cathode: hydrogen **(1)**

e Substance which breaks down / decomposes when electric current passed **(1)**; idea of ionic nature of the electrolyte **(1)**.

Question 10.02

a Zinc is more reactive than silver **(1)** so loses electrons more readily than silver **(1)**.

b negative electrode: $Zn \rightarrow Zn^{2+} + 2e^-$ **(1)**

positive electrode: $2H^+ + 2e^- \rightarrow H_2$ **(1)**

c From the zinc to the copper / from the negative to the positive **(1)**. The electrons move from where they are released (at negative electrode) to where they are taken up (at positive electrode) **(1)**.

d The two electrodes lose electrons to the same extent **(1)**.

e Magnesium becomes the negative electrode **(1)** because it is more reactive than zinc **(1)**.

Question 10.03

a Cathode: $2H^+ + 2e^- \rightarrow H_2$ **(2)** (**1 mark** for reactants and products, **1 mark** for balance)

Anode: $2I^- \rightarrow I_2 + 2e^-$ **(1)** and $4OH^- \rightarrow O_2 + 2H_2O + 4e^-$ **(2)** (**1 mark** for reactants and products, **1 mark** for correct balance)

b Hydrogen ions lower in discharge series than sodium **(1)**. Ions lower in series are more likely to be discharged **(1)**.

c Iodine lower than oxygen in discharge series **(1)**. Ions lower in discharge series more likely to be discharged **(1)**.

d Ions remaining in solution / not discharged are OH^- (from water) and Na^+ **(1)**.

Sodium hydroxide is alkaline / OH^- ions are alkaline **(1)**.

e Discharge also depends on concentration of the solution **(1)**. Dilute solutions produce more oxygen than concentrated solutions **(1)**.

f Solid does not conduct because the ions cannot move **(1)**. In solution the ions can move **(1)**.

Unit 11

Question 11.01

a exothermic **(1)**

b Vertical axis labelled energy **(1)**.

Magnesium and hydrochloric acid / reactants on left and products on the right **(1)**.

Line from product level to reactant level **(1)** labelled energy released **(1)**.

c i magnesium chloride **(1)**

ii Doubling the amount of sodium fluoride doubles the temperature rise **(1)**. Doubling the amount of water halves the temperature rise **(1)** because the energy is spread out over twice the volume **(1)**.

Question 11.02

a Does not pollute **(1)**; more energy per kg **(1)**.

b Needs pressurised tanks / potentially flammable or explosive / expensive to produce / production uses fossil fuels **(1)**.

c Vertical axis labelled energy and horizontal labelled reaction pathway / course of reaction **(1)**.

Hydrogen and oxygen on left and water on the right **(1)**.

Reactant level above product level **(1)**.

Arrow pointing downwards from product level to reactant level (labelled energy released) **(1)**.

d i $H_2(g) \rightarrow 2H^+(aq) + 2e^-$ **(1)**

ii Hydrogen releases electrons by reaction at negative electrode **(1)** oxygen accepts electrons by reaction at positive electrode **(1)** electrons go round external circuit because of imbalance of electrons on electrodes **(1)**.

e

bonds broken in reactants (endothermic, ΔH positive)	bonds formed in products (exothermic, ΔH negative)
E(H–H) = 436	2 × E(H–Cl) = 2 × 432
E(Cl–Cl) = 243	
Total = +679 kJ	Total = –864 kJ

The exothermic change is greater by 854 – 679 kJ = 175 kJ

So the heat energy change is – 175 kJ.

(**1 mark** for total energy in reactants, **1 mark** for energy in products, **1 mark** for correct answer)

Question 11.03

a vertical axis labelled energy and horizontal labelled reaction pathway / course of reaction **(1)**

nitrogen and hydrogen / reactants on left and ammonia / products on the right **(1)**

reactant level above product level **(1)**

arrow pointing downwards from product level to reactant level (labelled energy released) **(1)**

b The energy needed to break a mole of covalent bonds between two particular atoms **(1)**

c

bonds broken in reactants	bonds formed in products
(endothermic, ΔH positive)	(exothermic, ΔH negative)
$E(N \equiv N) = 945$	$2 \times 3 \times E(N–H) = 6 \times 391.$
$3 \times E(H–H) = 3 \times 436$	
Total = +2253 kJ	Total = −2346 kJ

The exothermic change is greater by 2346 − 2253 kJ = 93 kJ

So the heat energy change is −93 kJ.

(**1 mark** for total energy in reactants, **1 mark** for energy in products, **1 mark** for correct answer)

d i $O_2 + 2H_2O + 4e^- \rightarrow 4OH^-$

(**1 mark** for correct reactants and products, **1 mark** for correct balance)

ii Advantages: **1 mark** each for any two of: non-polluting / produce more energy per kg of fuel burned / higher efficiency of energy transmission / can be used in restricted places.

Disadvantage: **1 mark** for one of: materials used to make them are expensive / high pressure tanks are needed / do not work well at low temperatures / hydrogen is expensive to produce / production of hydrogen uses energy sources which pollute the atmosphere.

Unit 12

Question 12.01

a i 120 s **(1)** ALLOW: 110-120 s

ii 50 cm^3 **(1)** ALLOW: 48-52 cm^3

iii A **(1)**: has the biggest gradient / biggest slope **(1)**

b Initial gradient / slope for the 1 mol/dm^3 acid is greater than the initial gradient / slope for the 0.5 mol/dm^3 acid (or compare gradients at any other specified time) **(1)**.

c Reaction would be faster with smaller pieces **(1)**. Greater surface area with smaller particles **(1)**. So more particles exposed to the acid / comment on increasing frequency of collisions **(1)**.

Question 12.02

a Rate is inversely proportional to time / proportional to 1/time **(1)**.

b Rate doubles for each 10 °C rise in temperature **(2)**.

BUT if this is not scored, a mark is gained for: rate increases as temperature increases **(1)**.

c The higher the temperature, the faster the particles move / particles have more energy the higher the temperature **(1)**.

The higher the temperature, the greater the number of collisions per unit time **(1)**.

The higher the temperature, the more particles there are with energy greater than the activation energy **(1)**.

d Particles become more spread out / fewer particles per unit volume **(1)**

Collision rate decreases / frequency of collisions decreases **(1)** so the rate of reaction decreases **(1)**.

e When the concentration is higher, the particles are closer together / there are more particles per cm^3 / per unit volume **(1)**. The collision frequency increases / the rate of collisions increases **(1)**.

Unit 13

Question 13.01

a $CoCl_2.6H_2O \rightarrow CoCl_2 + 6H_2O$ (**1 mark** for correct reactants and products, **1 mark** for balance)

b Add water to the anhydrous cobalt chloride **(1)**.

c $CoCl_2 + 6H_2O \rightarrow CoCl_2.6H_2O$ **(1)**

d anhydrous **(1)**

Question 13.02

a Deeper red-brown colour **(1)** because the position of equilibrium shifts to the left **(1)** to oppose the increase in H^+ ion concentration **(1)**.

b Lighter red-brown colour **(1)** because the position of equilibrium shifts to the right **(1)** because the alkali removes the H^+ ions **(1)** and lowers the H^+ concentration **(1)**.

c Deeper red-brown colour **(1)** because position of equilibrium shifts to the left **(1)** to oppose the increase in the Br^- ion concentration **(1)**.

Question 13.03

a At first the rate of forward reaction is greater than the rate of the back reaction **(1)** because of the higher concentration of hydrogen and iodine **(1)**. After a time the back reaction increases in reaction rate **(1)** because the concentration of hydrogen iodide is increasing **(1)**. At equilibrium, the rate of the forward reaction is the same as that of the backward reaction **(1)**.

b No effect **(1)** because there are equal numbers of molecules of gas in the stoichiometric equation on each side of the equation **(1)**.

c Position of equilibrium shifts to the right **(1)** to oppose the increase in concentration of hydrogen **(1)**.

d Exothermic **(1)**. Heat is given out in exothermic reaction, so reaction goes in the direction to oppose this change / the endothermic direction / direction of lower concentration of hydrogen iodide **(1)**.

Question 13.04

a The position of equilibrium moves to the left **(1)** to try to lower the concentration of HCl **(1)**. The amount of white solid decreases / solution appears less white **(1)**.

b The position of equilibrium moves to the right **(1)** to try to lower the concentration of water **(1)**. The amount of white solid increases / solution appears whiter **(1)**.

c The position of equilibrium moves to the left **(1)** to try to reduce the amount of added white solid **(1)**. The amount of white solid decreases / solution appears less white after initial addition **(1)**.

Unit 14

Question 14.01

a Oxidation state **(1)** shows how oxidised a substance is **(1)**.

b $PbO + C \rightarrow Pb + CO$ **(1)**

c Lead oxide **(1)** as loses oxygen **(1)**.

d redox **(1)**

Question 14.02

a $Cl_2 + 2e^- \rightarrow 2Cl^-$ **(1)** reduction **(1)** $2I^- \rightarrow I_2 + 2e^-$ **(1)** oxidation **(1)**.

b Reducing agent is I^- / iodide ions **(1)** because they give electrons to the chlorine.

Oxidising agent is chlorine / Cl_2 **(1)** because accepts electrons from iodide ions **(1)**.

c Add potassium iodide to hydrogen peroxide. Iodide goes from colourless **(1)** to brown **(1)** due to iodine being formed **(1)**.

d Fe^{2+} oxidised **(1)** because increases oxidation state to 3+ **(1)**.

H_2O_2 reduced **(1)** as decreases oxidation state (from −1 to −2) **(1)**.

Question 14.03

a $Fe \rightarrow Fe^{2+} + 2e^-$ **(1)** $2H^+ + 2e^- \rightarrow H_2$ **(1)**

b Oxidising agent is H^+ ions **(1)** because takes electrons from iron **(1)**.

Reducing agent is Fe **(1)** because gives electrons to H^+ ions **(1)**.

c Aluminium / Al **(1)**. Increase in oxidation number / + charge on aluminium in Al_2O_3 but not on Al **(1)** reference to 0 → +3 for charge or oxidation number **(1)**.

Unit 15

Question 15.01

a proton acceptor / substance which has pH above 7 **(1)**

b pH between 8–14 **(1)**; turns red litmus blue **(1)**

c i calcium hydroxide + hydrochloric acid → calcium chloride + water **(1)**

ii calcium oxide + nitric acid → calcium nitrate + water **(1)**

d To neutralise **(1)** excess acidity in the soil **(1)** so that plants can grow well **(1)**.

e $CaCO_3 + 2HCl \rightarrow CaCl_2 + CO_2 + H_2O$
(**1 mark** for $CaCl_2$, **1 mark** for $CO_2 + H_2O$, **1 mark** for balance)

Question 15.02

a i Blue litmus turns red (1).

ii Bubbles / effervescence (1) because carbon dioxide is given off (1).

b sulfuric acid + ammonia → ammonium sulfate (1)

sulfuric acid + sodium hydroxide → sodium sulfate + water (1)

c $Fe + H_2SO_4 \rightarrow FeSO_4 + H_2$ (**2 marks** for correct equation, **1 mark** if one error)

d Add litmus / methyl orange / other named indicator other than universal (1) to the potassium hydroxide (1). Add acid drop by drop / gradually (1) until the indicator changes colour (1).

e Add a drop of solution to universal indicator (1): compare the colour with a pH colour chart for universal indicator (1).

Question 15.03

a $HCl \rightarrow H^+ + Cl^-$ (1)

$CH_3CH_2COOH \rightleftharpoons H^+(aq) + CH_3CH_2COO^-$
(**1 mark** for equation, **1 mark** for equilibrium sign)

b Hydrochloric acid is completely ionised in water (1), propanoic acid is partially ionised in water (1). Concentration of H^+ ions is higher in hydrochloric acid / lower in propanoic acid (1) so the frequency of collisions is higher in hydrochloric acid, lower in propanoic acid (1).

c Take pH (1): hydrochloric acid will have a higher pH / propanoic acid will have a lower pH (1). Compare electrical conductivity (1): hydrochloric acid will have a higher conductivity / propanoic acid will have a lower conductivity (1) because (for both pH and conductivity there is a higher concentration of H^+ ions in HCl / lower concentration in propanoic acid (1).

d $2CH_3CH_2COOH + 2Na \rightarrow 2CH_3CH_2COONa + H_2$

(**1 mark** for equation, **1 mark** for correct balance)

Unit 16

Question 16.01

a sulfuric acid (1)

b cobalt carbonate + sulfuric acid → cobalt sulfate + water + carbon dioxide

(**2 marks** if all correct, **1 mark** if one of products omitted or incorrect)

c So that all the sulfuric acid is used up (1).

d Filter off the excess cobalt carbonate (1).

e Evaporate filtrate to crystallisation point (1). Leave to crystallise (1). Filter off crystals (1). Dry crystals between filter papers (1).

Question 16.02

a Add excess lead oxide (1) to dilute nitric acid (1). Warm / heat (and stir) (1). Filter off the excess lead oxide (1). Evaporate filtrate to crystallisation point (1). Leave to crystallise (1). Filter off crystals (1). Dry crystals between filter papers (1).

b Add (aqueous) lead nitrate (1) to named (aqueous) iodide, e.g. potassium iodide (1); idea that both are aqueous solutions (1). Filter off the precipitate of lead iodide (1). Wash precipitate with water (1). Dry precipitate with filter papers / warm in oven to dry (1).

$Pb^{2+}(aq) + I^-(aq) \rightarrow PbI(s)$

(**1 mark** for correct reactants and product, **1 mark** for correct state symbols.)

Unit 17

Question 17.01

a Turns blue litmus red (1) because the pH is 1.5 / it is acidic (1).

b Bubble gas through limewater / hold drop of limewater above the gas (1), limewater (1) turns milky (1).

c Cu^{2+} / copper(II) (1)

d copper carbonate (1)

e Add aqueous sodium hydroxide and aluminium foil (1), warm gently (1), test gas evolved with damp red litmus paper (1), red litmus turns blue (1).

Question 17.02

a chlorine (1)

b Add dilute acid (1), warm gently (1), test gas evolved with acidified aqueous potassium manganate(VII) (1), potassium manganate(VII) turns from purple to colourless (1).

c barium sulfate (1)

d To aqueous solution of chloride ions add dilute nitric acid **(1)** then aqueous silver nitrate **(1)**; white precipitate formed **(1)**.

e Similarities: both form light blue **(1)** precipitate **(1)** when sodium hydroxide or ammonia not in excess **(1)**.

Differences: when in excess the precipitate with sodium hydroxide is insoluble **(1)**; when in excess the precipitate with aqueous ammonia dissolves **(1)** to form a dark blue solution **(1)**.

Unit 18

Question 18.01

a Period 3 and Group V **(1)**

b Use litmus / universal indicator paper **(1)**. Solution from oxide of calcium turns red litmus blue **(1)** because it is a basic oxide / oxide of a metal **(1)**. Solution from oxide of phosphorus turns blue litmus red **(1)** because it is an acidic oxide / oxide of a non-metal **(1)**.

c Magnesium **(1)** because it is a metal **(1)**.

d Magnesium **(1)**. Most metallic structures have high melting points **(1)**; phosphorus has simple molecular structure / is a non-metal **(1)** so has low melting point **(1)**.

e Phosphorus is a solid **(1)**. The elements above and below it in the Group have melting points above room temperature **(1)**.

f From non-metals at the top of the Group to metals at the bottom **(1)**. Nitrogen is a non-metal and antimony and bismuth are metals **(1)**.

Question 18.02

a Giant structure **(1)** of covalent bonds **(1)**; tetrahedral structure / every Ge atom linked to four others **(1)**.

b Takes a high temperature / a lot of energy to break the bonds **(1)** because there is strong covalent bonding in three dimensions / lots of covalent bonds **(1)**.

c Conductivity increases down Group IV **(1)**. The elements get more metallic down the Group **(1)**.

d Accept values between 210 and 290 (actual value = 222) **(1)**.

e i Covalent **(1)** because chlorine is a non-metal and germanium has some non-metallic character / elements at the top of Group IV tend to be non-metals **(1)**.

ii $GeCl_4$

Unit 19

Question 19.01

a Any four of:

Lithium reacts slowly, potassium reacts violently **(1)**.

Lithium produces bubbles slowly, potassium produces bubbles rapidly **(1)**.

Lithium moves slowly on the surface of the water but potassium moves rapidly **(1)**.

Lithium fizzes a little, potassium fizzes a lot **(1)** (remember that fizzing is sound).

Lithium does not form a ball / does not melt but potassium melts / forms a ball **(1)**.

Lithium does not catch fire but potassium does **(1)** with a lilac flame **(1)**.

Lithium does not explode but potassium does **(1)**.

b Add red litmus **(1)**; (red litmus) turns blue **(1)**.

OR add universal indicator **(1)**; turns blue **(1)**.

c potassium + water → potassium hydroxide + hydrogen (**2 marks** if all correct, **1 mark** if one error).

Question 19.02

a Any two of: conducts electricity (or heat) / fresh surface if shiny (lustrous / silvery) / malleable / ductile **(2)**.

b Any three of: forms coloured compounds / high melting (or boiling) point / high densities / stronger (or harder) **(3)**.

c Any two of: increases strength / increases hardness / increases corrosion resistance **(2)**.

d $2Na + 2H_2O \rightarrow 2NaOH + H_2$ (**1 mark** for H_2, **1 mark** for 2(NaOH))

Question 19.03

An alloy is a mixture **(1)** of a metal with another metal **(1)** or non-metal **(1)**. Properties in common of zinc and copper (any four of): lustrous / shiny surface (NOT silvery surface because copper is pink-brown colour) / conduct heat / conduct electricity / malleable / ductile (**1 mark** each).

(High melting point, high density, sonorous and strong are not included because Group I elements have low melting points etc.)

Unit 20

Question 20.01

a They have a complete outer electron shell (1) so cannot gain, lose or share electrons (1).

b Has only one atom (1).

c less dense than air (1) it is unreactive / does not burn (1)

d It is unreactive / inert (1) so does not allow filament to burn out / prevents reaction of components of bulb with oxygen (1).

e 2,8,8

f ALLOW: Between 1.6 and 3.3 (actual 2.15) (1).

Question 20.02

a Has two atoms in a molecule (1).

b Chlorine is more reactive than iodine (1).

c Colourless solution (1) turns brown (1).

d chlorine + potassium iodide → potassium chloride + bromine (1)

e $Cl_2(aq) + 2KI(aq) \rightarrow 2KCl(aq) + I_2(aq)$

(**1 mark** for correct formulae, **1 mark** for correct balance)

f $Cl_2(aq) + 2I^-(aq) \rightarrow 2Cl^-(aq) + I_2(aq)$

(**1 mark** for correct formulae, **1 mark** for correct balance, **1 mark** for correct state symbols.)

Chlorine is the oxidising agent and bromide ions are the reducing agent (1) because chlorine is accepting electrons (1) and iodide ions are donating electrons (1).

Unit 21

Question 21.01

a Magnesium: very few bubbles / hardly any bubbles (1). Sodium: moves around on the surface (1), lots of bubbles / effervescence (1), fizzes (1), melts into a ball (1). Zinc: no reaction with cold water (1).

b Sodium produces bubbles at a faster rate than magnesium (1). Magnesium reacts with water as well as steam but zinc only reacts with steam (1). So order of reactivity is sodium > magnesium > zinc (1).

c zinc + water / steam → zinc oxide + hydrogen

d $2Na + 2H_2O \rightarrow 2NaOH + H_2$ (**1 mark** for H_2 and **1 mark** for balance)

e Zinc is below carbon in the reactivity series (1) so carbon can remove the oxygen from zinc oxide (1). Magnesium is above carbon in the reactivity series (1).

f i Add the same amount of metal to the same concentration of acid (1); compare the rate at which bubbles are given off (1). The faster the rate of bubbling, the more reactive is the metal (1).

ii The reaction is explosive (1).

Question 21.02

a zinc > cobalt > nickel > tin (1)

b Any two of: Tin loses electrons less readily than zinc (1) because it is lower in the reactivity series (1). Tin forms ions less readily than zinc (1). Zinc is a better reducing agent than tin (1).

c Ni^{2+}, Sn^{2+} (1)

d Sn^{2+} (1)

e i Breakdown of a compound into two or more substances by heating (1).

ii $Sn(OH)_2 \rightarrow SnO + H_2O$ (1)

Question 21.03

On heating, black mixture turns brownish (1). Reducing agent is magnesium (1), oxidising agent is copper oxide (1). Magnesium loses electrons more readily than copper (1) so forms positive ions more readily than copper (1).

Unit 22

Question 22.01

a i A rock which contains a compound from which a metal can be extracted (1).

ii haematite (1)

b i S

ii U

iii X (**1 mark** each)

c i $Fe_2O_3 + 3CO \rightarrow 2Fe + 3CO_2$ (1 **mark** for 2(Fe), 1 **mark** for 3CO and $3CO_2$.)

ii Iron oxide + carbon monoxide → iron + carbon dioxide (1)

iii Oxygen is removed from the iron oxide (1).

d $Fe_2O_3 + 3C \rightarrow 2Fe + 3CO$ (1)

e To remove impurities / to remove silicon dioxide (1) by decomposing to form calcium oxide (1) which reacts with silicon dioxide (1) to form slag (1).

f It is not strong enough / it is brittle (1).

g i car bodies / making machines (1)

ii surgical instruments / cutlery / chemical plant (1)

Question 22.02

a i Covering a metal with a coating of zinc (1).

ii A mixture of a metal with another metal or non-metal (1).

b i zinc blende (1)

ii $2ZnS + 3O_2 \rightarrow 2ZnO + 2SO_2$ (1 **mark** for correct reactants and products, 1 **mark** for balance)

c zinc blende or zinc ore / coke / air (**2 marks** for all three correct; 1 **mark** for two correct)

d $2C + O_2 \rightarrow 2CO$ (1 **mark** for correct reactants and products, 1 **mark** for balance.)

OR two equations: $C + O_2 \rightarrow CO_2$ (1) and $C + CO_2 \rightarrow 2CO$ (1)

$ZnO + CO \rightarrow Zn + CO_2$ (1)

e i Graphite (1) as they conduct electricity (1) and are inert (1).

ii $Zn^{2+} + 2e^- \rightarrow Zn$ (1)

Unit 23

Question 23.01

a 1 **mark** each for any two uses, e.g. coolant / raw material for specified reaction, e.g. making ethanol.

b i To remove harmful substances / so that it is fit to drink (1).

ii Filtration (1). Larger particles of insoluble substances get trapped in the filter (1) Water particles drain through (1).

iii To kill bacteria (1).

c Anhydrous copper sulfate (1) turns from white (1) to blue (1).

OR

Anhydrous cobalt chloride (1) turns from blue (1) to pink (1).

d Oiling or greasing (1): the oil remains on the surface of the moving parts / paint / plastic would rub off (1).

Question 23.02

a nitrogen = 78% (1) oxygen = 21% (1) noble gases / argon = 1% (1)

b carbon dioxide (1)

c Fractional distillation (1) due to differences in boiling points (1). Nitrogen has a lower boiling point than oxygen (1) so nitrogen boils off first (1).

d Zinc coating on the iron (1). If the coating is broken zinc corrodes instead of iron (1) because zinc is more reactive than iron (1).

Unit 24

Question 24.01

a Nitrogen and oxygen combine (1) as a result of lightning activity / in high temperature furnaces / in car engines (1) nitrogen dioxide dissolves in rainwater (1) (to form an acidic solution).

b (Moist) sulfur dioxide is acidic (1). Sulfur dioxide reacts with added calcium oxide (1), which neutralises the acid (1) to form calcium sulfite (1).

Question 24.02

a Any two sources (1 **mark** each): car exhausts or high temperature furnaces / bacterial action in soil / lightning (combining N and O in atmosphere).

b i Combustion when oxygen is limiting (1).

ii $C_{10}H_{22} + 10\frac{1}{2}O_2 \rightarrow 10CO + 11H_2O$ (1 **mark** for correct reactants and products, 1 **mark** for correct balance)

c Nitrogen oxides are passed over catalysts (1). Carbon monoxide is converted to carbon dioxide (1). Nitrogen oxides converted to nitrogen (1) by redox reactions (1).

Carbon monoxide is a reducing agent / nitrogen oxides are oxidising agents (1).

Unit 25

Question 25.01

a i Breakdown of a substance using heat (1).

ii Any two of: respiration / action of acids on carbonates / combustion of carbon-containing substances (**1 mark** each).

b i Lime is a base (1): it reacts with acids in the soil (1) to neutralise them / raise the pH (1).

ii Lime reacts with water in the soil (1) to form a strongly alkaline solution / makes the soil too alkaline (1) so plants cannot grow (1). Calcium carbonate is insoluble in water (1) and not alkaline (1). So the pH cannot go (much) above neutral (1).

c base + acid → salt + water (**1 mark** for salt, **1 mark** for water)

d i sulfuric acid (1) calcium hydroxide (1)

ii neutralisation (1)

Question 25.01

a Acid that is completely ionised (in aqueous solution) (1).

b $Mg(s) + H_2SO_4(aq) \rightarrow MgSO_4(aq) + H_2(g)$

(**1 mark** for equation, **1 mark** for state symbols.)

c Sulfuric acid has two moles of hydrogen that can ionise (1) to form hydrogen ions (1). Hydrochloric acid has only one mole of ionisable hydrogen (1). So the concentration of hydrogen ions in sulfuric acid is greater (1).

d **1 mark** each for any two of: car battery acid / as acid drain cleaner / for making phosphate fertilisers / for making detergents / making paints (or dyes).

Unit 26

Question 26.01

a A 'family' of similar compounds with similar chemical properties (1) due to the same functional group (1).

b alkene(s) (1)

c C_2H_4 (1)

```
H   H (1)
|   |
C = C
|   |
H   H
```

d i

```
    H     H
    |     |      //O
H — C — C = C — C
    |   |        \
    H   O—H       O—H
```

(**1 mark** for each correct)

ii alcohol (1)

iii An atom or group of atoms which is responsible for the characteristic reactions of a particular homologous series (1).

e

```
    H
    |
H — C — H (1)
    |
    H
```

Question 26.02

a Can be represented by a general formula (1).

Differs from the members immediately before or after by a CH_2 group (1).

Shows a gradual change in physical properties as the number of carbon atoms in the increases (1).

b i C_4H_8 (1) C_5H_{10} (1)

ii

```
    H   H   H (1)
    |   |   |
H — C — C = C
    |       |
    H       H
```

c Compounds which have the same molecular formula but different structural formulae (1).

d $CH_3—CH_2—CH_2—Br$ (1)

```
      Br
      |
CH3 — CH — CH3 (1)
```

Unit 27

Question 27.01

a The breakdown / thermal decomposition of long chain alkanes **(1)** into shorter chain alkanes and alkenes (and perhaps hydrogen) **(1)**.

Cracking is used to convert fractions containing larger hydrocarbons into smaller hydrocarbons **(1)**. Larger hydrocarbons are less useful but smaller ones are more useful **(1)**.

b Catalyst **(1)** of silicon(IV) oxide and aluminium oxide **(1)** at 400–500 °C **(1)**.

c $C_{12}H_{26} \rightarrow C_{10}H_{22} + C_2H_4$ (**1 mark** for C_2H_4 and **1 mark** for rest of equation correct)

d i Saturated hydrocarbons have no double bonds **(1)** unsaturated hydrocarbons have (carbon–carbon) double bonds **(1)**.

ii Aqueous bromine / bromine water **(1)**. With saturated hydrocarbon there is no colour change / no reaction **(1)**. With unsaturated hydrocarbon the bromine water turns from orange to colourless **(1)**.

e $2C_2H_6 + 7\underline{O_2} \rightarrow \underline{4}CO_2 + \underline{6}H_2O$ (**1 mark** for each of the underlined)

Question 27.02

a $2C_4H_{10} + 13O_2 \rightarrow 8CO_2 + 10H_2O$ (**1 mark** for correct reactants and products, **1 mark** for balance)

b

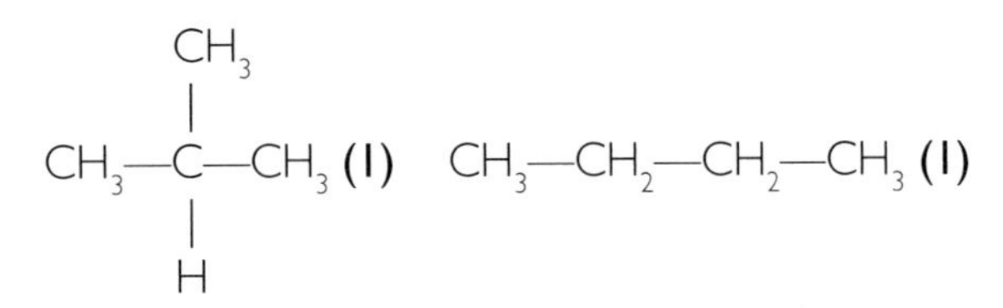

Any two of: (**1 mark** each)

$CH_3—CH_2—CH{=}CH_2$ $CH_3—CH{=}CH—CH_3$ $CH_2{=}C(CH_3)—CH_3$

c $C_2H_6 + Cl_2 \rightarrow C_2H_5Cl + HCl$

(**2 marks** for equation correct, allow **1 mark** for C_2H_5Cl)

ultraviolet light / sunlight needed **(1)**

d i H–C(H)(H)–C(Br)(H)–C(Br)(H)–H **(1)** (structure: H Br Br on top; H–C–C–C–H; H H H below)

ii H–C(H)(H)–C(H)(H)–C(H)(H)–H **(1)** (structure: H H H on top; H–C–C–C–H; H H H below)

e $C_{16}H_{34} \rightarrow C_4H_8 + C_{12}H_{24} + H_2$

(**3 marks** for correct equation. If not scored **1 mark** for C_4H_8 and **1 mark** for $C_{12}H_{24}$.)

Unit 28

Question 28.01

a H–C(H)(H)–C(H)(H)–O–H **(2)** (**1 mark** if OH instead of O–H) (structure: H H on top; H—C—C—O—H; H H below)

b anaerobic / no oxygen present **(1)** yeast **(1)** room temperature **(1)**

ALLOW: presence of water / pH (near) 7

c steam / water **(1)**. Catalyst of concentrated phosphoric acid **(1)**. Temperature 330 °C **(1)**. Pressure 60–70 atmospheres **(1)**.

d i Add universal indicator / blue litmus to the acid (or vice versa) **(1)**: indicator turns red **(1)**. OR Dip pH electrode in acid **(1)**: record pH to see if below pH 7 **(1)**.

ii carboxylic acids **(1)**

iii sodium + ethanoic acid → sodium ethanoate + hydrogen

(**2 marks** for correct equation, **1 mark** if one error or omission)

iv CH_3COOH on left **(1)** 2 (CH_3COOK) **(1)** CO_2 **(1)** H_2O **(1)**

Question 28.02

a Ethanoic acid is a weak acid and hydrochloric acid is a strong acid **(1)**. Ethanoic acid is only partially ionised **(1)** and hydrochloric acid is fully ionised **(1)**. So concentration of H^+ ions in ethanoic acid is lower (than in HCl) **(1)**. So the frequency of collisions (of H^+ ions with magnesium) is lower in ethanoic acid **(1)**.

b $MgCO_3 + 2CH_3COOH \rightarrow (CH_3COO)_2Mg + H_2O + CO_2$

(**1 mark** for correct reactants, **1 mark** for correct products, **1 mark** for balance)

c i butyl ethanoate **(1)**

ii

```
     H  O       H  H  H  H
     |  ||      |  |  |  |
  H— C— C— O— C— C— C— C—H
     |          |  |  |  |
     H          H  H  H  H
```

(**2 marks** for correct structure, **1 mark** if ethyl butanoate drawn)

iii (sulfuric) acid catalyst **(1)** warm / heat / reflux **(1)**

d potassium manganate(VII) **(1)** sulfuric acid catalyst **(1)** heat / reflux **(1)**

Unit 29

Question 29.01

a long chain molecule / macromolecule **(1)** covalent bonding **(1)**

b i

```
H   H
|   |
C = C
|   |
H   H (1)
```

ii The C=C double bond **(1)**.

c Any three of (**1 mark** each):

blocking drains / blocking airways or gullets of animals / birds and fish getting trapped in plastic netting / when burned they may produce toxic gases / toxic additives leach out of plastic dumped on the soil.

Question 29.02

a A reaction where two organic molecules join together **(1)** with the elimination of a small molecule **(1)** to form the repeating units of a polymer **(1)**.

b HOOC—□—COOH **(1)** H_2N—□—NH_2 **(1)**

c i Three units drawn **(1)**. Units and bonding correct **(1)** continuation bonds present **(1)** amide link correctly ringed **(1)**.

```
   H      O           O                 O
   |      ||          ||                ||
 —N— X— C— N— X— C— O— N— X— C—
              |               |
              H               H
```

ii water **(1)**

Question 29.03

a **1 mark** each for any two of: Addition polymerisation only produces the polymer and condensation polymerisation involves elimination of a small molecule **(1)**. Addition polymerisation (usually) involves one monomer and condensation polymerisation (usually) involves two **(1)**. Addition polymerisation involves C=C double bonds joining and condensation polymerisation involves two different functional groups reacting **(1)**.

b i There no other group to react with the NH_2 group / C=C can't react with NH_2 group **(1)**

ii

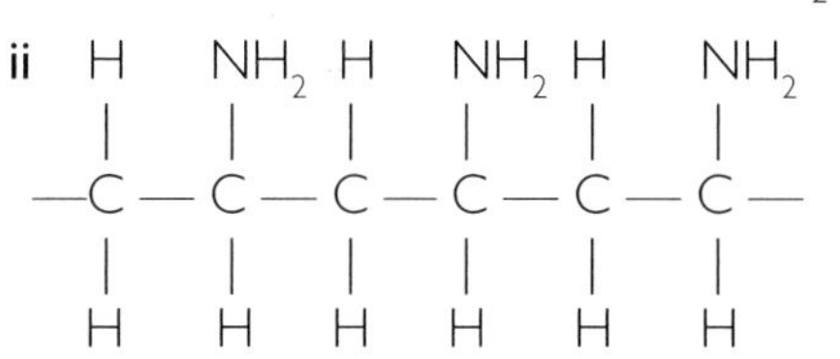

(**1 mark** for correct structure, **1 mark** for continuation bonds.)

Unit 30

Question 30.01

a $-NH_2$ **(1)** –COOH **(1)**

b amide linkage / peptide linkage **(1)** –CO–NH– **(1)**

c i Breaking down a compound by water / acid / alkali. **(1)**

ii Heat **(1)** with concentrated hydrochloric acid **(2)** (**1 mark** if concentrated left out).

iii To neutralise the (hydrochloric) acid **(1)**.

d Because the R_f values of some amino acids are very similar **(1)**. These amino acids may move different distances in another solvent **(1)**.

e Amino acids are colourless **(1)**: locating agent reacts with the amino acid **(1)** to form a coloured compound **(1)**.

Question 30.02

a HO—□—OH **(1)**

b i proteins **(1)**

ii They are catalysts **(1)**: they lower the activation energy of the reaction **(1)**.

c To prevent loss of vapour **(1)** of the acid **(1)**.

d Spray the chromatogram with a locating agent **(1)**. Measure the distance spot has run from the baseline and the distance of the solvent front from the baseline **(1)**.

Calculate the R_f value **(1)**.

Compare with known R_f values in a data book **(1)**.

Glossary

Acid: A proton donor. It reacts with bases to form a salt and water.

Acidic oxide: An oxide that reacts with a base to form a salt and water.

Acid rain: Rain with pH below about pH 5.6 due to the reaction of acidic gases with rainwater.

Activation energy: The minimum energy needed for particles to react when they collide.

Addition polymerisation: A reaction in which monomers containing C=C double bonds react to form polymers without any other substance being formed.

Addition reaction: In organic chemistry, a reaction where two molecules combine to give a single product.

Alcohol: A compound containing the –OH functional group, having the general formula $C_nH_{2n+1}OH$.

Alkali: A soluble base.

Alkali metals: The elements in Group I of the Periodic Table.

Alkane: A hydrocarbon which contains only single bonds.

Alkene: A hydrocarbon which contains one or more double bonds.

Alloy: A mixture of a metal with another element or elements.

Amphoteric oxide: An oxide that reacts with both acids and bases to form a salt and water.

Anhydrous (salts): Salts without any water of crystallisation.

Anion: A negative ion.

Anode: The positive electrode.

Anomalous result: A result or piece of data which does not fit the pattern of the rest of the data.

Atom: The smallest part of an element that can take part in a chemical change.

Avogadro constant: The number of defined particles (atoms, molecules or ions) in one mole of those particles.

Base: A proton acceptor. It reacts with acids to form a salt and water.

Basic oxide: An oxide that reacts with an acid to form a salt and water.

Bond energy: The energy needed to break one mole of covalent bonds between two particular atoms.

Brownian motion: The random movement of small visible particles in a suspension caused by the unequal random bombardment of molecules of liquid or gas on the visible particles.

Carbohydrates: The general name for simple and complex sugars having the general formula $C_x(H_2O)_y$.

Carbon cycle: The cycle of processes which move carbon compounds between the Earth, atmosphere and oceans and keeps the amount of carbon dioxide in the atmosphere fairly constant.

Carboxylic acid: A compound containing the –COOH functional group, having the general formula $C_nH_{2n+1}COOH$.

Catalyst: A substance which speeds up a chemical reaction without being used up or chemically changed after the reaction.

Catalytic converter: A device fitted onto a car exhaust system to reduce the emission of oxides of nitrogen and carbon monoxide.

Cathode: The negative electrode.

Cation: A positive ion.

Chemical change: How elements or compounds react with other substances.

Chromatography: see **Paper chromatography**.

Chlorination: The addition of chlorine to the water supply to kill bacteria and other microorganisms.

Complex carbohydrates (polysaccharides): Polymers formed from simple carbohydrate monomers, e.g. starch.

Compound: A substance made up of two or more different atoms (or ions) bonded together.

Concentration (of solution): The amount of solute dissolved in a defined volume of solution (usually mol/dm^3 or g/dm^3).

Condensation polymerisation: A reaction where two organic molecules join together in a condensation reaction to form the repeating units of a polymer.

Condensation reaction: A reaction where two organic molecules join together with the elimination of a small molecule such as water.

Condensing: The changing of state when a gas changes to a liquid.

Contact process: The industrial process of making sulfuric acid using a vanadium(V) oxide catalyst and high temperature.

Corrosion: The 'eating away' of the surface of a metal by chemical reaction.

Covalent bond: A bond formed by sharing a pair of electrons between two atoms.

Cracking: The process by which large, less useful hydrocarbon molecules are broken down into smaller, more useful alkanes and alkenes.

Delocalised electrons: Electrons which are not associated with any particular atom and are able to move between atoms or ions.

Diatomic: Molecule containing two atoms.

Diffusion: The spreading movement of one substance into another due to the random motion of the particles.

Displacement reaction: A reaction in which one atom or ion (usually a metal) replaces another atom or ion in a compound.

Ductile: Can be drawn into wires by a pulling force.

Electrochemical cell: A source of electrical energy where two metals of different reactivity dipping into an electrolyte are connected via an external circuit.

Electrode: A rod of metal or graphite which leads an electric current to or from an electrolyte.

Electrolysis: The breakdown of an ionic compound (molten or in aqueous solution) by the passage of electricity.

Electrolyte: An ionic compound which conducts electricity when molten or dissolved in water.

Electron shells: The regions at different distances from the nucleus where one or more electrons are found.

Electronic arrangement (electronic structure): The number and arrangement of electrons in the electron shells of an atom.

Electroplating: The coating of the surface of one metal with a layer of another by an electrolytic reaction.

Element: A substance containing only one type of atom.

Empirical formula: A formula showing the simplest whole number ratio of atoms or ions in a compound.

Endothermic: A reaction or process which absorbs energy.

Energy level diagram: A diagram showing the energy change of the reaction on the vertical axis and the reactants and products on the horizontal axis.

Enzyme: A biological (protein) catalyst.

Equilibrium reaction: A reaction which does not go to completion and in which reactants and products are present in fixed concentration ratios at a particular temperature and pressure.

Ester: A substance with the general formula RCOOR', formed by the reaction between a carboxylic acid and an alcohol.

Esterification: The reaction between a carboxylic acid and an alcohol to produce an ester and water.

Evaporation: The change in state from the liquid phase to the gaseous phase below the boiling point of the liquid.

Exothermic: A reaction or process which releases energy.

Fair test: An experiment where the independent variable affects the dependent variable and all other variables are controlled.

Fermentation: The breakdown of organic substances by microorganisms with effervescence and release of energy. It commonly refers to the breakdown of sugars by enzymes in yeast in the absence of oxygen to form ethanol.

Filtrate: In filtration, the liquid which goes through the filter paper.

Flame test: A test for particular metal ions by heating a sample of a compound containing the ion in a blue Bunsen flame. Characteristic colours are produced.

Flue gas desulfurisation: The removal of sulfur dioxide from waste gases produced when fossil fuels containing sulfur are burned.

Fraction: A group of molecules with a defined boiling point range which distils off at the same place during fractional distillation.

Fractional distillation: The separation of liquids with different boiling points from a mixture of liquids by evaporation and condensation in a long column.

Fuel cell: A type of electrochemical cell where hydrogen and oxygen undergo a reaction to produce electrical energy.

Functional group: An atom or group of atoms which is responsible for the characteristic reactions of a particular homologous series.

Galvanising: Coating a metal, usually iron, with a layer of zinc.

General formula: A formula which applies to all members of a particular homologous series.

Giant covalent structures (macromolecular structures): Structures with a lattice (network) of covalent bonds which repeats throughout the whole structure.

Global warming: The warming of the atmosphere due to greenhouse gases trapping infrared radiation emitted from the Earth's surface.

Greenhouse gases: A gas which absorbs infrared radiation emitted from the Earth's surface.

Group: A vertical column of elements in the Periodic Table.

Haber process: The industrial process for making ammonia from hydrogen and nitrogen using a catalyst of iron.

Halide: Compound containing an ion formed when a halogen atom has gained an electron to complete its outer shell of electrons.

Halogens: The elements in Group VII of the Periodic Table.

Homologous Series: A 'family' of similar compounds with similar chemical properties due to the same functional group.

Hydrated (salts): Salts containing water of crystallisation.

Hydration: An addition reaction in which water is combined with another compound.

Hydrocarbon: A compound containing only carbon and hydrogen atoms.

Hydrolysis: The breakdown of a compound by water (often catalysed by acid or alkali).

Indicator: A substance that changes colour over a narrow range of pH to show the end point of a titration.

Intermolecular forces: The weak forces between molecules.

Ion: An electrically charged particle formed from an atom or group of atoms by loss or gain of electrons.

Ionic bond is a strong bond formed by the electrostatic attraction between positive and negative ions in an ionic structure.

Ionic equation: An equation showing only those ions and molecules taking part in the reaction.

Ionic half equation: An equation balanced by electrons which shows either oxidation or reduction.

Isotopes: Atoms of the same element which have the same proton number but a different nucleon number.

Kinetic theory: Particles in solids, liquids and gases behave as hard spheres which are constantly moving from place to place (in liquids and gases) or vibrating (in solids).

Lattice: A continuous regular arrangement of particles which repeats itself throughout the structure.

Limiting reactant: The reactant that is not in excess.

Locating agent: A substance that reacts with colourless spots on a chromatogram to make them visible as coloured spots.

Lone pair: A pair of electrons in the outer shell of an atom in a molecule which does not form a covalent bond.

Lustrous: Shiny like a mirror.

Macromolecular structure: see **Giant covalent structures.**

Malleable: Can be beaten into different shapes.

Metallic bonding: A lattice of metal ions surrounded by a 'sea' of delocalised electrons.

Mixture: An impure substance which contains two or more different components.

Mole: The amount of substance which has the same number of specific particles (atoms, molecules or ions) as there are atoms in exactly 12 g of the isotope carbon-12.

Molecular formula: The formula showing the number and type of atom of each element in a molecule.

Molecule: A particle having two or more atoms joined by covalent bonds.

Monatomic: Existing in the natural state as single atoms.

Monomer: A small reactive molecule that reacts and joins together with itself or another molecule to form the repeating units of a polymer.

Neutral oxide: An oxide that does not react with an acid or base.

Neutral solution: Solution with a pH of 7.

Noble gas structure: The electronic structure of ions or atoms with a complete outer shell of electrons.

Noble gases: Elements in Group VIII of the Periodic Table.

Non-biodegradable: Unable to be broken down in the environment by microorganisms or other living things.

NPK fertilisers: Fertilisers containing the elements nitrogen, phosphorus and potassium in various compounds.

Nucleon number (mass number): The total number of protons and neutrons in the nucleus of an atom.

Ore: A rock containing a metal or metal compound in sufficient quantity that a metal can be extracted from it.

Oxidation: The gain of oxygen, loss of electrons or increase in oxidation state.

Oxidation state: A number describing the degree of oxidation (or reduction) of an atom in a compound.

Oxidising agent: A substance that oxidises another substance by accepting electrons.

Paper chromatography: The separation of a mixture of soluble compounds using chromatography paper and a solvent.

Period: A horizontal row of elements in the Periodic Table.

Periodic Table: An arrangement of elements in order of increasing proton number so that elements with the same number of electrons in their outer shell fall in the same vertical column.

Photochemical reaction: A reaction which is catalysed by light or dependent on light for the reaction to occur.

Photosynthesis: The process of producing glucose and oxygen from carbon dioxide and water in plants in the presence of chlorophyll and light.

Physical change: A change in a physical property, e.g. melting, boiling.

Pollution: The introduction of substances into the environment which should not normally be present.

Polyamide: A polymer whose monomers are bonded to each other by the amide link, –CONH–.

Polyester: A polymer whose monomers are bonded to each other by the ester link, –COO–.

Polymer: A large molecule built up from smaller units called monomers.

Polymerisation: The chemical reaction where monomers react to form a polymer.

Position of equilibrium: The relationship between the equilibrium concentrations of reactants and products.

Precipitation reaction: A reaction in which a solid is formed when two solutions are mixed.

Proteins: Condensation polymers formed from amino acids which are joined by peptide (amide) groups.

Proton number (atomic number): The number of protons in the nucleus of an atom.

Rate of reaction: The rate at which reactants are used up or the rate at which products are formed usually calculated as change in concentration in mol/dm^3 divided by time.

Reactivity series: A list of elements (usually metals) in order of their reactivity, with the most reactive first.

Recycling: The processing of used materials into new products.

Redox reaction: A reaction where oxidation and reduction occur together.

Reducing agent: A substance that reduces another substance by donating (adding) electrons.

Reduction: The loss of oxygen, gain of electrons or decrease in oxidation state.

Relative atomic mass, A_r: The average mass of naturally occurring atoms of an element on a scale where a carbon-12 atom has a mass of exactly 12-units.

Relative molecular mass, M_r: The sum of the relative atomic masses of all the atoms in a molecule.

Residue: In filtration, the solid that is trapped on the filter paper.

Respiration: The production of energy in living things by a series of reactions involving the conversion of glucose and oxygen to carbon dioxide and water.

Reversible reaction: A reaction in which the reactants combine to form products and the products can also react together to form the original reactants.

***R_f*:** In chromatography, the distance moved by a particular substance from the base line divided by the distance moved by the solvent front from the base line.

Rusting: The reaction of iron with water and oxygen to form hydrated iron(III) oxide (rust).

Sacrificial protection: When a more reactive metal is placed in contact with a less reactive metal, the more reactive metal corrodes first, preventing the corrosion of the less reactive metal.

Salt: A compound formed when the hydrogen in an acid is replaced by a metal or ammonium ion.

Saturated hydrocarbon: A hydrocarbon which has only single carbon–carbon bonds.

Separating funnel: A piece of glassware used to separate two immiscible liquids or a solute which is more soluble in one liquid than another.

Simple distillation: A method of separating a volatile from a non-volatile substance by evaporation and condensation.

Spectator ions: Ions which do not take part in a reaction.

Stoichiometry: The (mole) ratios of the reactants and products shown in the balanced equation.

Strong (acid/base): an acid or base which ionises completely in solution.

Structural formula: A formula showing how the atoms are bonded in a molecule.

Structural isomers: Compounds with the same molecular formula but different structural formulae.

Sublimation: The change in state directly from solid to gas and/or gas to solid without the liquid state being formed.

Substitution reaction: A reaction in which one atom or group of atoms in a compound is replaced by another.

Thermal decomposition: The breakdown of a compound into two or more products by heating.

Titration: A method for finding out the amount of substance (in moles) in a solution of unknown concentration (usually by using a burette and indicator).

Transition element: A block of metals in the middle of the Periodic Table which have characteristic properties such as high melting points and the formation of coloured compounds.

Universal indicator: A mixture of indicators used to measure pH values.

Unsaturated hydrocarbon: A hydrocarbon which has one or more carbon–carbon double (or triple) bonds.

Volatile: Easily changed to a vapour. Volatile substances have low boiling points.

Weak (acid/base): An acid or base which ionises partially in solution.

Index

D

E

F

G

H

I

K

L

M

N

O

P

R

S

T

U

W

Z